CISM COURSES AND LECTURES

The series presents lecture notes, monographs, edited works and proceedings in the field of Mechanics, Engineering, Computer Science and Applied Mathematics.
Purpose of the series is to make known in the international scientific and technical community results obtained in some of the activities organized by CISM, the International Centre for Mechanical Sciences.

INTERNATIONAL CENTRE FOR MECHANICAL SCIENCES

COURSES AND LECTURES - No. 368

MECHANICS OF SOLIDS WITH PHASE CHANGES

EDITED BY

M. BERVEILLER
UNIVERSITY OF METZ

F.D. FISCHER
LEOBEN UNIVERSITY

Le spese di stampa di questo volume sono in parte coperte da contributi del Consiglio Nazionale delle Ricerche.

This volume contains 131 illustrations

Printed in Italy

In order to make this volume available as economically and as rapidly as possible the authors' typescripts have been reproduced in their original forms. This method unfortunately has its typographical limitations but it is hoped that they in no way distract the reader.

ISBN 3-211-82904-0 Springer-Verlag Wien New York

PREFACE

This book contains the lectures delivered at the International Centre for Mechanical Sciences, Udine, Italy, in the session "Mechanics of Solids with Phase Changes", September 1995.

The increased interest in mechanics of materials has raised the motivation to organize such a session. Continuum mechanics has strongly influenced the research in classical plasticity of inhomogeneous materials and in composites and multiphase materials. However, all these materials possess one common feature: the more or less, spatially and temporally, fixed distribution of the various phases. This is, however, not the case for materials with solid-solid phase changes. The thermomechanical loading is responsible for the development of one or more new phases in a parent material. Therefore an interdisciplinary research by involving branches such as material sciences, physical chemistry and thermodynamics and continuum mechanics has proved to be necessary to understand the transformation behavior and the corresponding thermomechanical deformation behavior of such phase changing materials. This matter makes the research more difficult, but also more exciting. To the opinion of the authors, no current books exist that present solid-solid phase transformations in elastic and elastic-plastic materials in the light of mechanics of materials and describe the tools how to understand solid-solid phase transformations.

Therefore, the editors assume that this book will close a certain gap.

The book may be of use for both material scientists and people working in continuum mechanics. The editors tried hard to keep the presentation of the contents on such a mathematical level that post-graduate students and also practitioners with an interest in mechanics of materials can use this book as a source of better understanding as well as a tool for further development. In addition, we expect that researchers in the field of phase change phenomena may find some new information and hints for problem solutions.

The book is mainly devoted to diffusive and displacive phase transformations with and without plastic behavior of the parent and product phases. Mainly metals, both non ferrous and ferrous alloys are dealt with. Additionally an interesting introduction of phase change mechanics into geological problems is included.

Thermoelastic materials like shape memory alloys are treated both with respect to experiments and to mechanics of materials related aspects. Practical applications of theoretical concepts how to control the phase transformation and, therefore, the thermomechanical behavior of the phases are outlined. The phenomenon of transformation induced plasticity (TRIP) as an enhanced plastic deformation of phase changing materials is dealt with in detail mainly for ferrous alloys under diffusive and displacive (martensitic) phase transformations.

The first two chapters supply an introduction to phase change phenomena and classical thermodynamics within the framework of physical metallurgy and solid state physics. The third chapter presents an overview on the displacive (martensitic) transformation with and without plasticity. An overview on the experimental work of the last decades forms an important part of this chapter. The following chapter is outlined in the same way; however, it deals with the diffusive transformation in solids. The next two chapters are mainly devoted to the modelling and simulation of the deformation behavior of shape memory alloys and ferrous alloys. Both micromechanical and numerical concepts are introduced and tested in various experiments. The concept of simulation may be used also with respect to material design if one thinks of low alloyed TRIP - Steel or shape memory alloys designed for several specific purposes.

Chapter 7 reports on a geological application and shows how non hydrostatic thermodynamics in conjunction with continuum mechanics may help to understand geological phenomena like wet compacting sediments.

The last block of three chapters is devoted to some examples of the exploitation of solid-solid phase changes like the heat treatment of metals or the design of smart structures from shape memory alloys.

Finally it is hoped that the reader will gain an overview on phase change phenomena from elementary physics via experiments and continuum mechanical formulations to applications of concepts in technical fields.

The editors are deeply indebted to the CISM Staff for all organization facilities, in particular to Prof. S. Kaliszky who represents the scientific committee and to Prof. C. Tasso for his patience as the CISM Editor.

M. Berveiller
F. D. Fischer

CONTENTS

Page

THERMODYNAMICS AND KINETICS OF PHASE TRANSITIONS

AN INTRODUCTION

J. Ortin

University of Barcelona, Barcelona, Spain

Abstract

This work presents an introduction to the thermodynamics and kinetics of first-order and continuous phase transitions. In the first three sections both types of transitions are described in the framework of classical equilibrium thermodynamics, as well as using a relatively simple version of Landau's phenomenological approach. Section 4 introduces the statistical mechanics approach to phase transitions, making use of reticular models. Finally, in section 5, reticular models are used to study two different non-equilibrium aspects of phase transitions: the dynamics of domain growth and the dynamics of avalanches and hysteresis in first-order phase transitions.

1 Preface

In this chapter, I plan to review some important topics in the theory of first-order and continuous phase transitions, using the powerful and well established tools of thermodynamics and statistical mechanics. The idea is to focus on a few simple (idealized) problems, generic enough to be applicable to a variety of systems, and for which some rigorous results can be derived. The level is adequate for last year undergraduate students in physics, and for researchers on materials science that have an interest on the relationship between phase transitions and the behaviour of materials. Actually, my own perspective is that of a physicist interested in the behaviour of materials.

The review starts with a traditional presentation of the classical thermodynamic approach to first-order and continuous phase transitions in homogeneous systems, based on references [1], [2] and [3]. I include here a very simple version of the phenomenological Landau approach to phase transitions, which I have felt that could be of interest to materials scientists. Then I turn the attention to spin models, particularly the Ising model, which I use to introduce the classical techniques of equilibrium statistical mechanics, and to show how to perform explicit calculations of thermodynamic properties near phase transitions. This part is based on references [4], [5] and [6]. In the last part of the review, I introduce nonequilibrium aspects of phase transitions through two subjects of current interest: the kinetics of domain growth [7] and avalanches [8] in first-order phase transitions.

I am indebted to M.Berveiller and F.D.Fischer for their invitation to participate in the CISM course, and for their patient encouragement in the completion of this project. My research on materials science and nonequilibrium phenomena at the University of Barcelona is supported by projects MAT92-0884 (CICyT, Spain) and PB93-0054-C02-01 (DGICyT, Spain).

1.1 Basic concepts in classical equilibrium thermodynamics

Consider, in the framework of classical equilibrium thermodynamics, a closed hydrostatic system (e.g. a system that does not exchange mass with its environment and is subjected only to hydrostatic external forces). Such a system can modify its internal energy U in a reversible way by either doing mechanical work or exchanging heat, reversibly, with its environment. This is summarized by the fundamental thermodynamic identity for closed systems:

$$dU = -pdV + TdS \tag{1}$$

where p is the pressure of the system, V its volume, T its absolute temperature and S its entropy. It follows that V and S are the natural variables of the internal energy $U(V, S)$. Alternatively, thermodynamic potentials with natural variables more adequate to be controlled experimentally can be defined through Legendre transforms. These are:

$$\begin{aligned} \text{Enthalpy:} \quad & H \equiv U + pV \\ \text{Helmholtz free energy:} \quad & F \equiv U - TS \\ \text{Gibbs free energy:} \quad & G \equiv H - TS \end{aligned} \tag{2}$$

and therefore:

$$\begin{aligned} dH &= Vdp + TdS & \text{so that} \quad & H = H(p, S). \\ dF &= -pdV - SdT & \text{so that} \quad & F = F(V, T). \\ dG &= Vdp - SdT & \text{so that} \quad & G = G(p, T). \end{aligned} \tag{3}$$

This last potential, the Gibbs free energy, is particularly relevant in our subsequent discussion of phase transitions because its natural variables are p and T, and these quantities remain constant at a phase transition. It can be proved that, for a system kept at constant pressure and temperature, the state of equilibrium is the state of minimum Gibbs free energy. In addition, stability conditions applied to the thermodynamic potentials lead to a definite (positive) sign of the response functions C_p (heat capacity at constant pressure) and κ_T (isothermal compressibility):

$$Cp \equiv T\left(\frac{\partial S}{\partial T}\right)_p = -T\left(\frac{\partial^2 G}{\partial T^2}\right)_p \geq 0 \tag{4}$$

$$\kappa_T \equiv -\frac{1}{V}\left(\frac{\partial V}{\partial p}\right)_T = -\frac{1}{V}\left(\frac{\partial^2 G}{\partial p^2}\right)_T \geq 0 \tag{5}$$

If a magnetic system is considered, the fundamental thermodynamic identity for the internal energy U reads:

$$dU = HdM + TdS \qquad \text{and} \qquad U = U(M,S) \tag{6}$$

where M is the magnetization of the system and H the external field applied to it. Again, a family of thermodynamic potentials can be defined through Legendre transforms:

$$\begin{aligned} \text{Energy:}\quad & U' \equiv U - HM \\ \text{Helmholtz free energy:}\quad & F \equiv U - TS \\ \text{Free energy:}\quad & F' \equiv U' - TS \end{aligned} \tag{7}$$

and then:

$$\begin{aligned} dU' &= -MdH + TdS \qquad &\text{so that} \qquad U' &= U'(H,S). \\ dF &= HdM - SdT \qquad &\text{so that} \qquad F &= F(M,T). \\ dF' &= -MdH - SdT \qquad &\text{so that} \qquad F' &= F'(H,T). \end{aligned} \tag{8}$$

Concerning magnetic systems, however, there is no general *consensus* on the names of the potentials and the symbols used for them. In the statistical mechanical description of magnetic phase transitions, for example, the potential whose natural variables are the *intensive* variables H and T is usually written F. This potential plays a role equivalent to that of G in hydrostatic systems. A system at constant temperature and external magnetic field is in equilibrium if its free energy F' is at a minimum. The stability conditions determine again the sign of the response functions C_H (heat capacity at constant magnetic field) and χ_T (isothermal magnetic susceptibility):

$$C_H \equiv T\left(\frac{\partial S}{\partial T}\right)_H = -T\left(\frac{\partial^2 F'}{\partial T^2}\right)_H \geq 0 \tag{9}$$

$$\chi_T \equiv \left(\frac{\partial M}{\partial H}\right)_T = -\left(\frac{\partial^2 F'}{\partial H^2}\right)_T \geq 0 \tag{10}$$

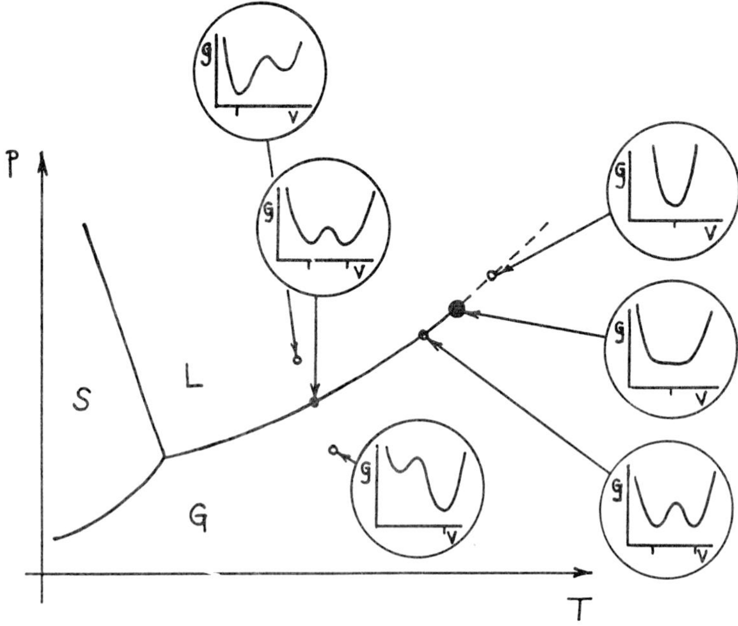

Figure 1: Schematic phase diagram of water.

1.2 Phase transitions

A phase transition is the result of a failure in the thermodynamic stability of the system under consideration.

For a qualitative understanding of the thermodynamic behaviour of a system in the neighbourhood of a phase transition, consider figure 1, which shows a portion of the $p - T$ phase diagram of pure water. Three *coexistence* lines define three different regions, corresponding to the solid (S), liquid (L) and gas (G) phase. The three lines merge at a single point, the triple point, defined by the pressure and temperature at which the three phases coexist simultaneously in equilibrium ($p_{tr} = 6.026 \times 10^{-3} atm$ and $T_{tr} = 273.16K$ for pure water). The liquid-gas coexistence line is also limited at the other end by the so-called critical point ($p_c = 218.5 atm$ and $T_c = 647.4K$ for pure water); beyond this point, the two fluid phases (liquid and gas) cannot be distinctly defined anymore.

The thermodynamic behaviour of the system at given p and T can be discussed in terms of a functional $\mathcal{G}$, representing the Gibbs free energy of the system as a function of the different available configurations compatible with p and T. The configurations are characterized by the values of the extensive quantities V and S, conjugate to p and T, but we will consider only the projections $\mathcal{G}(V)$ for simplicity. The absolute minima of $\mathcal{G}(V)$ define the true equilibrium states of the system, with Gibbs free energy $G = \mathcal{G}(V)$ and volume V.

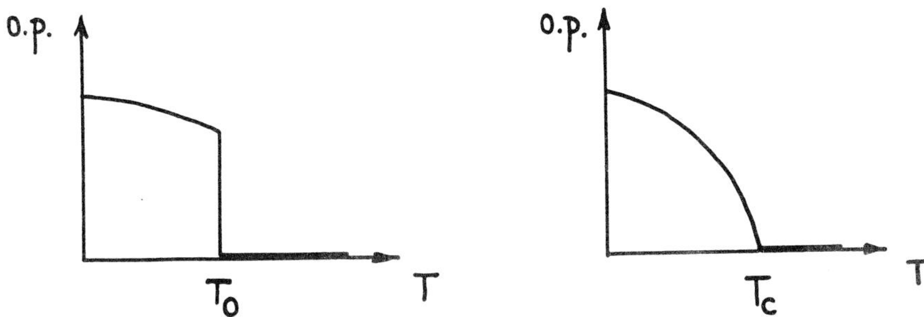

Figure 2: Order parameter as a function of temperature. Left: first-order phase transition. Right: continuous phase transition.

In the liquid phase $\mathcal{G}(V)$ displays two different minima, but the absolute of these two minima identifies the phase of small V(i.e. the most condensed phase or liquid) as the equilibrium state. On going from the liquid to the gas phase $\mathcal{G}(V)$ changes progressively, until a point is reached where the two minima take the same value. At this point in the phase diagram, the liquid (most condensed) and gas (less condensed) phases can coexist in equilibrium, and hence this point lies on the liquid-gas coexistence line. Beyond this point, the absolute minima of $\mathcal{G}(V)$ identifies a state of large V (gas) as the true equilibrium state. It is important to note that in the coexisting state the two phases have different equilibrium values of V . In other words, on going from the liquid to the gas phase through the coexistence line there is a *discontinuity* in the extensive variable used to characterize the behaviour of the system at the phase transition.

It is costumary to introduce a new quantity, the *order parameter* of the system, defined as a monotonous function of this extensive variable and chosen in such a way that it is 0 above the phase transition and positive below. In the case considered the order parameter could be $(V - V_L)$, where V_L would represent the volume of the liquid phase. The discontinuity experienced by the order parameter at the phase transition is represented schematically in figure 2 as a function of temperature.

The behaviour is considerably different when the phase transition takes place at the critical point. Consider, in figure 1, an evolution from a fluid state to a state in the liquid-gas coexistence line, through the critical point. In the fluid phase $\mathcal{G}(V)$ has a single minimum, associated with a single phase. This minimum becomes rather flat at the critical point, showing that large fluctuations in volume can take place there, and splits into two minima of equal depth (representing the liquid phase and the gas phase coexisting in equilibrium) as we get into the coexistence line. This time the order parameter goes continuously from 0 above T_c to a positive value below T_c, as shown in figure 2.

The two types of phase transitions described provide examples of first-order and continuous transitions, respectively. Both are discussed in detail in the next two sections.

2 First-order phase transitions

The $p - V - T$ diagram of a pure substance is presented in figure 3, together with its $p - V$ and $p - T$ projections. The lines of constant T (isotherms) in the $p - V$ projection display *plateaus* of constant p in the regions where two phases coexist: the coresponding phase transitions take place at constant p and T. The extent in V of the liquid-gas coexistence region decreases with temperature until it disappears at (p_c, T_c), where the isotherm displays an inflection point. Above this temperature there is no distinction between liquid and gas. This is not the case for the solid-liquid coexistence region, which extents into the range of very large pressures without any indication of the presence of a critical point.

2.1 Gibbs free energy at a first-order phase transition

We focus now on one of these transitions taking place at constant p and T. In particular, we consider the transition on crossing the liquid-gas coexistence line in the $p - T$ projection of the phase diagram (figure 3), and the behaviour of the Gibbs free energy $G(p, T)$.

Suppose first that the system is kept at constant pressure p_0 and its temperature raised starting from the liquid phase. Below T_0, defined by the coexistence line, we have $G_{liquid} < G_{gas}$ and hence the liquid phase is the stable phase. When the temperature reaches T_0, the two phases are equally stable and the transition from liquid to gas takes place in equilibrium. Finally, above T_0, $G_{gas} < G_{liquid}$ and all the liquid has turned into the stable gaseous phase. Note that the slope of $G(p_0, T)$ of the equilibrium phase suffers a discontinuity at the phase transition; since $S = -\left(\partial G/\partial T\right)_p$, we conclude that the entropy of the system is discontinuous across the transition.

Suppose now that the system remains at constant temperature T_0 and its pressure decreases, starting again from the liquid phase. Above p_0, defined by the coexistence line, we have $G_{liquid} < G_{gas}$ and the liquid phase is the stable phase. When the pressure reaches p_0, the two phases are equally stable and the transition from liquid to gas takes place in equilibrium. Finally, below p_0, $G_{gas} < G_{liquid}$ and all the liquid has turned into the stable gaseous phase. Note that this time it is the slope of $G(p, T_0)$ of the equilibrium phase that suffers a discontinuity at the phase transition; since $V = (\partial G/\partial p)_T$, we conclude that the volume of the system is also discontinuous across the transition.

This is the reason for calling these transitions first-order: a thermodynamic system undergoes a first-order phase transition when the *first* derivatives of the Gibbs free energy experience a discontinuity. Other interesting consequences of the discontinuous change of S and V at a first-order transition are discussed next.

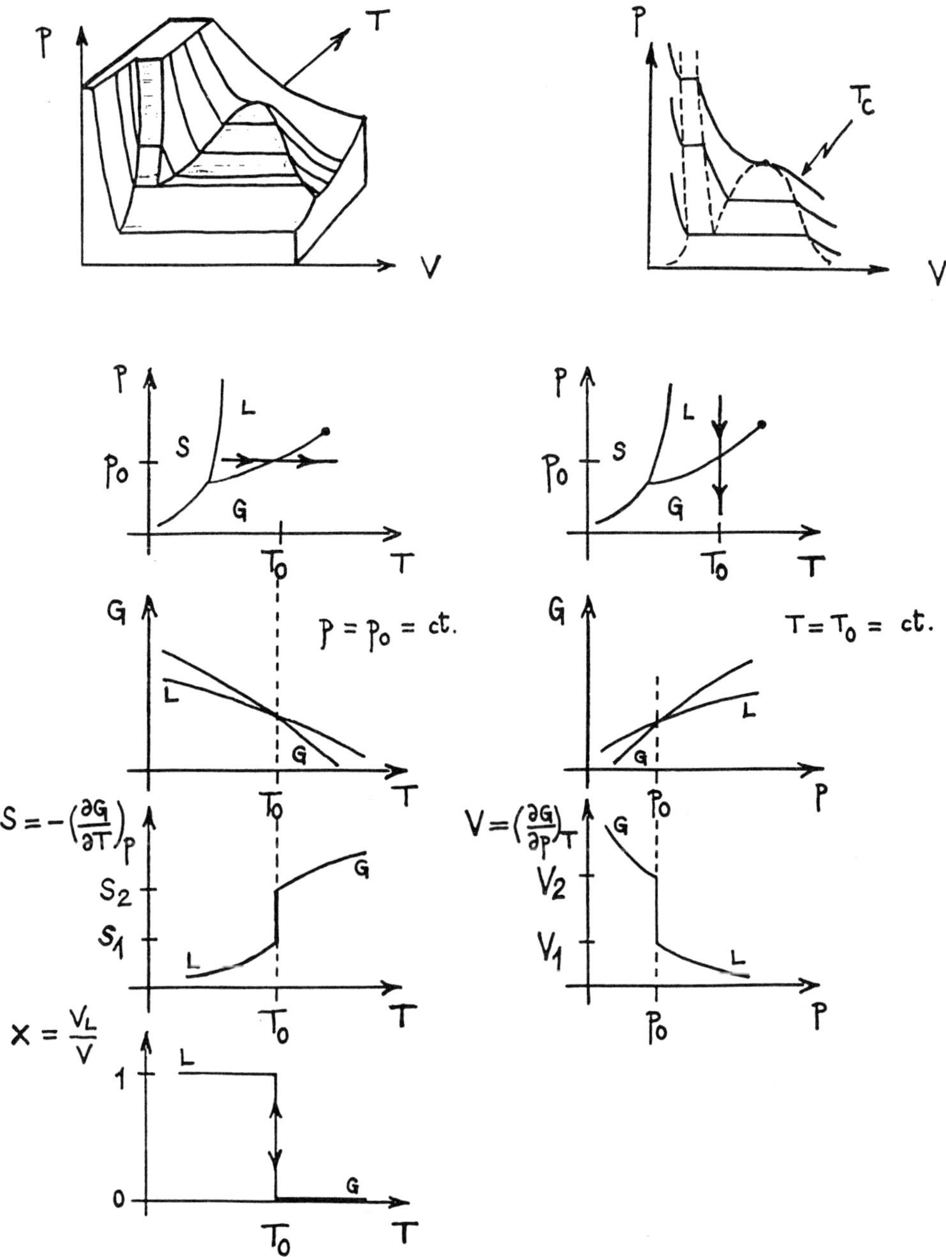

Figure 3: $p - V - T$ diagram of a pure substance, $p - V$ and $p - T$ projections, and schematic behaviour of $G(p, T)$ and its first derivatives at a first-order phase transition.

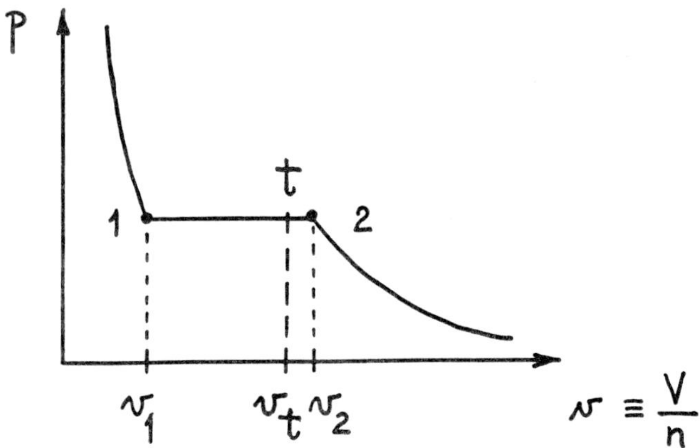

Figure 4: $p - V$ isotherm in the two-phase region.

2.2 Lever rule, latent heat and the Clausius-Clapeyron equation

The lever rule gives the amount of each phase present in equilibrium when the system is in the coexistence region. Consider the isotherm that crosses the liquid-gas region of the $p - V$ phase diagram shown in figure 4.

Point 1, at the left end of the *plateau*, represents the vapor-saturated liquid phase; point 2, at the other end, the liquid-saturated vapor. Let t be the actual state of the system where the two phases, liquid and gas, coexist in equilibrium and let x_1, x_2 denote the molar fraction of each phase in the two-phase state. We have:

$$\left.\begin{array}{rr} \text{Average volume of the system:} & V_t = x_1 V_1 + x_2 V_2 \\ \text{Sum of molar fractions:} & x_1 + x_2 = 1 \end{array}\right\} \quad V_t = V_1 + x_2(V_2 - V_1)$$

and therefore:

$$x_2 = \frac{V_t - V_1}{V_2 - V_1} = \frac{\overline{1t}}{\overline{12}} \tag{11}$$

This is the lever rule.

Since the entropy of the system is discontinuous at a first-order phase transition, the system absorbs or releases a given amount of heat at the transition. This is the latent heat of transition, l, given by:

$$0 = \Delta G = \Delta H - T_0 \Delta S \quad \rightarrow \quad l \equiv \Delta H = T_0 \Delta S \tag{12}$$

The latent heat is absorbed by the system when it transforms from a phase of low entropy to one of high entropy ($\Delta S > 0$) and released in the opposite sense ($\Delta S < 0$).

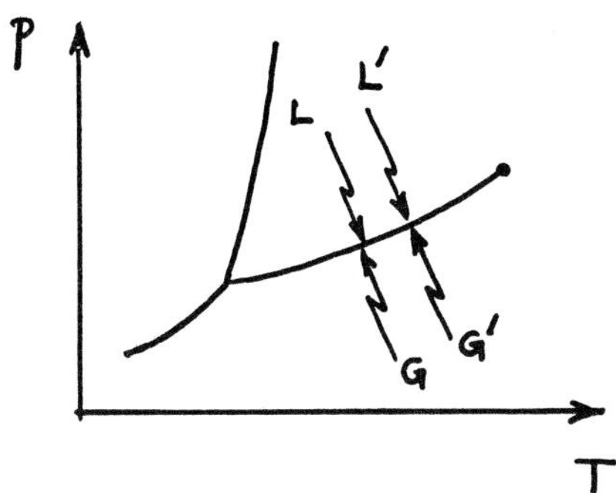

Figure 5: $p - T$ diagram, showing two points of liquid-gas phase equilibrium.

Actually, the amount of heat exchanged by the system controls the fraction transformed.

The discontinuities in S and V determine the slope of the coexistence line at the pressure and temperature of the transition. To see this, consider two points along the coexistence line, as indicated in figure 5. The phase equilibrium condition for these two points reads:

$$\left.\begin{array}{l} G_l = G_g \\ G_l' = G_g' \end{array}\right\} \quad G_l' - G_l = G_g' - G_g$$

or, since the two points considered can be arbitrarily close to each other:

$$dG_l = dG_g \quad \rightarrow \quad -S_l dT + V_l dp = -S_g dT + V_g dp$$

and we arrive to the Clausius-Clapeyron equation:

$$\left(\frac{dp}{dT}\right)_{coex} = \frac{S_l - S_g}{V_l - V_g} \tag{13}$$

For the liquid-gas phase transition $S_l < S_g$ and $V_l < V_g$. Therefore $(dp/dT)_{coex} > 0$, i.e. the liquid-gas coexistence line has a positive slope in the (p, T) diagram. Usually, the same is true for the solid-liquid and solid-gas transitions. A notable exception is the solid-liquid coexistence line of pure water, which has a negative slope because $V_s > V_l$.

Finally, the fact that the equilibrium transition takes place at constant p and T, while the conjugate variables V and S experience a discontinuity, makes the three response functions to diverge at a first-order phase transition:

$$Cp = T\left(\frac{\partial S}{\partial T}\right)_p \rightarrow \infty, \quad \beta = \frac{1}{V}\left(\frac{\partial V}{\partial T}\right)_p \rightarrow \infty, \quad \kappa_T = -\frac{1}{V}\left(\frac{\partial V}{\partial p}\right)_T \rightarrow \infty \tag{14}$$

2.3 Metastability and hysteresis

One important aspect of first-order phase transitions, of basic as well as practical relevance, is the possibility that a system crossing a coexistence line, instead of transforming to the new phase at the equilibrium point (p_0, T_0), gets trapped for some time in a a metastable state. This possibility arises from the fact that the two phases are separated by an energy barrier at the transition point, as shown in figure 6. Consider the sequence of $\mathcal{G}(V)$ curves in the figure, obtained by increasing the pressure of the system at constant temperature T_0. The minimum at point A indicates that gas is the stable equilibrium phase at low pressures. Increasing pressure makes an inflection point to appear at point M, which turns into a second minimum N at higher pressures. In these conditions $\mathcal{G}(V)$ has two minima: C (stable, gas phase) and N (metastable, liquid phase), and one maximum in between: L (unstable, unphysical). Going up in pressure, we find next that the two minima become equal (D and O): this means that p_0 (pressure for transformation in equilibrium at T_0) has been reached and hence it should be possible to observe an equilibrium transition from the gas to the liquid phase, according to strictly thermodynamic considerations. There is, however, an energy barrier to be surmounted for the system to find the equilibrium state corresponding to the liquid phase. It may well be that the system remains in the gas phase even if we keep increasing its pressure: the corresponding state E (gas) is metastable with respect to the stable state Q (liquid) but corresponds to a true minimum of $\mathcal{G}(V)$ and hence it is physically possible. The limit of metastability for the gas phase is reached when the associated minimum becomes an inflection point, such as F. From then on, upon increasing pressure, the only equilibrium state is in the liquid phase (point S, for example).

We could equally well consider an evolution from the liquid phase to the gas phase, reducing the pressure of the system at constant temperature, T_0. This corresponds to consider the sequence of $\mathcal{G}(V)$ curves in figure 6 in ascending order. It is easy to see that this time the liquid phase can exist as a metastable phase in a range of pressures for which gas is the true stable phase (sequence of minima O, N and M).

The energy barrier between the two minima, in summary, makes the isothermal transformation from gas to liquid to occur at some pressure above the equilibrium pressure, and the reverse one from liquid to gas at some pressure below the equilibrium pressure. This behaviour, characteristic of first-order phase transitions, is called *hysteresis* (i.e. delay in responding to the external driving).

For each value of the pressure p, we can take the values of V for which $\mathcal{G}(V)$ displays a local maximum or minimum and plot them in a $p-V$ diagram. The resulting isotherm is shown in figure 6. The segments ABCD (gas) and OQRS (liquid) correspond to stable states, and the points D and O are the *limits of the two-phase region* at (p_0, T_0). The segments DEF and ONM are the metastable continuations of the stable isoterm for the gas and liquid phase, upon increasing and decreasing pressure, respectively. The metastable continuations terminate at F and M, the *limits of metastability* at this temperature. Hence, F and M define the maximum possible extension of the hysteresis cycle for the isotherm considered, as shown in figure 6. Finally, the segment FJKLM represents states that are unstable at T_0 (note that κ_T is negative in this portion of the isotherm) and cannot physically exist as such.

In the $p-V$ diagram (figure 7) the boundary of the two-phase region is defined

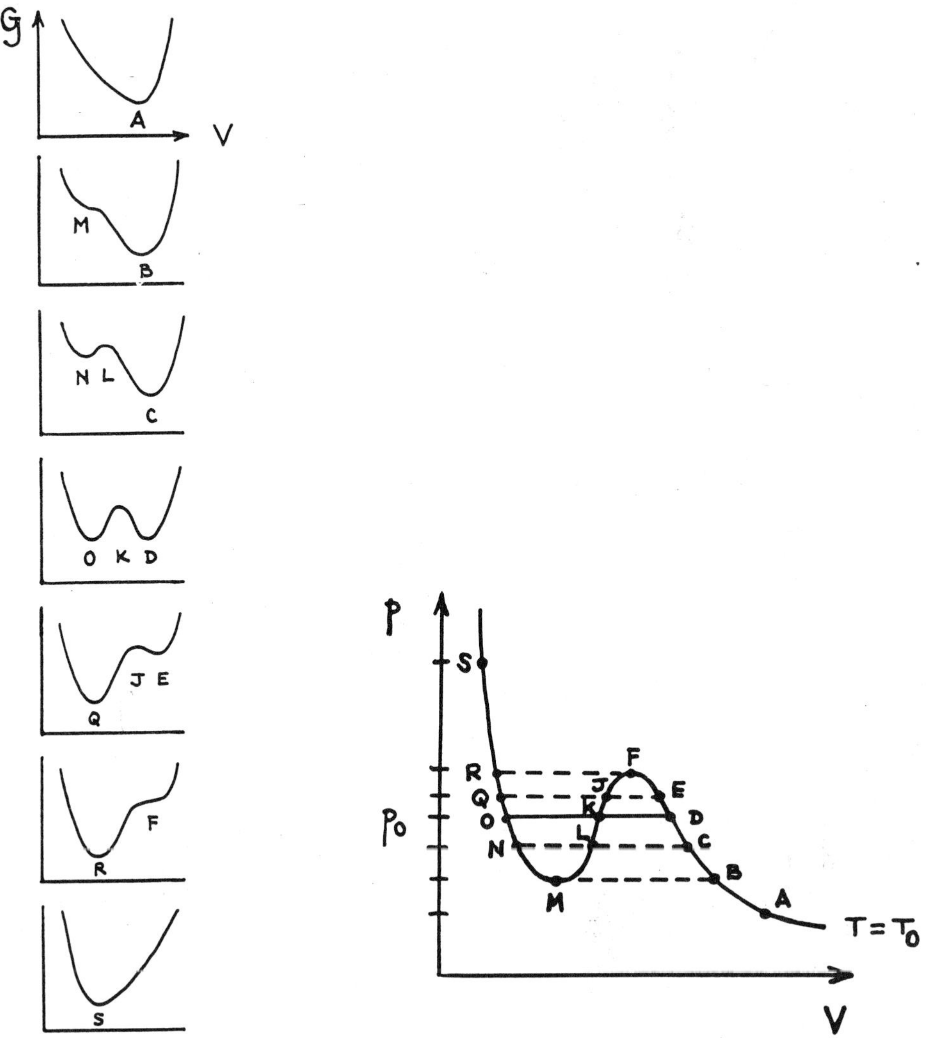

Figure 6: Sequence of $\mathcal{G}(V)$ curves on going from a less condensed to a more condensed phase, at constant temperature (left), and the corresponding isotherm in a $p - V$ diagram (right).

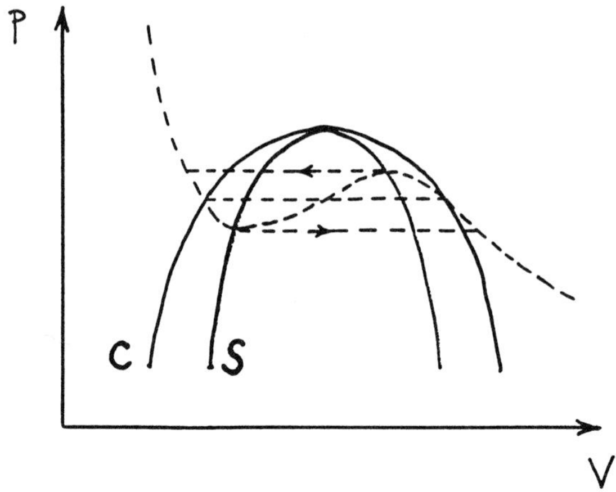

Figure 7: Coexistence line (C) and spinodal line (S).

by the location of the points of transformation in equilibrium (such as D and O) for the different isotherms. Within the two-phase region, on its turn, the line that goes through the limits of metastability of the different isotherms and separates metastable from unstable states is called the *spinodal.*

The equilibrium isotherm can be obtained from the general isotherm taking the branches corresponding to the pure phases outside the two-phase region and replacing the metastable and unstable portions by a horizontal segment that restores equilibrium. The actual location of this horizontal segment of the equilibrium isotherm (i.e. the pressure p_0 corresponding to points D and O) can be calculated following an argument due to Maxwell. In general:

$$dG = -SdT + Vdp$$

If we integrate this equation from point D to point O along the equilibrium isotherm:

$$\left.\begin{array}{ll} \int_D^O dG = G_O - G_D = 0 & \text{(equilibrium)} \\ dT = 0 & \text{(isotherm)} \end{array}\right\} \quad \int_D^O V dp = 0$$

or, equivalently,

$$\int_D^F V dp + \int_F^K V dp + \int_K^M V dp + \int_M^O V dp = 0$$

and rearranging the integrals:

$$\int_D^F V dp - \int_K^F V dp = \int_M^K V dp - \int_M^O V dp \tag{15}$$

The left-hand side of this equation represents the area enclosed between the isotherm and the horizontal equilibrium line from K to D, and the right-hand side the area enclosed between the isotherm and the horizontal equilibrium line from O to K. The equation states that the equilibrium line intersects the isotherm T_0 at a pressure p_0

such that the two areas are identical. This result is known as Maxwell's equal area construction.

2.4 First-order phase transitions in multicomponent systems: Gibbs phase rule and phase diagrams

Up to now, our presentation has been limited to a single component system (one atomic species) existing in one or several phases. Let us extend the problem of thermodynamic equilibrium, now, to a generic system of C components and P phases. For this system the Gibbs free energy must be generalized to account for mass transfer between the different phases, as a result of either chemical reactions or phase transitions:

$$dG(p,T,n_i^k) = -SdT + Vdp + \sum_{i=1}^{C}\sum_{k=1}^{P} \mu_i^k dn_i^k \tag{16}$$

where n_i^k is the number of moles and μ_i^k the molar partial potential of the i component in the k phase.

The condition of thermodynamic equilibrium at p and T reads:

$$\left.\begin{array}{l} \mu_A^\alpha = \mu_A^\beta = \mu_A^\gamma = \cdots \\ \mu_B^\alpha = \mu_B^\beta = \mu_B^\gamma = \cdots \\ \mu_C^\alpha = \mu_C^\beta = \mu_C^\gamma = \cdots \\ \quad\vdots \end{array}\right\} \quad \text{where} \quad \begin{array}{l} A, B, C, \ldots\text{: components} \\ \alpha, \beta, \gamma, \ldots\text{: phases} \\ \\ C(P-1) \text{ equations} \end{array}$$

For each phase:

$$x_i^k = \frac{n_i^k}{\sum_i n_i^k}$$

and therefore:

$$\left.\begin{array}{l} x_A^\alpha + x_B^\alpha + x_C^\alpha + \cdots = 1 \\ x_A^\beta + x_B^\beta + x_C^\beta + \cdots = 1 \\ \quad\vdots \end{array}\right\} \quad P \text{ equations}$$

Since the system in equilibrium is completely defined by the value of its intensive variables:

$$p, T, \left\{\begin{array}{l} x_A^\alpha, x_A^\beta, x_A^\gamma, \cdots \\ x_B^\alpha, x_B^\beta, x_B^\gamma, \cdots \\ x_C^\alpha, x_C^\beta, x_C^\gamma, \cdots \\ \quad\vdots \end{array}\right\} \quad 2 + CP \text{ variables}$$

Therefore, a multicomponent phase equilibrium has a number of degrees of freedom F (number of intensive variables that can be externally fixed) given by $2 + CP - C(P - 1) - P$. This result is known as the Gibbs phase rule:

$$F = C - P + 2 \tag{17}$$

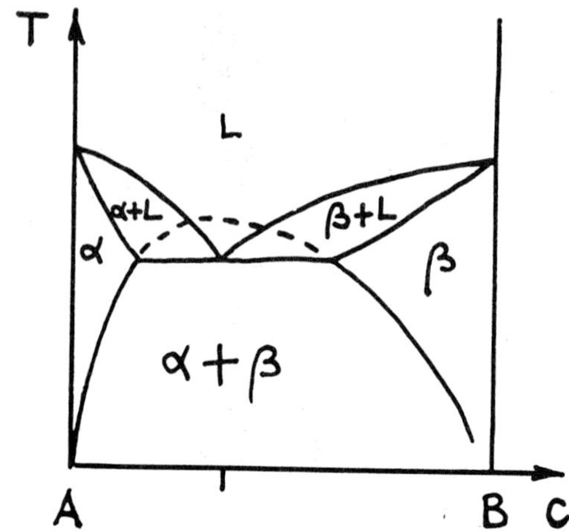

Figure 8: Phase diagram of an eutectic two-component system.

Example 1: Consider the simplest case of a pure substance (a single component system). For this system $x_A^\alpha = x_A^\beta = x_A^\gamma = \cdots = 1$.
(a) Three phases can simultaneously coexist in equilibrium only at fixed p and T (e.g. at the triple point shown in figure 3), because:

$$\left.\begin{array}{c} C = 1 \\ P = 3 \end{array}\right\} \quad F = 0 \Rightarrow p, T \text{ fixed}$$

(b) Two phases coexisting in equilibrium have only one intensive variable free, because:

$$\left.\begin{array}{c} C = 1 \\ P = 2 \end{array}\right\} \quad F = 1 \Rightarrow p = p(T)$$

The relations $p = p(T)$ define the three coexistence lines in the $p - T$ phase diagram. The slopes of these lines are given by equation (13).
(c) If only one phase is present, it can be found in a whole region of the $p - T$ diagram. Indeed:

$$\left.\begin{array}{c} C = 1 \\ P = 1 \end{array}\right\} \quad F = 2 \Rightarrow p, T \text{ free}$$

Example 2: Consider now a two-component system of atomic species A and B, which are completely miscible in the liquid state but not in the solid state. The phase diagram is usually presented in a plot of temperature T as a function of composition, at constant pressure. A typical example for a simple eutectic is shown in figure 8. L represents the liquid phase (the melt) and α, β two different solid phases; at all temperatures, α is richer than β in the atomic species A and viceversa.

(a) If only one phase is present:

$$\left.\begin{array}{l} C=2 \\ P=1 \end{array}\right\} \quad F=3$$

Since p has been fixed in constructing the diagram, one degree of freedom has been used already. The other two can be used for temperature and composition. Hence, single phases are found in regions of variable temperature and composition in the phase diagram.
(b) When two phases coexist in equilibrium:

$$\left.\begin{array}{l} C=2 \\ P=2 \end{array}\right\} \quad F=2$$

Again, p is already fixed, and hence there is only one remaining intensive variable that can be fixed arbitrarily: either T or the composition of one of the two phases. Consider, for example, the $\alpha + L$ region of the phase diagram (figure 8). If we choose T to be fixed at some value T_1, a horizontal line at constant T_1 (called tie-line) cuts the limits of the two-phase region at two points α_1 (on the left) and L_1(on the right). It follows from the phase rule that all points in this tie-line, within the $\alpha + L$ region, are mixtures in different proportions of an L phase and an α phase of fixed composition. Since L_1 represents the limit where the presence of solid phase in the mixture is negligible and α_1 represents this same limit for the liquid, and the two points belong to the tie-line, we conclude that the fixed compositions of the liquid and solid phases of all two-phase mixtures on the tie-line must be precisely L_1 and α_1.
(c) Finally, consider the coexistence of three phases in equilibrium. According to Gibbs' phase rule:

$$\left.\begin{array}{l} C=2 \\ P=3 \end{array}\right\} \quad F=1$$

The only freedom available has been used to select p. This means that the three phases coexist only at a single temperature (that of the eutectic line, the horizontal line in the phase diagram) and have fixed composition: the composition of the α phase is the composition of the left end of the eutectic line, where it reaches the α region; the composition of the L phase is the composition of the eutectic point, where the tie-line meets the L region; and finally the composition of the β phase is the composition of the right end of the tie-line, where it reaches the β phase.

2.5 Landau theory of a first-order phase transition

Let us consider a phase transition described by a single, scalar order parameter e (an uniaxial strain, say), such that $e = 0$ corresponds to the high temperature phase and $e = \pm e_s$ to two variants of the low temperature phase. The transition can be driven by temperature T and by an external field σ (an applied stress, say). We will use this particular problem to show the basics of a Landau theory of a first-order phase

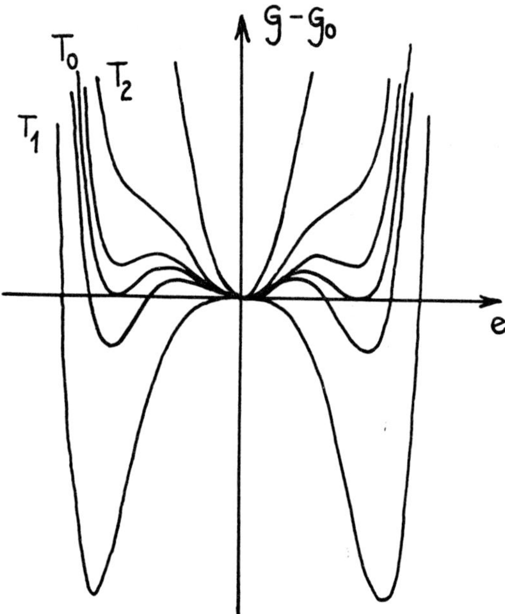

Figure 9: $\mathcal{G} - \mathcal{G}_0$ as a function of e, the order parameter, at different temperatures.

transition and the conclusions that can be derived from this kind of analysis. This section is largely based in [9]; for the interested reader, a complete presentation of Landau theory is given in reference [10].

2.5.1 In the absence of applied field ($\sigma = 0$)

A Gibbs free energy functional $\mathcal{G}(p,T,e)$ can be constructed as a power series in the order parameter. In this series, the odd terms are discarded for symmetry reasons (when $\sigma = 0$, replacing e for $-e$ should leave $\mathcal{G}$ invariant) and the signs are chosen to provide for three different minima. The functional reads:

$$\mathcal{G}(p,T,e) = \mathcal{G}_0(p,T) + Ae^2 - Be^4 + Ce^6 \tag{18}$$

where A carries the only temperature dependence, in the form $A(p,T) = a(T - T_1)$, and a, B, C, T_1 are positive constants. This choice leads to the family of curves shown in figure 9, in which four different temperature regions can be considered: (a) Above T_2 there is only one minimum, at $e = 0$; this is the only stable phase. (b) Between T_2 and T_0, the two other minima at $\pm e_s$ indicate that the low temperature phase can exist as a metastable phase. (c) Lowering the temperature between T_0 and T_1 makes the two minima at $\pm e_s$ deeper than the minimum at $e = 0$, so that $\pm e_s$ is the stable phase in this temperature region and $e = 0$ is metastable. (d) Below T_1, finally, the minimum at $e = 0$ becomes a maximum and this phase becomes unstable.

The equilibrium phase transition takes place at T_0. At this temperature the order parameter experiences a discontinuity given by $\pm\sqrt{B/2C}$, showing the first order character of the transition. From the discontinuity of the order parameter, the latent heat of transition $T_0\Delta S$ can be computed from the difference in entropy between the two phases at the transition temperature:

$$\left.\begin{array}{ll} T \geq T_0, & S = -\left(\frac{\partial G_0}{\partial T}\right)_p \\ T \leq T_0, & S = -\left(\frac{\partial G_0}{\partial T}\right)_p - ae^2 \end{array}\right\} \quad T = T_0, \quad \Delta S = -ae^2(T_0) = -a\frac{B}{2C} \tag{19}$$

Finally, the temperatures T_1 and T_2 define the upper and lower limits of metastability.

2.5.2 With applied field ($\sigma \neq 0$)

The simplest approach is to couple the external field linearly to the order parameter. The free energy functional reads now:

$$\mathcal{G}_\sigma(p,T,\sigma) = \mathcal{G}(p,T,e) - \sigma e \tag{20}$$

and the condition for a transition in equilibrium between two states e_a and e_b is given by:

$$\mathcal{G}_\sigma(p,T,\sigma)|_{e_a} = \mathcal{G}_\sigma(p,T,\sigma)|_{e_b} \tag{21}$$

This means that the equilibrium states are given by a double tangent construction on $\mathcal{G}(p,T,e)$ with slope σ.

The phase diagram of the thermally-induced transitions ($\sigma =$ constant $> 0, T$ variable) is shown in figure 10. The first noticeable role of σ is to break the $\pm e$ symmetry. In addition, there is a critical value σ_c such that, when $\sigma < \sigma_c$ (including $\sigma = 0$) the transition is still first-order and exhibits thermal hysteresis, when $\sigma = \sigma_c$ the transition is continuous and takes place at $T = T_c$, and when $\sigma > \sigma_c$ there is no thermally-induced transition.

In the presence of σ, the phase transitions can take place at constant temperature and be driven by the external field. The associated $\sigma - e$ trajectories can be obtained from a double tangent construction on $\mathcal{G}(p,T,e)$, as shown in figure 11 for some temperature T between T_0 and T_2. Since the relative stability between phases depends on temperature, the $\sigma - e$ trajectories (representing the response of the system to the external field) look very different at different temperatures. This is shown in figure 12. The $\sigma - e$ trajectories remind the stress-strain curves of a shape-memory crystal subjected to a tension-compression test at different temperatures [11]. Actually, the parallelism is not incidental: pseudoelasticity and shape-memory can be demonstrated in the same alloy, at different temperatures, because the thermomechanical properties of the alloy depend on the relative stability between the two phases involved in the structural transition (austenite and martensite).

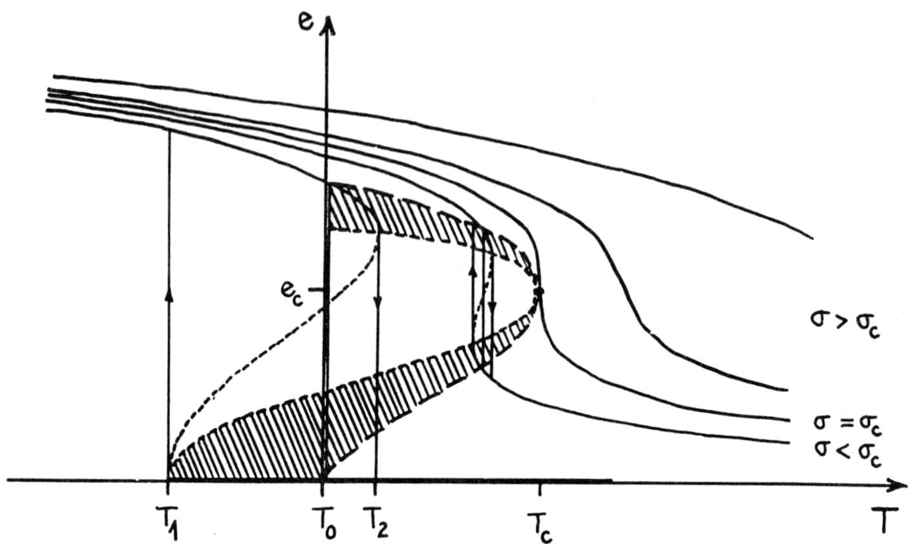

Figure 10: Phase diagram of the thermally-induced transitions at constant applied field.

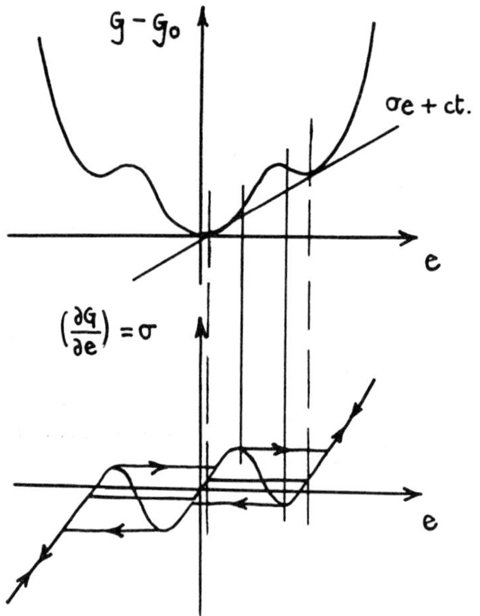

Figure 11: Derivation of $\sigma - e$ curves from a double-tangent construction on $\mathcal{G}$. Here $T_0 < T < T_2$.

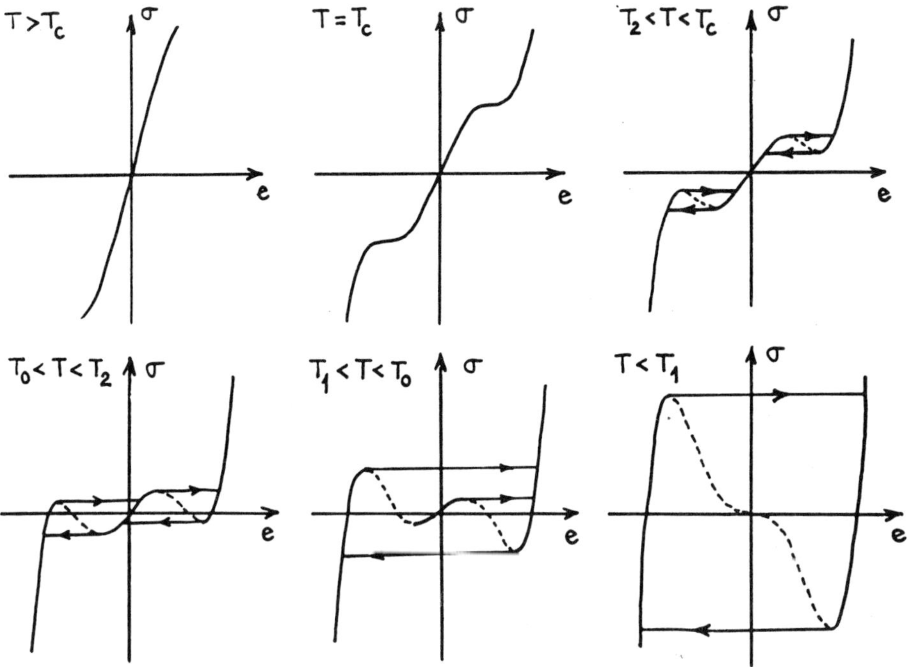

Figure 12: $\sigma - e$ trajectories at different temperatures.

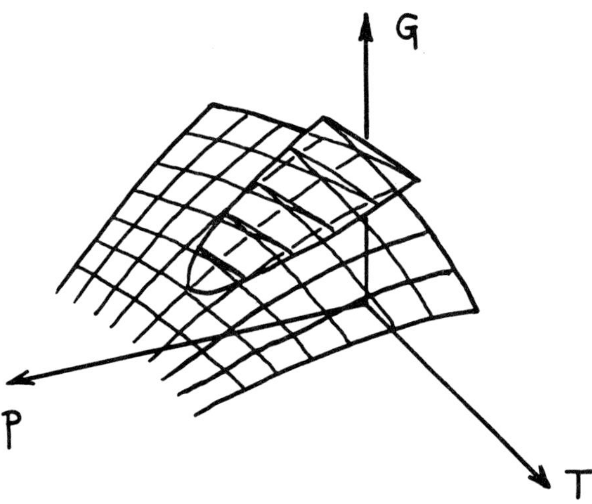

Figure 13: The Gibbs free energy of a pure substance as a function of pressure and temperature (schematic).

3 Continuous phase transitions

3.1 Gibbs free energy at a continuous phase transition

In this section we focus on the behaviour of a thermodynamic system in the neighbourhood of a critical point. A critical point is defined as a point in the $p - V - T$ (or equivalent) surface of the system where a phase transition takes place with no discontinuity in the first derivatives of the Gibbs free energy. For this reason, phase transitions at critical points are called continuous. Figure 13 shows schematically the Gibbs free energy surfaces (given by the different minima of the functional $\mathcal{G}$) corresponding to two different phases of a pure substance; the two surfaces intersect at a line of first-order phase transitions, which ends precisely at the critical point, where they merge continuously into each other.

3.2 Critical points and critical exponents

Response functions and other properties of the system exhibit divergences and interesting behaviour near a critical point, which are the subject of this section.

Consider first the different projections of the phase diagram of a hydrostatic system (a system described by $p - v - T$ variables), such as a pure substance. Instead of v, we choose this time the density ρ as extensive variable ($\rho \sim 1/v$). The projections are presented in figure 14, where p_c, T_c and ρ_c are the state variables at the critical point. Note that the critical isotherm displays an inflection at the critical point.

We have chosen ρ as a state variable to show the strong resemblance of the resulting

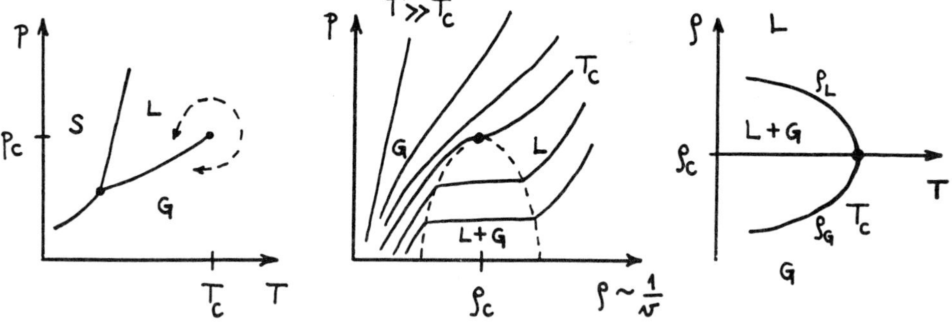

Figure 14: Family of phase diagrams of a hydrostatic system.

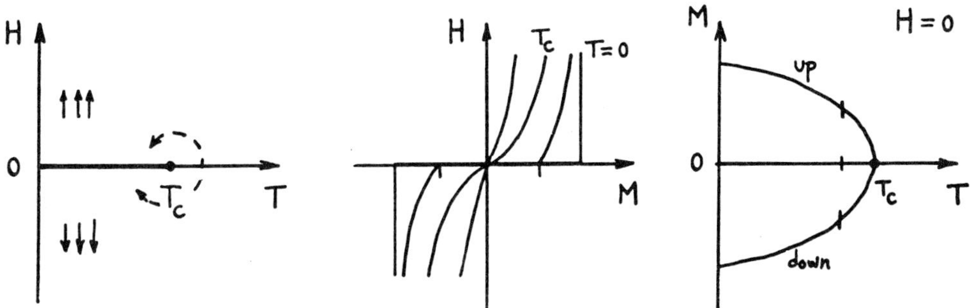

Figure 15: Family of phase diagrams of a magnetic system.

projections of the phase diagram with the equivalent projections of a magnetic system that undergoes a paramagnetic-ferromagnetic transition. Those are shown in figure 15. For $H = 0$, the stable phase above T_c (known as Curie point) is paramagnetic, i.e. there is no net alignment of the magnetic moments of the ions. At T_c, however, as a result of mutual interactions between the magnetic moments, the system starts to magnetize spontaneously and becomes ferromagnetic. Lowering the temperature below T_c makes the magnetization to increase, until it reaches its saturation value at $T = 0$, when thermal agitation ceases. If the system presents a strong magnetic anisotropy, there are two opposite directions of easy magnetization (up or $\uparrow$ and down or $\downarrow$); the system finally magnetizes along one or the other depending on some random fluctuation at the beginning of the transition. Hence, a symmetry of the system is spontaneously broken at the transition. The presence of an external magnetic field, $H \neq 0$, eliminates the phase transition. Now the system tends to magnetize in the direction of the field at all temperatures.

The thermodynamic behaviour at a critical point is characterized quantitatively by the precise values of several exponents. These *critical exponents* describe the nature of the singularities in some selected measurable magnitudes at a continuous transition. Thus, the critical exponent λ for the magnitude F is obtained from the relation:

$$F(t) \sim |t|^{\lambda} \qquad \text{as} \qquad t \to 0 \tag{22}$$

where t is a reduced temperature defined by:

$$t \equiv \frac{T}{T_c} - 1 \tag{23}$$

The symbol $\sim$ in equation (22) means that F, in addition to a regular part that remains finite, has a singular part that either diverges or has divergent derivatives and is proportional to a power λ (generally fractional) of t.

The first four critical exponents for a fluid system are defined as follows:

$$\begin{aligned} \text{Heat capacity at constant volume } v_c\text{:} &\quad C_v \sim |t|^{-\alpha} \\ \text{Liquid-gas density difference:} &\quad (\rho_L - \rho_G) \sim (-t)^{\beta} \\ \text{Isothermal compressibility:} &\quad \kappa_T \sim |t|^{-\gamma} \\ \text{Critical isotherm } (t=0)\text{:} &\quad p - p_c \sim |\rho_L - \rho_G|^{\delta}\, sgn(\rho_L - \rho_G) \end{aligned} \tag{24}$$

Their equivalent for a magnetic system are:

$$\begin{aligned} \text{Zero-field heat capacity:} &\quad C_H \sim |t|^{-\alpha} \\ \text{Zero-field magnetization:} &\quad M \sim (-t)^{\beta} \\ \text{Zero-field isothermal susceptibility:} &\quad \chi_T \sim |t|^{-\gamma} \\ \text{Critical isotherm } (t=0)\text{:} &\quad H \sim |M|^{\delta}\, sgn M \end{aligned} \tag{25}$$

The exponent β describes the behaviour of the order parameter in the neighbourhood of the transition point and the exponent δ the shape of the critical isotherm. α and γ, on their hand, describe the behaviour of the response functions.

To be complete but avoid details that would go beyond the scope of this introductory review, I should mention that there are two other exponents, ν and η, defined in the same way for fluid and magnetic systems, which concern a quantity called the correlation function. The reader interested in learning more about this may refer to [4].

3.3 Exponent inequalities

Thermodynamics restricts the possible values of the critical exponents for a given system. These restrictions are expressed as exponent inequalities. They are derived from the convexity properties of the free energy for a physical system and from reasonable assumptions about the behaviour of the thermodynamic variables or correlation functions. A complete derivation is given in [3].

The first inequality (derived here for a magnetic system) follows from the thermodynamic relation:

$$\chi_T \left(C_H - C_M \right) = T \left(\frac{\partial M}{\partial T} \right)_H^2$$

Since $C_M \geq 0$ and (by definition) $C_H \sim (-t)^{-\alpha}$, $\chi_T \sim (-t)^{-\gamma}$ and $(\partial M/\partial T)_H \sim (-t)^{\beta-1}$ when $H = 0$ and $t \to 0^-$, it follows that:

$$\alpha + 2\beta + \gamma \geq 2 \tag{26}$$

which is known as Rushbrooke's inequality. The other three inequalities, due to Widom, Fisher and Josephson respectively, read:

$$\gamma \geq \beta\,(\delta - 1) \tag{27}$$

$$\gamma \leq \nu\,(2 - \eta) \tag{28}$$

$$d\nu \geq 2 - \alpha \tag{29}$$

Note that the last inequality involves the space dimensionality d.

Experimental data, and models that can be solved exactly or numerically, show that the inequalities actually hold as equalities. A theoretical explanation of this remarkable observation was provided by the Renormalization Group approach to critical phenomena, for which K.G.Wilson was awarded the Nobel Prize in Physics in 1982 [12].

3.4 Universality

Systems exhibiting critical behaviour can be grouped in universality classes, each class characterized by a set of values of critical exponents. From the large amount of experimental data available and from extensive work on statistical models of continuous phase transitions, it is found that systems of a very different nature can be grouped in a reduced number of universality classes. The classes differ only in:

- the space dimensionality d
- the symmetry of the order parameter
- the range of interaction.

Once different systems are grouped by classes, it is customary to use the simplest statistical model in each class to label the class and to compute the critical exponents for all the systems within the class. Universality classes and the values of their critical exponents are compiled in Table 1.

The exponent β for three-dimensional systems governed by an scalar order parameter provides a nice example of universality. In 1945 Guggenheim [13] showed already that the coexistence curve of many different fluid systems, plotted in reduced units (i.e. reescaled by their values at the critical point), collapse into a universal curve (figure 16) and that, in the neighbourhood of the critical point, the behaviour is described by a critical exponent $\beta = 1/3$. Heller and Benedek [14] in 1962 measured the magnetization of MnF_2, a magnet with uniaxial magnetic anisotropy, and obtained $\beta = 0.335$. Finally, Thompson and Rice [15] in 1964 studied the phase separation of the binary fluid mixture $CCl_4 + C_7F_{16}$ and found $\beta = 0.33$. Numerical estimates of β for the $d = 3$ Ising model, representative of the same universality class, give $\beta = 0.33$ (Table 1).

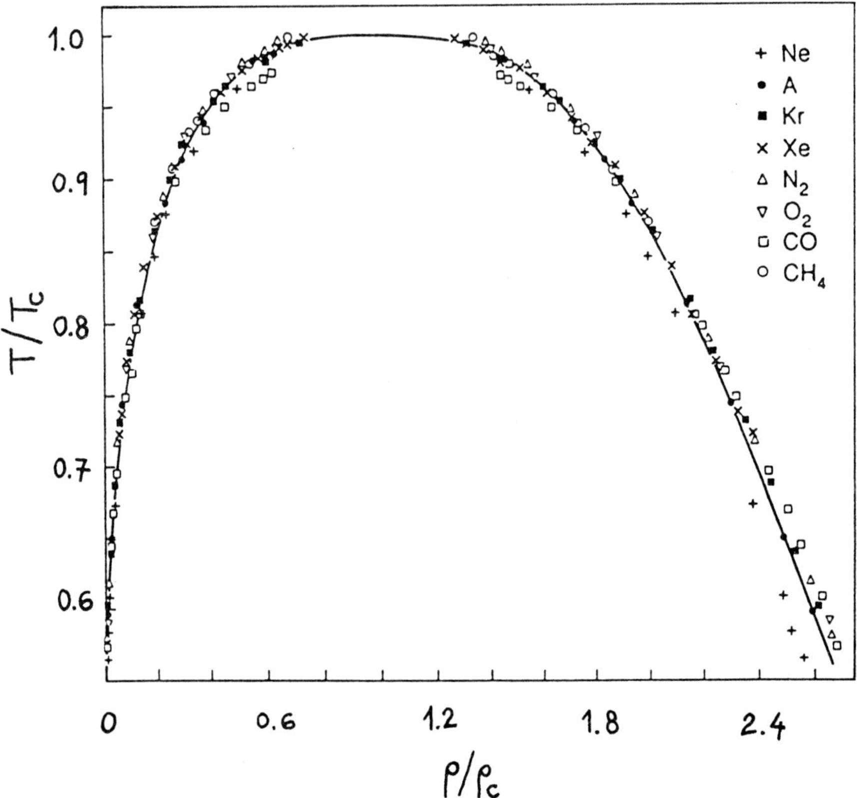

Figure 16: Guggenheim's plot of the reduced temperature vs. reduced density in the gas-liquid coexistence region, for eight different substances.

Universality class and physical examples	Symmetry of order parameter	α	β	γ	δ	ν	η
2-d Ising Adsorbed monolayers, e.g. H on Fe	2-compt scalar	0 (log)	1/8	7/4	15	1	1/4
3-d Ising Phase separation and order-disorder trans.	2-compt scalar	0.10	0.33	1.24	4.8	0.63	0.04
3-d X-Y Superfluids and superconductors	2-dim vector	0.01	0.34	1.30	4.8	0.66	0.04
3-d Heisenberg Isotropic magnets	3-dim vector	−0.12	0.36	1.39	4.8	0.71	0.04
Mean-Field		0 (dis)	1/2	1	3	1/2	0
2-d Potts, $q = 3$	q-compt scalar	1/3	1/9	13/9	14	5/6	4/15
2-d Potts, $q = 4$ Adsorbed monolayers, e.g. Kr on graphite	q-compt scalar	2/3	1/12	7/6	15	2/3	1/4

Table 1: Exponents and universality classes

3.5 Landau theory of a continuous phase transition

We want to describe a continuous, paramagnetic-ferromagnetic transition, in the framework of a simple Landau theory. We consider a lattice of magnetic ions whose magnetic moments can be found in two possible orientations, ↑ (up) and ↓ (down). At high temperatures the magnetic moments can be found indistinctly in any of the two orientations because of thermal agitation; the average magnetic moment in the system (the magnetization) is $M = 0$, and the system is paramagnetic. At low temperatures, mutual interaction between magnetic moments prevails over thermal disorder, making the moments to align preferably in one of the two possible orientations; then $M > 0$ (for orientation up) or $M < 0$ (for orientation down), and the system is ferromagnetic. The phase transition between the paramagnetic and the ferromagnetic phase takes place at some intermediate temperature T_c at which the first spontaneous magnetization sets in. Notice again that the spontaneous magnetization of the system is accompanied by spontaneous breaking of the up-down symmetry.

3.5.1 In the absence of external field ($H = 0$)

The simplest Landau free-energy functional that accounts for a continuou phase transition between a paramagnetic phase ($M = 0$) and a degenerated ferromagnetic phase ($\pm M \neq 0$) can be written as:

$$\mathcal{F} = \mathcal{F}_0 + a_2 m^2 + a_4 m^4 \tag{30}$$

where m is the magnetization density ($m = M/V$), a_2 is a function of the reduced temperature t of the form $a_2 = \tilde{a}_2 t$, and $\tilde{a}_2$, a_4 are positive constants. Odd powers of m have been eliminated to preserve the $\pm$ symmetry of the order parameter m. At temperatures such that $a_2 > 0$ the functional F has the shape of a single-well potential centered around $m = 0$. On lowering the temperature we reach $a_2 = 0$, for which the minimum of the well becomes very flat. Thereafter, for $a_2 < 0$, the minimum at $m = 0$ turns continuously into a maximum and two symmetric minima appear at $\pm m$.

The magnetization density m in the vicinity of the transition point, as a function of the reduced temperature t, is given by the minima of $\mathcal{F}$:

$$\frac{\partial \mathcal{F}}{\partial m} = 2\tilde{a}_2 t m + 4 a_4 m^3 = 0$$

For $t < 0$ we have $m \sim (-t)^{1/2}$. We conclude that the exponent for the magnetization density (the order parameter) is $\beta = 1/2$.

Above the transition point, $t > 0$, $m = 0$ and $\mathcal{F} = \mathcal{F}_0$. Below the transition point, $t < 0$, $m^2 = -\tilde{a}_2 t / 2a_4$ and thus $\mathcal{F} = \mathcal{F}_0 - \tilde{a}_2^2 t^2 / 4a_4$. The specific heat, therefore, experiences a discontinuity at $t = 0$, which implies that the critical exponent α is $\alpha = 0$.

3.5.2 In the presence of external field ($H \neq 0$)

In the simplest approximation, the external field couples linearly to the order parameter. The Landau functional reads then:

$$\mathcal{F} = \mathcal{F}_0 - Hm + \tilde{a}_2 t m^2 + a_4 m^4 \tag{31}$$

The magnetic susceptibility at constant temperature, χ_T, can be derived from the equilibrium values of m:

$$\frac{\partial \mathcal{F}}{\partial m} = 0 = -H + 2\tilde{a}_2 t m + 4 a_4 m^3 \quad \rightarrow \quad 0 = -dH + 2\tilde{a}_2 t\, dm + 12 a_4 m^2 dm \quad \rightarrow$$

$$\chi_T = \left(\frac{\partial m}{\partial H}\right)_t = \frac{1}{2\tilde{a}_2 t + 12 a_4 m^2} \quad \rightarrow \quad h = 0 \begin{cases} t > 0, & m^2 = 0, \quad \chi_T = \frac{1}{2\tilde{a}_2 t} \\ t < 0, & m^2 = -\frac{\tilde{a}_2 t}{2a_4}, \quad \chi_T = -\frac{1}{4\tilde{a}_2 t} \end{cases}$$

so that:

$$\frac{\chi_T(t \rightarrow 0^+)}{\chi_T(t \rightarrow 0^-)} = 2$$

and the critical exponent is $\gamma = 1$.

Finally, from the isotherms given by the state equation:

$$\frac{\partial \mathcal{F}}{\partial m} = 0 = -H + 2\tilde{a}_2 t m + 4a_4 m^3$$

we derive the behaviour of the critical isotherm $t = 0$:

$$4a_4 m^3 = H$$

and obtain the exponent $\delta = 3$.

Comparing these results with the values of the exponents in Table 1 we conclude that the Landau theory belongs to the universality class of Mean-Field theories.

4 Spin models of systems undergoing phase transitions

Universality classes of critical phase transitions are represented by reticular models. A reticular model is a system of interacting particles on a lattice, well suited to be studied by the methods of statistical mechanics, both theoretical and numerical. In defining the model one looks for the simplest system that properly takes into consideration the distinctive properties of the universality class: space dimensionality, symmetry of the order parameter and range of particle interaction. Reticular models are usually defined in the language of magnetism: the interacting particles represent the magnetic moments (spins) of magnetic ions on a periodic lattice, the order parameter is the magnetization, and the system can interact with an external magnetic field. Nevertheless, we will see that spin models can map other phase transitions, such as order-disorder and phase separation.

Since the equilibrium properties of reticular models have been studied in great detail and are in many cases well understood, it is natural to start from these same models to study non-equilibrium phenomena, such as domain growth and avalanches, near a phase transition. This is why, in the last two decades, spin models have been increasingly used to study the kinetics of first-order phase transitions. The last section of these notes deals with this kind of problems.

4.1 The Ising model

The Ising model is one of the simplest spin models that can be formulated. Historically, Onsager's exact solution of the model in two dimensions (2-d) [16] started a whole revision of our understanding of continuous phase transitions: it showed that the behaviour near a critical point was not well described by existing mean-field theories (the Van der Waals theory for fluids and the molecular field theory due to Curie and Weiss for magnetic systems) and, even more surprising, this behaviour changed with system dimensionality. Actually, it was very discouraging for Ising himself to find out that his model has no phase transition in 1-d. Nevertheless, in 2-d and 3-d (a case for which there is not yet an exact solution) the model presents a continuous phase transition with different sets of exponents in each case. Interestingly, if we go to 4-d and higher dimensions the model still presents a continuous phase transition but the critical behaviour falls in the mean-field universality class.

To define the model, consider the N lattice sites of a periodic lattice in d-dimensions. On each site there is a magnetic moment or spin, which can be in any of two orientations: $\uparrow$ (up) or $\downarrow$ (down). To each lattice site we assign an index i and a spin variable S_i, which takes the value $+1$ when the spin points upwards and -1 when it points downwards. A microscopic configuration of the system is given by the set of N numbers $\{S_i\}$. Every spin interacts with its q nearest-neighbours and, eventually, with an external field H. The energy of a configuration is given by:

$$\mathcal{H} = -J \sum_{<ij>} S_i S_j - H \sum_i S_i \tag{32}$$

Here, $< ij >$ is a convention to indicate that the summation extends only to nearest-neighbours and thus contains $qN/2$ terms. J is the spin-spin interaction energy, assumed isotropic. The interaction can be chosen to be ferromagnetic ($J > 0$), favouring that all spins take the same orientation, or antiferromagnetic ($J < 0$), favouring that all spins take the orientation opposite to its nearest neighbours.

The free energy of the system $F'(H,T)$ can be written as:

$$F'(H,T) = -k_B T \cdot \ln Q(H,T) \tag{33}$$

where k_B is Boltzmann's constant and $Q(H,T)$ is the classical partition function, given by:

$$Q(H,T) = \sum_{S_1} \sum_{S_2} \ldots \sum_{S_N} e^{-\mathcal{H}\{S_i\}/k_B T} \tag{34}$$

Other thermodynamic properties like the internal energy U, the heat capacity C_H and the state equation $M(H,T)$, follow from $F'(H,T)$:

$$U = -k_B T^2 \frac{\partial}{\partial T} \left(\frac{F'}{k_B T} \right)_H \tag{35}$$

$$C_H = \left(\frac{\partial U}{\partial T} \right)_H \tag{36}$$

$$M(H,T) = -\left(\frac{\partial F'}{\partial H} \right)_T \tag{37}$$

It is not difficult, using the last equation, to realize that the macroscopic magnetization of the system is the statistical average of $\sum_i S_i$ over all possible microscopic configurations compatible with H and T, i.e. $M(H,T) =< \sum_i S_i >$ with $<>$ representing an ensemble average. $M(0,T)$ is the spontaneous magnetization of the system.

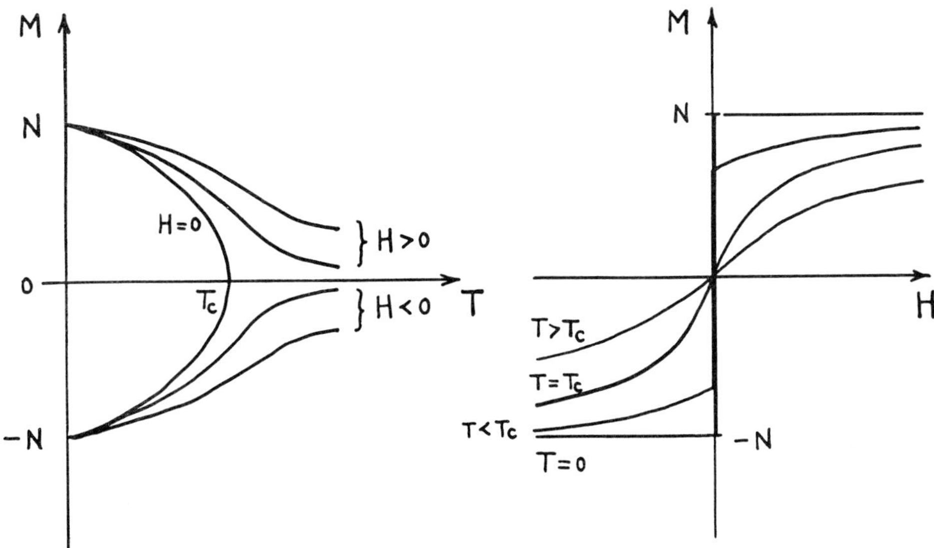

Figure 17: Phase diagrams of the ferromagnetic Ising model.

4.2 Phase diagrams of the Ising model

The $M-T$ and $M-H$ phase diagrams of the 2-d and 3-d ferromagnetic Ising model are reproduced qualitatively in figure 17. Since these diagrams are very similar to the phase diagrams of a real magnetic system (figure 15), we conclude that the model reproduces adequately the phenomenology of a real paramagnetic-ferromagnetic transition.

Consider first the $M-T$ diagram. In the absence of an applied magnetic field ($H=0$), the spontaneous magnetization $M(0,T)=0$ above the critical temperature (paramagnetic phase) and becomes $M(0,T)\neq 0$ below the critical temperature (ferromagnetic phase), reaching a saturation value $+N$ or $-N$ at $T=0$, at which all the spins align in the same direction. The phase diagram is symmetric around the horizontal axis because the two directions of space (up and down) are equivalent when $H=0$. An applied field $H\neq 0$ has a profound effect on the system of spins: it breaks the spatial symmetry, favouring the spin alignment in the direction of the field; in these conditions the system does not magnetize spontaneously and the phase transition disappears.

The $M-H$ diagram shows the isotherms of the model, i.e. the evolution of the macroscopic magnetization when the applied field is ramped from $-\infty$ to $+\infty$ at constant temperature. Above T_c the isotherms are typical for a paramagnet: M is proportional to H at moderate fields (Curie's law) and goes to the magnetization saturation value ($-N$ and $+N$) for extreme fields. The isotherm for the critical temperature T_c presents an inflection point at $H=0$ and, as a consequence, the isothermal susceptibility diverges, a signature of the paramagnetic-ferromagnetic continuous phase transition that takes place at this point. Below T_c, the magnetization curves behave rather differently: increasing the field from below, the system gradually loses its (negative) magnetization, up to $H = 0$. At this point the preferred spatial orientation dictated by the applied field disappears, and the sign of the magnetization becomes irrelevant: the system transits in equilibrium from negative to positive magnetization. It is worth noting that the transition at $H = 0$ (for all $T < T_c$) is first-order because the order parameter M experiences a discontinuity, but takes place without latent heat because the two phases involved have the same entropy. Finally, increasing again the field, from $H = 0$, results in a gradual increase of the magnetization up to its saturation value.

Before closing this section, let us remark that the model exhibits both a first-order and a continuous phase transition. A first-order transition is found for any temperature below T_c, at $H = 0$, on ramping the field from $-\infty$ to $+\infty$. A continuous phase transition takes place at $T = T_c$ when the temperature of the system is ramped with $H = 0$. Table 2 presents numerical estimates of T_c and the critical exponent β, for 3-d Ising models built on different lattice structures. Note that the critical temperature is not of universal character.

Crystal lattice	$k_B T_c/J$	β
Simple cubic	0.2216	0.327
bcc	0.1574	0.327
fcc	0.1021	0.327

Table 2: Some numerical estimates for the 3-d Ising model.

4.3 Mean-field solution of the Ising model

In spite of its apparent simplicity, the Ising model has only been solved exactly in $d = 2$. The difficulties lie in the summation over microscopic configurations involved in the calculation of $Q(H,T)$, given by equation (34). These difficulties can be overcome by a rather crude approximation in which the interaction of every magnetic moment with its nearest neighbours is replaced by the interaction with an average magnetic moment resulting from the overall magnetization of the lattice (molecular field). This general procedure is called the mean-field approximation.

4.3.1 Paramagnet ($J = 0$)

As a first step, consider a paramagnetic system, i.e. a system of non-interacting magnetic moments ($J = 0$). The Hamiltonian of the system reads:

$$\mathcal{H} = -H \sum_i S_i \tag{38}$$

and the partition function is given by:

$$Q(H,T) = \sum_{S_1} \sum_{S_2} \dots \sum_{S_N} \exp\left[+\beta H \sum_i S_i\right] \tag{39}$$

where $\beta \equiv (k_B T)^{-1}$. The behaviour of the paramagnet is controlled by the competition between the magnetic field H, which tends to align the spins in one direction, and the thermal disorder represented by $k_B T$.

Since the spins do not experience mutual interactions, the total energy of the paramagnet is simply the sum of energies of individual spins. This allows an exact solution of the problem. We have:

$$Q(H,T) = \sum_{S_1} \sum_{S_2} \cdots \sum_{S_N} \prod_i \exp[\beta H S_i] = \prod_i \left[\sum_{S_i = \pm 1} \exp[\beta H S_i] \right] \equiv \prod_i [Q_1(H,T)]$$

where the single spin partition function is simply:

$$Q_1(H,T) = \sum_{S_i=\pm 1} \exp[\beta H S_i] = e^{+\beta H} + e^{-\beta H} = 2\cosh[\beta H]$$

The free energy of the paramagnet is now:

$$F'(H,T) = -k_B T \ln Q(H,T) = -N k_B T \ln Q_1(H,T)$$

and hence the magnetization can be computed as:

$$M(H,T) = -\left(\frac{\partial F'}{\partial H}\right)_T = N k_B T \frac{\partial}{\partial H} \ln\cosh[\beta H] = N k_B T \frac{\sinh[\beta H]}{\cosh[\beta H]} \beta$$

resulting in:

$$M(H,T) = N \tanh \frac{H}{k_B T} \tag{40}$$

This is the state equation of the paramagnet. In the limit $T \to 0$ (equivalent to $H \to \pm\infty$) the magnetization saturates to $\pm N$. In the opposite limit, $T \to \infty$ (equivalent to $H \to 0$), the magnetization follows Curie's law of paramagnetism: $M \sim N \frac{H}{k_B T}$.

4.3.2 Ferromagnet (mean field)

The exact solution of the paramagnet turns out to be very helpful to derive a solution of the ferromagnetic Ising model in the mean-field approximation. In this approximation the Hamiltonian of the model, given by equation (32), can be rewritten as:

$$\mathcal{H} = -J \sum_i S_i \sum_{\substack{j \\ j|<ij>}} S_j - H \sum_i S_i \simeq -J \sum_i S_i \gamma \frac{M}{N} - H \sum_i S_i = -\left[J\gamma \frac{M}{N} + H\right] \sum_i S_i \tag{41}$$

where the notation $j \; |< ij >$ means that the sum for j is carried over the γ nearest neighbours of the spin i. This sum represents the total magnetization of the γ nearest neighbours. The mean field approximation is taken in the second step, when this magnetization (which depends on the particular configuration of the spin system) is replaced by γ times M/N, the average magnetization per spin. Within this approximation, it turns out that the Hamiltonian given by equation (41) looks the same as that of a paramagnet, subjected to an effective magnetic field given by $\left[J\gamma\frac{M}{N}+H\right]$. Thus, using the results obtained for the paramagnet:

$$M(H,T) = N\tanh\left[\frac{J\gamma\frac{M}{N}+H}{k_BT}\right] \tag{42}$$

This is the state equation of the ferromagnetic Ising model in the mean-field approximation.

In the absence of applied field ($H=0$), the system is paramagnetic at high temperatures and becomes ferromagnetic (i.e. magnetizes spontaneously) when its temperature reaches a critical value T_c. To see this, consider independently the two members of the state equation when $H=0$:

$$y_1 = M$$

$$y_2 = N\tanh\left[\frac{J\gamma\frac{M}{N}}{k_BT}\right]$$

At high temperatures these two functions intersect only at the origin $M=0$. Decreasing T, the slope of y_2 at the origin approaches that of y_1, till both become the same when:

$$T_c = \frac{J\gamma}{k_B} \tag{43}$$

Below T_c the two functions intersect not only at $M=0$ (which actually becomes an unstable state) but also at two other values $\pm M_s$ (the two equivalent stable states with spontaneous magnetization). Concerning the critical temperature, it should be noticed that $J\to 0$ (the non-interacting limit) leads correctly to $T_c\to 0$, and also that the critical temperature depends on the coordination number γ.

The critical behaviour is best characterized by the values of the critical exponents. Since $M\sim 0$ near T_c, we can use

$$\tanh x \underset{x\sim 0}{\sim} x - \frac{1}{3}x^3 + \cdots$$

to find that

$$M \sim NJ\gamma\frac{M}{N}\frac{1}{k_BT} - \frac{1}{3}N\left(J\gamma\frac{M}{N}\frac{1}{k_BT}\right)^3 = \frac{T_c}{T}M - \frac{1}{3}\left(\frac{T_c}{T}\right)^3\frac{M^3}{N^2}$$

and finally

$$\frac{M}{N} \sim \sqrt{3}\left(\frac{T}{T_c}\right)\left(1-\frac{T}{T_c}\right)^{\frac{1}{2}}$$

which results in an exponent $\beta=\frac{1}{2}$ for the magnetization. The behaviour of the magnetic susceptibility near T_c is obtained by taking an implicit derivative of equation

(42) with respect to the field:

$$\left(\frac{\partial M}{\partial H}\right)_T \sim N\frac{\frac{J\gamma}{N}\left(\frac{\partial M}{\partial H}\right)_T + 1}{k_B T}$$

so that

$$\left(\frac{\partial M}{\partial H}\right)_T \sim \frac{\frac{N}{k_B T}}{1 - \frac{J\gamma}{k_B T}} = \frac{N}{k_B(T - T_c)}$$

and therefore $\gamma = 1$. Finally, the critical behaviour of the heat capacity at constant field can be derived from

$$C_H = T\frac{1}{\chi_T}\left(\frac{\partial M}{\partial T}\right)_H^2$$

For $H = 0$ we have

$$\begin{array}{ll} T > T_c & C_H = 0 \\ T < T_c & C_H = \frac{3}{2}Nk_B\left\{1 - \mathcal{O}\left(1 - \frac{T}{T_c}\right) + \cdots\right\} \end{array}$$

which means that C_H experiences a discontinuity at the transition point and the associated exponent is $\alpha = 0$. Presumably you have already noticed that these exponents coincide with those for the Landau theory of a continuous phase transition presented before. Actually it can be shown that a mean field approximation of the Ising model, in a continuous limit, leads to a polynomial free energy functional of the Landau kind [5].

The mean-field approximation is not only elegant and simple. It reveals already that the collective behaviour at the phase transition originates in the mutual interaction between magnetic moments. Yet, its quantitative predictions (transition temperature and critical exponents) do not agree with the ones obtained experimentally in magnetic systems and fluids, or derived from numerical simulations of the Ising model (see Table 1). The problem originates in the fact that a mean-field approximation disregards correlations in the fluctuations of the magnetic moments, though these correlations diverge at the transition. This aspect of the problem is properly taken into consideration by the techniques of the Renormalization Group [6].

4.4 Mapping of an order-disorder transition to the Ising model

Let us focus on the order-disorder transition in stoichiometric β -brass, a Cu-Zn alloy with a bcc structure (figure 18). We denote with I each of the eight lattice sites at the corners of the cubic unit cell, and with II the site at the center of the cell. The ordering transition at T_c can be characterized by the probability that an I site is occupied by a Zn atom:

$T > T_c$	$\text{Prob}[I = \text{Zn}] = \frac{1}{2}$	Disordered state
$T < T_c$	$\text{Prob}[I = \text{Zn}] > \frac{1}{2}$	Partial B2 ordering
$T = 0K$	$\text{Prob}[I = \text{Zn}] = 1$	Complete B2 ordering

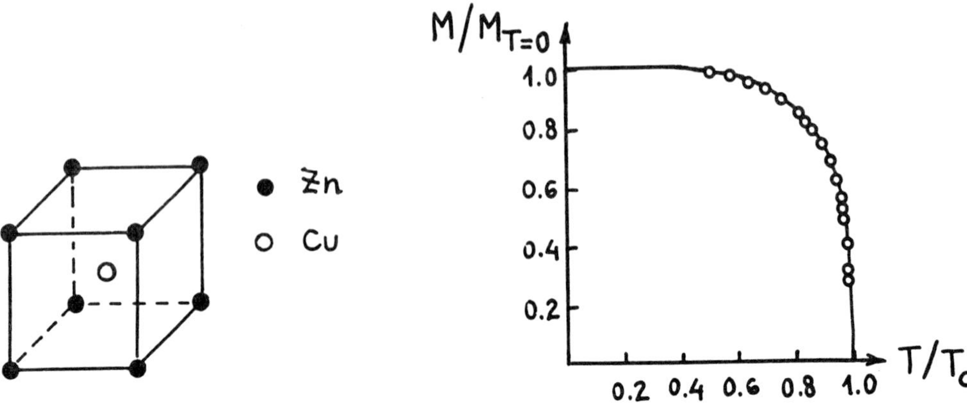

Figure 18: Unit cell of a bcc CuZn alloy (left), and the behaviour of the order parameter with temperature in this alloy (right).

The disorder is substitutional, not topological (the lattice remains). The order parameter can be defined as the difference in the number of Cu and Zn atoms on the sublattice II. Figure 18 shows neutron scattering measurements of the order parameter $M/M_{T=0}$ as a function of the reduced temperature T/T_c (dots) with $T_c = 733K$, compared to the prediction of a compressible Ising model (solid line).

The Ising model in a cubic lattice can be mapped to the order-disorder transition in β-brass in the following way: assign $S_i = +1$ to a site i occupied by a Cu atom, and $S_i = -1$ to a site i occupied by a Zn atom. Then the energy of the system reads

$$\mathcal{H} = \frac{1}{4} \sum_{<ij>} J_{CuCu}(1 + S_i)(1 + S_j) + \frac{1}{4} \sum_{<ij>} J_{ZnZn}(1 - S_i)(1 - S_j)$$

$$+ \frac{1}{4} \sum_{<ij>} J_{CuZn} \left\{(1 + S_i)(1 - S_j) + (1 - S_i)(1 + S_j)\right\} \tag{44}$$

which is identical to the Ising hamiltonian

$$\mathcal{H} = -J \sum_{<ij>} S_i S_j - H \sum_i S_i + C$$

with $J = \frac{1}{4}(J_{CuCu} + J_{ZnZn} - 2J_{CuZn})$, $\sum_i S_i = 0$ (stoichiometry), and C a spin independent term.

The use of an Ising variable to describe the system is justified from the fact that each lattice site is strictly in one of two states (Cu or Zn) when we neglect the presence of vacancies or impurities. The model reproduces the actual critical properties of the system accurately, as shown in table 3. Other properties may depend on details of the atomic interactions, and may lead to the need of considering further-neighbour

	β	γ
Experimental	0.305 ± 0.005	1.24 ± 0.015
3d-Ising	0.33	1.24

Table 3: Universal properties of the order-disorder transition in stoichiometric β-brass.

interactions or multi-spin terms of the kind $S_iS_jS_k$. For β -brass, in particular, it has been found necessary to incorporate a variation of J with T, which results from the thermal expansion of the lattice.

5 Phase transitions in solids: dynamic aspects

Up to this point, we have been considering the equilibrium properties of physical systems at phase transitions. Very often, however, real systems undergo phase transitions in non-equilibrium conditions, either during their preparation (ordering and precipitation following a quench from a stable region of the phase diagram, for example) or during their operation (formation of magnetic domains in a magnetic recording material by application of external magnetic fields, for example). The dynamics associated with these non-equilibrium processes is technologicaly relevant: it is determinant of the microstructure of the material, responsible for many of its technological properties.

The dynamics of phase transitions in solids is one of the most interesting topics of a more general discipline, the physics of non-equilibrium phenomena in complex systems. Here, we restrict ourselves to review two different problems:

- First, the kinetics of domain growth of a new phase that follows a quench through a first-order phase transition. Usually the domain growth is assumed to take place not too far from thermodynamic equilibrium, and the dynamics is governed by thermal fluctuations.

- Second (on the other extreme) the dynamics of systems driven by an external field, when they present a multiplicity of metastable states separated by high energy barriers. The evolution is practically insensitive to thermal fluctuations, and takes place by avalanches, far from equilibrium.

Both problems have been the subject of numerous studies in the last years. My aim here is simply to review the main concepts and models used in these studies, and present some of the most important results. Most of this section is based on reference [7] for the subject of domain growth, and on reference [8] for the subject of hysteresis and avalanches. You should refer to these works for additional information and an extensive list of references.

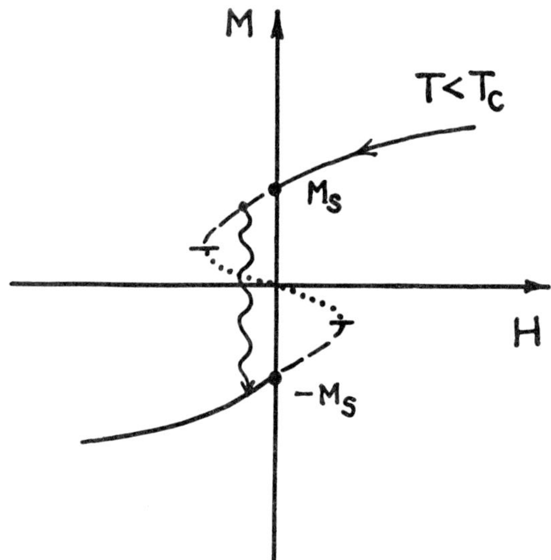

Figure 19: Non-equilibrium evolution of a ferromagnetic system (wiggling line) following a sudden decrease of magnetic field at constant temperature.

5.1 Domain growth

The generic problem to be considered is the following: a system is in thermodynamic equilibrium with its environment when, suddenly, the environmental conditions change and the system finds itself out of equilibrium. While it approaches the new equilibrium state, the system undergoes a first-order phase transition. The focus is on the rate at which the phase transition will take place, and the morphologies of the resulting microstructure.

5.1.1 Systems with non-conserved order parameter

The description corresponds to the kind of process undergone by a ferromagnetic system which, initially, is in a ferromagnetic state at $T < T_c$, subjected to $H > 0$, such that most of the magnetic moments point in the direction of the field and the magnetization is $M > 0$. The field is suddenly reversed to $H < 0$. The state $M > 0$ becomes metastable, and the system makes a non-equilibrium isothermal evolution to the true stable state $M < 0$, as shown schematically in figure 19.

Alternatively, the ferromagnetic system can be prepared in a uniform paramagnetic state at $(H = 0, T > T_c)$, and then suddenly quenched to a temperature $T < T_c$ (figure 20). Since the paramagnetic state is unstable in this region of the phase diagram, the system experiences a non-equilibrium evolution to the new stable state at $(H = 0, T)$.

Equivalent routes may be formulated for any problem that can be mapped to the ferromagnetic transition, such as the growth of ordered domains in systems undergoing

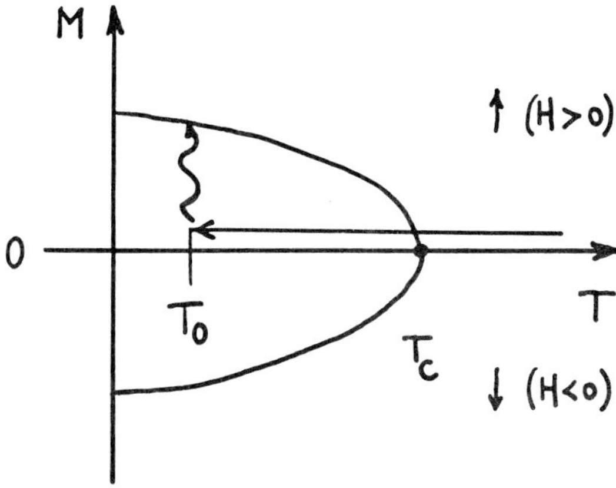

Figure 20: Non-equilibrium evolution of a paramagnet (wiggling line) following a sudden quench at zero field into the ferromagnetic phase.

an order-disorder transition, or the grain growth of polycrystalline materials. A common feature to these problems, of important kinetic consequences, is that all of them are described by an order parameter which is *not conserved* in the phase transition. The magnetization $M = \sum_i S_i$ (for example) is not conserved, in the sense that a spin S_i can flip from -1 to $+1$ without another spin having to flip simultaneously in the opposite direction.

5.1.2 Systems with conserved order parameter

The system to consider is a binary solid (or liquid) solution of atomic species A and B, which undergoes a phase separation into an α phase (rich in A) and a β phase (rich in B). A typical phase diagram for this system has been already presented in figure 8. There is a variety of non-equilibrium situations that can be considered in this diagram: phase separation in the $\alpha+L$ or $\beta+L$ regions (solid+liquid) leads to problems of cellular growth and dendritic growth, while phase separation in the $\alpha+\beta$ region (solid+solid) leads to problems of Ostwald ripening. The order parameter suitable to describe the process of phase separation is the scalar c, the composition, given by the average concentration of B atoms. Note that the composition is *conserved*, already at a local level.

To study the problem from a theoretical point of view, the real phase diagram is usually simplified in the form shown in figure 21. The solid line is the coexistence curve, and the broken line (which separates metastable and unstable states) is the spinodal. Initially the system is prepared at a temperature T, far above the temperature of the eutectic, as a uniformly mixed solution of given concentration c_0. The system is then instantaneously quenched to a temperature T_0. Since the point (c_0, T_0) lies

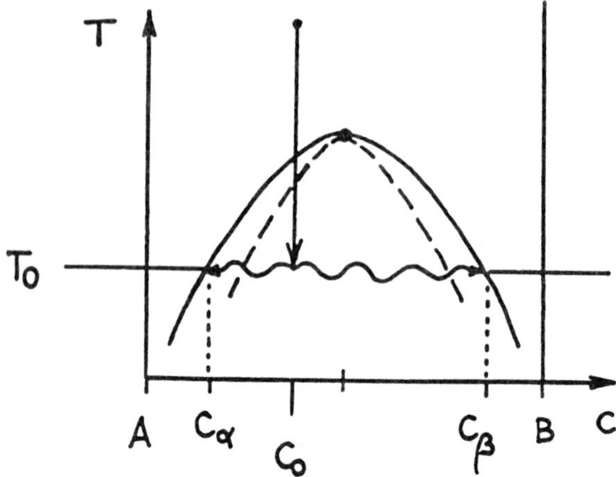

Figure 21: Simplified version of an eutectic phase diagram.

inside the two-phase region, the uniformly mixed solution is not stable, and the system undergoes a non-equilibrium phase separation, which ultimately will take it to the true equilibrium state at T_0: a mixture of coexisting regions of the α phase (with equilibrium concentration c_α) and of the β phase (with equilibrium concentration c_β).

The kinetics of phase separation depends strongly on whether the system is quenched in the metastable region ($c_0 \simeq c_\alpha$) or it is quenched in the unstable region (c_0 near the eutectic). The different behaviour is shown schematically in the one-dimensional cuts of the microstructure presented in figure 22 [17]. The phase separation following a quench in the metastable region proceeds by nucleation and growth: droplets of the β-phase precipitate slowly out of the slightly supersaturated α-phase. The phase separation following a quench in the unstable region is a relatively fast process, called spinodal decomposition, in which α-like and β-like regions appear in approximately equal proportions.

5.1.3 A continuum model of the Ising ferromagnet (non-conserved order parameter)

Our purpose here is to derive a continuum model of the Ising ferromagnet which (i) bridges the gap between the atomic level description provided by the model and its macroscopic behviour in equilibrium, and (ii) can be extended naturally to situations not too far from equilibrium and thus can be used to study the kinetics of phase transitions.

We evaluate the partition function of the Ising model using the following *coarse-graining* procedure.

First, as shown in figure 23, we divide the system into blocks of linear size $l \gg a$ (where a is the original lattice spacing), centered at positions $\vec{r}$. The order parameter of

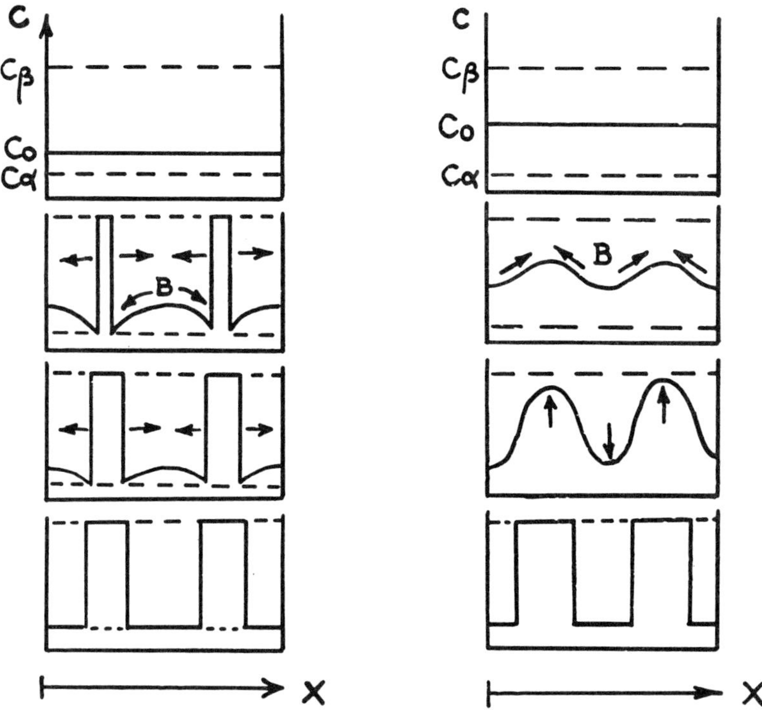

Figure 22: Temporal evolution of the concentration profile, from top to bottom, after a quench in the metastable region (left) and in the unstable region (right) of the phase diagram. Schematic.

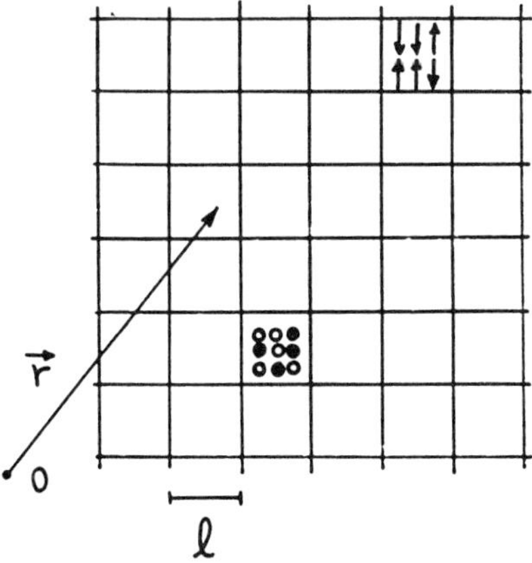

Figure 23: Scheme of the coarse-graining procedure.

the ferromagnet, now a function of position $\vec{r}$ and time t, is the magnetization density

$$m(\vec{r},t) = \frac{1}{l^d} \sum_{i \in \text{block at } \vec{r}} S_i \tag{45}$$

where d is the lattice dimensionality. For the procedure to make sense, we assume that (i) $m(\vec{r},t)$ varies smoothly on scales smaller than l, and (ii) equilibration of the system on length scales smaller than l is much faster than the processes that we wish to describe. Based on this second assumption, we can compute the partial partition function

$$Z_l(\{m\}, H) = \sum_{\{S\}} e^{-E\{S\}/k_BT} \tag{46}$$

with the sum constrained to spin configurations $\{S\}$ consistent with $m(\vec{r},t)$. The free energy of the coarse-grained system, given by

$$F_l(\{m\}, H) = -k_BT \ln Z_l(\{m\}, H), \tag{47}$$

adopts a Ginzburg-Landau form:

$$F_l(\{m\}, H) = \int d\vec{r} \left[\frac{1}{2} K (\nabla m)^2 + f_l(m) - Hm + \cdots \right] \tag{48}$$

where $f_l(m)$ is a Landau free energy functional of the local order parameter m.

Second, the thermodynamic free energy is computed as the functional integral

$$\overline{f}(m) = -k_BT \lim_{V \to \infty} \frac{1}{V} \ln \int \mathcal{D}_m e^{-F_l\{m\}/k_BT} \tag{49}$$

taken over all the remaining configurations $m(\vec{r})$ consistent with a macroscopic magnetization

$$m = \frac{1}{V} \int d\vec{r} m(\vec{r}).$$

Figure 24 shows schematically the form of $f_l(m)$ and $\overline{f}$ at $H = 0$. While the former is computed at a mesoscopic scale l and does not need to be convex, the latter results from taking the thermodynamic limit in equation (49) and therefore must be a convex function of m.

This approach admits a natural extension to situations not too far from thermodynamic equilibrium. The aim is to write a nonequilibrium equation of motion for the local order parameter, in the limit of a purely dissipative system for which inertial effects can be neglected. From a purely phenomenological point of view, the local rate of displacement of the order parameter can be taken proportional to the local thermodynamic force in this limit, so that:

$$\frac{\partial m(\vec{r})}{\partial t} = -\Gamma \frac{\delta F}{\delta m(\vec{r})}. \tag{50}$$

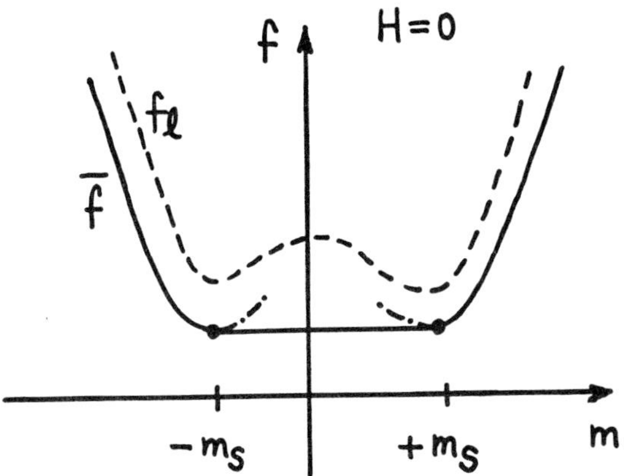

Figure 24: The coarse-grained free energy $f_l(m)$ and the thermodynamic limit $\overline{f}(m)$ for an Ising model.

Γ is a response coefficient which defines a time-scale for the system. Using for F (the local coarse-grained free energy) the form given by the integrand in equation (48), we finally obtain:

$$\frac{\partial m}{\partial t} = -\Gamma \left[-K\nabla^2 m + \frac{df}{dm} - H \right] . \tag{51}$$

We apply now this equation to study the motion of an spherical droplet of spins $\uparrow$ in a sea of spins $\downarrow$, in the presence of an applied magnetic field which favours the $\uparrow$ phase. The equation reads:

$$\frac{\partial m}{\partial t} = \Gamma \left[K \frac{\partial^2 m}{\partial r^2} + K \frac{d-1}{r} \frac{\partial m}{\partial r} - \frac{df}{dm} + H \right] .$$

We skip all details of the calculation, which can be found in reference [7], and give directly the solution:

$$v = \frac{dR}{dt} = \Gamma K \, (d-1) \left(\frac{1}{R^*} - \frac{1}{R(t)} \right) . \tag{52}$$

R^* is the critical radius for nucleation of the droplet, given by:

$$R^* = \frac{d-1}{2} \frac{\sigma}{m_s H}$$

where σ is the equilibrium surface tension and m_s is the magnetization of the $\uparrow$ phase. It follows from equation (52) that droplets of radius $R < R^*$ have $v < 0$ and collapse, while droplets of radius $R > R^*$ have $v > 0$ and grow. $R = R^*$ coresponds to the unstable stationary solution, with $v = 0$. Finally note that when $H = 0$ the critical radius $R^* \to \infty$, so that all droplets of finite size collapse in the absence of a favouring applied field.

The problem of domain coarsening can also be addressed using equation (51). We consider the quench of a magnetic system from the paramagnetic state to some temperature T below the critical temperature T_c, at $H = 0$ (so that phases $\uparrow$ and $\downarrow$ are equally likely to appear). Immediately after the quench the system will exhibit a pattern of interpenetrating domains of the two phases, which will coarsen in time. Since the two phases have the same energy, all the excess energy is located at the interfaces: the coarsening process is driven by the tendency to reduce the interfacial energy.

We will assume that, after some transient, the pattern consists of spherical domains of one phase embedded in a background of the other phase. Actually, although lines around domains do not cross, the domain pattern is more intrincate: domains are not spherical in general and there are domains inside domains. These details are not essential for our discussion, however, since experiments and simulations have shown that the coarsening process is scale invariant. We can associate the radius R of a spherical domain to the typical radius of curvature of the two-phase interface, which represents the characteristic length scale of the process.

For $H = 0$, equation (52) reads:

$$\frac{dR}{dt} = -\Gamma K \, (d-1) \, \frac{1}{R}$$

from which we conclude that:

- $dR/dt < 0$, ie all convex domains shrink.
- $dR/dt \sim 1/R$, so that the largest domains shrink most slowly.
- The only domains that remain in the system at time t are those with $R(t=0) > R_0$, when R_0 is given by:

$$\left. \begin{array}{l} R^2(t) = R_0^2 - 2\Gamma K(d-1)t \\ R(t) = 0 \end{array} \right\} \quad \rightarrow \quad R_0 = [2\Gamma K(d-1)t]^{\frac{1}{2}}$$

  This result implies that the characteristic size of domains follows a scaling law of the form $R_0 \sim t^{1/2}$. This law has been extensively confirmed by numerical simulations, and observed experimentally, for example, in the growth of antiphase domains in ordered Fe_3Al alloys [18]. It applies generically to systems with a scalar, non-conserved order parameter.

5.1.4 A continuum model of a binary solid solution (conserved order parameter)

Consider now a system with a conserved order parameter, typically a binary solid solution. To derive a continuum model for this system we use again a coarse-graining procedure, which as before starts by defining a mesoscopic scale of length l, much larger than the atomic spacing a. The scalar order parameter of the problem is the local concentration $c(\vec{r}, t)$, function of position $\vec{r}$ and time t, defined as the average

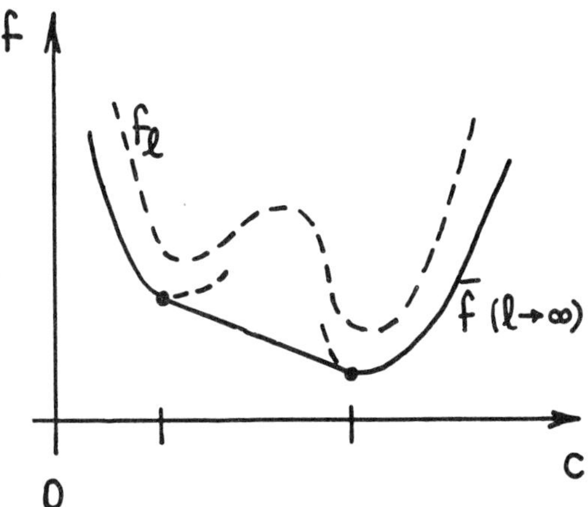

Figure 25: The coarse-grained free energy $f_l(c)$ and the thermodynamic limit $\overline{f}(c)$ for a binary alloy.

number of B atoms per unit volume in a block of length l. With the same assumptions than before, the partial partition function reads:

$$Z_l = \sum_{\text{constr.conf.}} e^{-E/k_BT} \tag{53}$$

where the sum is constrained to microscopic configurations consistent with $c(\vec{r})$. The coarse-grained free energy is given by:

$$F_l\{c\} = -k_BT \ln Z_l \tag{54}$$

and can be expected to have again the form:

$$F_l\{c\} = \int d\vec{r} \left[\frac{1}{2}K\,(\nabla c)^2 + f_l(c)\right] \tag{55}$$

where f_l is the free energy at scale l, which does not need to be symmetric (figure 25). Note also in equation (55) that there is no external field applied. Finally, the thermodynamic free energy $\overline{f}(c)$ is obtained as:

$$\overline{f}(c) \equiv -k_BT \lim_{V\to\infty} \frac{1}{V} \ln \int \mathcal{D}_c e^{-F_l\{c\}/k_BT} \tag{56}$$

where the functional integral is taken over all the remaining configurations $c(\vec{r})$ which are smooth on scale l and are consistent with a macroscopic concentration

$$c = \frac{1}{V}\int d\vec{r} c(\vec{r}).$$

Once the thermodynamics of the equilibrium states of the system has been written in the continuum, the approach can be extended to slightly non-equilibrium situations, such as the phase separation that follows a quench from a uniform state at some high temperature, outside the coexistence region, into the coexistence region. Our purpose is to derive a thermodynamic equation of motion for the order parameter $c(\vec{r},t)$. Since the concentration of A and B atoms is locally conserved, the transport mechanism to be considered is atomic diffusion. Let $\vec{j}\,(\vec{r})$ denote the flux of B atoms, which is subjected to the local conservation condition:

$$\frac{\partial c}{\partial t} = -\vec{\nabla} \cdot \vec{j}. \tag{57}$$

The flux $\vec{j}$ is driven by a thermodynamic force $\vec{\nabla}\mu$, where μ represents the difference in chemical potential between the two atomic species ($\mu \equiv \mu_B - \mu_A$), given by

$$\mu(\vec{r}) = \frac{\delta F}{\delta c(\vec{r})}. \tag{58}$$

When not too far from equilibrium, fluxes and forces can be assumed to be linearly related:

$$\vec{j}(\vec{r}) = -M(c)\vec{\nabla}\mu \tag{59}$$

where M is a mobility, that can be concentration dependent. Finally, from equations (55), (57), (58) and (59), we find the equation of motion for the concentration:

$$\frac{\partial c(\vec{r},t)}{\partial t} = \vec{\nabla} \cdot \left[M(c) \cdot \vec{\nabla} \left(-K\nabla^2 c + \frac{\partial f}{dc} \right) \right], \tag{60}$$

known as the Cahn-Hilliard equation. This equation becomes a simple diffusion equation in the limit where $M(c)$ is a constant, the term $-K\nabla^2 c$ is neglected, and the equation is linearized about some uniform concentration c_0.

Now, in analogy to the development of the preceding section, the Cahn-Hilliard equation can be used to describe the motion of an spherical droplet of the β phase immersed in supersaturated α phase, and study the corresponding domain coarsening problem. In the early stages of phase separation it is found that droplets grow independently, and the regime is characterized by a growth law of the form $R \sim t^{1/2}$. In the later stages of growth, the coarsening process is dominated by competition between droplets, and the scaling law changes to the Lifshitz-Slyozov-Wagner form $R \sim t^{1/3}$. The derivation of these important results is omitted here for reasons of conciseness.

5.2 Hysteresis and avalanches

A large variety of physical systems, particularly in the solid state, experience first-order phase transitions which happen to be nearly insensitive to thermal fluctuations. These *athermal* transitions are usually observed in systems with a complex free energy landscape, formed by a multiplicity of metastable states separated by energy barriers typically much larger than $k_B T$, the energy of thermal fluctuations. Since thermal fluctuations are not operative on reasonable time scales, the energy required to drive

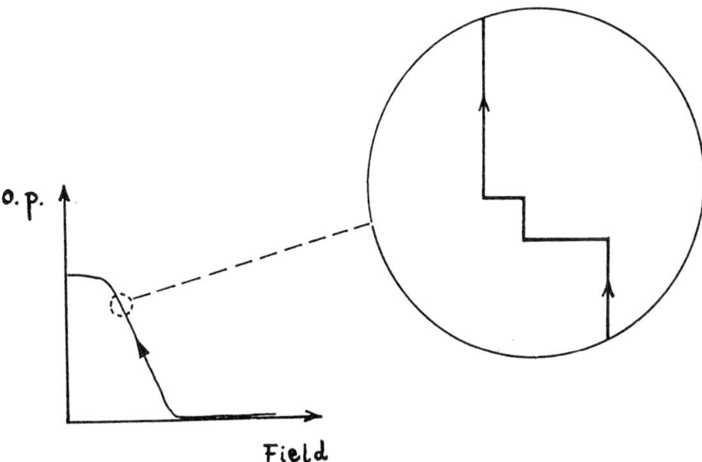

Figure 26: Schematic transformation path, obtained by plotting the order parameter as a function of the applied field. The magnification reveals an evolution by avalanches.

these systems through the phase transition must be provided externally, by increasing or decreasing the appropriate applied field.

Two important examples of athermal first-order phase transitions are provided by ferromagnetic systems driven by a magnetic field, and by martensitic materials driven by either temperature or a stress field. Actually, for example, the stability of magnetic recording relies on the fact that magnetic tapes are insensitive to thermal fluctuations at ordinary temperatures. The athermal dynamics is at the origin of many striking similarities observed in the behaviour of these two types of systems, particularly in the hysteresis cycle, obtained by plotting the order parameter against the driving field in the forward and reverse directions of transformation. The properties of the cycle are found to be independent of time (at least for moderate rates of driving) but strongly dependent on history.

A close-up examination of the hysteresis cycle (figure 26) reveals that, upon sweeping the applied field, the system goes from metastable to metastable state through avalanches. Before an avalanche starts, the system is trapped in a metastable state and the order parameter does not respond to the change of the field. At some point, however, the driving force provided by the field becomes sufficient to resume the phase transition in one part of the system. This local change, on its turn, may trigger new transformation events in other parts of the system, due to mutual interactions between different regions. The process goes on, as an avalanche, until the system gets locked in a new (long lived) metastable state, and a finite change of the field is again required to proceed with the transition. From a macroscopic point of view, every avalanche results in an abrupt change of the order parameter. In magnetic systems avalanches give rise to magnetization jumps, usually recorded as Barkhausen jumps. In martensitic materials avalanches release elastic waves that are detected as acoustic emissions (AE) in the ultrasonic range.

Remarkably, this kind of dynamics presents two distinct time scales: one associated with sweeping the field, and another associated with the avalanches. To all purposes, avalanches are instantaneous on the time scale of variation of the field, as shown schematically in figure 26.

5.2.1 The random field Ising model at $T = 0$

Rather recently, Sethna *et al.* [8] have shown that almost all relevant features of athermal first-order phase transitions are nicely captured by the random field Ising model (RFIM) at $T = 0$. The idea is to describe a first-order phase transition using an Ising model subjected to an external field H, and incorporate a modification in which every spin S_i experiences a static field h_i, distributed at random on the lattice. The random fields are introduced as a form of static disorder, and give rise to a multiplicity of metastable states. The energy of a spin configuration reads, for this model:

$$\mathcal{H} = -J \sum_{<ij>} S_i S_j - H \sum_i S_i - \sum_i h_i S_i \tag{61}$$

Since the transitions to describe are not influenced by thermal fluctuations, the model is simulated at $T = 0$.

The evolution of the spin configuration is dictated by a sequential minimization of the local energy. To start from an equilibrium configuration, we choose $H \ll 0$, for which all the spins point in the $\downarrow$ direction. Then we increase the field H, until the first *internal field*

$$F_i = J \sum_j S_j + H + h_i, \tag{62}$$

experienced by the spin S_i, changes its sign. At this moment S_i flips to the $\uparrow$ direction. Notice that the term $J \sum_j S_j$, which comes from the interaction of a spin with its nearest neighbours, enables a spin flip to trigger new spin flips in its neighbourhood without changing the field H, giving rise to a spreading avalanche at constant external field. When the avalanche arrests, the field is increased again until a new F_i changes sign and makes the corresponding S_i to flip, and eventually trigger a new avalanche. The process is iterated until all spins have flipped to the $\uparrow$ direction.

Usually, the random fields $\{h_i\}$ are chosen to follow a Gaussian distribution, with mean 0 and half-width R. A very narrow distribution of fields around 0 will make most spins to flip simultaneously, and the transformation to be nearly completed in a single, giant avalanche. Conversely, a wide distribution of random fields produces a transformation resolved in multiple, small scale avalanches. The transition from one kind of behaviour to the other can be mapped to a true critical phase transition, in the sense that at some critical value of R the model presents avalanches at all scales.

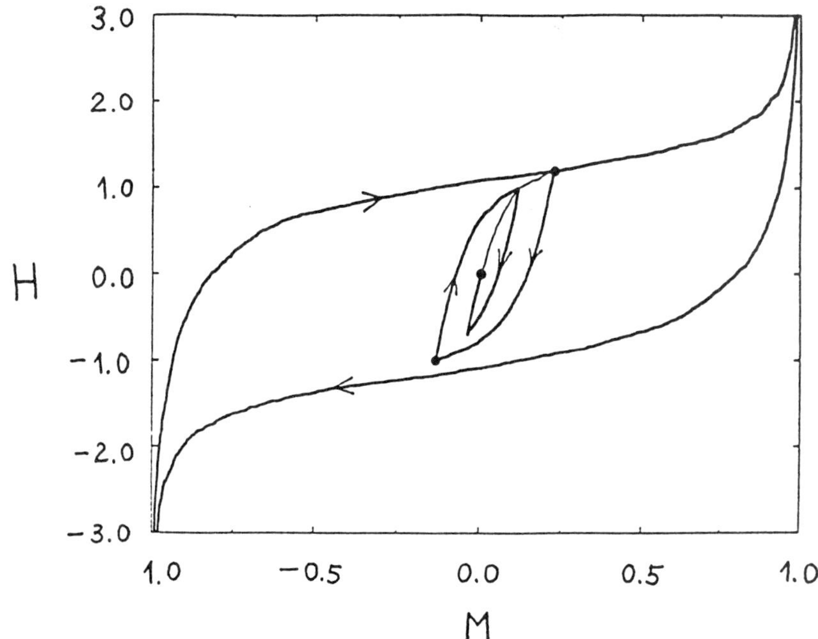

Figure 27: Hysteresis cycle of a $3-d$ RFIM of 30^3 spins, $J = 1$ and $R = 3.5J$, simulated at $T = 0$.

This kind of disorder-induced, non-equilibrium criticality of the RFIM at $T = 0$ has stimulated a large amount of research on the statistical distribution of avalanches, in both magnetic and martensitic systems [19].

It is worth noting that the dynamics described above, based on a criteria of *local* minimization, makes the system to follow a sequence of metastable spin configurations, instead of the sequence of equilibrium spin configurations that would result from an *overall* minimization of $\mathcal{H}$. As a consequence, when the field is swept back from $H \gg 0$ to $H \ll 0$ and the system follows the same dynamics, the order parameter does not retrace the same trajectory in the opposite direction. Instead, a hysteresis cycle is formed, as shown in figure 27 [8]. The area inside the cycle is a measure of the energy dissipated in the cyclic nonequilibrium evolution. In addition, the trajectories that can be traced inside the hysteresis cycle (resulting from partial cyclic excursions of the driving field H) reproduce the remarkable memory properties that are often observed in ferromagnets and martensitic materials: (i) the trajectories depend on the past history, through the values of the field H at which the direction was reversed (return-point memory), and (ii) the influence of a return-point vanishes when the trajectory rejoins this point, after an odd number of reversions (wipping out). Figure 27 presents examples of these two properties.

The return-point memory and the wipping-out property of the RFIM at $T = 0$ can be explained in full detail [8]. The explanation illuminates the intrinsically nonequilibrium dynamic behaviour of the model, and helps to understand the physical origin of the memory properties of the hysteresis cycle in real systems. It goes through the following steps:

- A partial ordering of spin configurations can be defined in the following way: let N be the number of lattice sites and let $\vec{s} = \{s_1, \ldots, s_N\}$, $\vec{r} = \{r_1, \ldots, r_N\}$ represent two different spin configurations. Then:

$$\vec{s} \geq \vec{r} \Leftrightarrow s_i \geq r_i \quad \forall i$$

  The ordering is only partial: many pairs of spin configurations do not have a definite relationship.

- Suppose that $\vec{s}(t)$ represents the sequence of configurations followed by the model under a driving $H_s(t)$, and $\vec{r}(t)$ the sequence under a driving $H_r(t)$. Then:

$$\left.\begin{array}{l} \vec{s}(0) \geq \vec{r}(0) \\ H_s(t) \geq H_r(t) \end{array}\right\} \Rightarrow \vec{s}(t) \geq \vec{r}(t)$$

  This means that the partial ordering is preserved by the dynamics. To prove it, remember that the internal field experienced by a spin S_i is given by $F_i = J \sum_j S_j + H + h_i$. For $\vec{r}(t) > \vec{s}(t)$, there should be a first time t for which $F_i^r > F_i^s$ for some spin S_i. But this is not possible, since $H_s(t) \geq H_r(t)$ and all the neighbours of S_i still preserve the ordering $\vec{s}(t) \geq \vec{r}(t)$.

- The dynamics are adiabatic: any monotonic evolution of $H(t)$ from H_A to H_B takes the system from a given configuration $\vec{s}(0) \equiv A$ to a configuration B, through a unique sequence of spin configurations. In other words, the path followed by the system is not influenced by the rate at which the system is being driven. Or, equivalently, avalanches are instantaneous on the time scale of the driving.

- The model exhibits return-point memory (figure 28), in the sense that:

$$\begin{array}{lll} t = 0, & \vec{s}(0) = A, & H(0) = H_A \text{ (initial state)} \\ & & H \text{ monotonically changes to } H(1) \\ t = 1, & \vec{s}(1) = B, & H(1) = H_B \text{ (final state)} \\ & & H \text{ varies back and forth, within the interval } (H_A, H_B) \\ t = T, & & H(T) = H_B \Rightarrow \vec{s}(T) = B \end{array}$$

  To prove this memory property, let $\vec{s}_{min}(t)$ represent the sequence of configurations of the model when it is driven by $H_{min}(t) = min_{t' \geq t} H(t')$, and $\vec{s}_{max}(t)$ the sequence when it is driven by $H_{max}(t) = max_{t' \leq t} H(t')$ (figure 29). Initially, let $\vec{s}_{min}(0) = \vec{s}_{max}(0) = A$. Since $H_{min}(t) \leq H(t) \leq H_{max}(t)$, and the dynamics preserve the partial ordering of configurations, we have $\vec{s}_{min}(t) \leq \vec{s}(t) \leq \vec{s}_{max}(t)$. But, since both $H_{min}(t)$ and $H_{max}(t)$ change monotonically between $t = 0$ and $t = T$, and the dynamics are adiabatic, we have $\vec{s}_{min}(T) = \vec{s}_{max}(T) = B$. In conclusion, $\vec{s}(T) = B$.

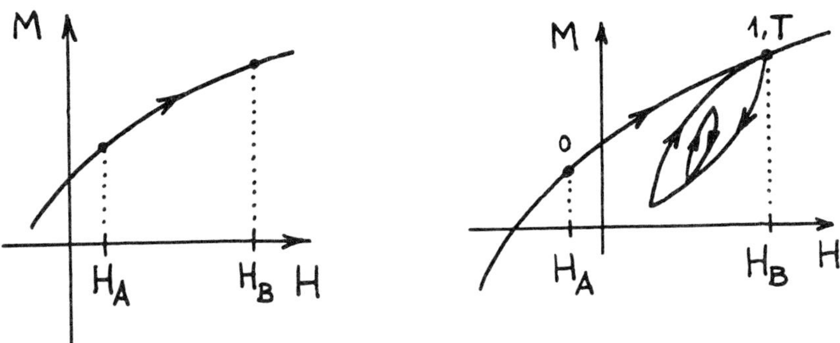

Figure 28: $M - H$ trajectories, displaying return-point memory.

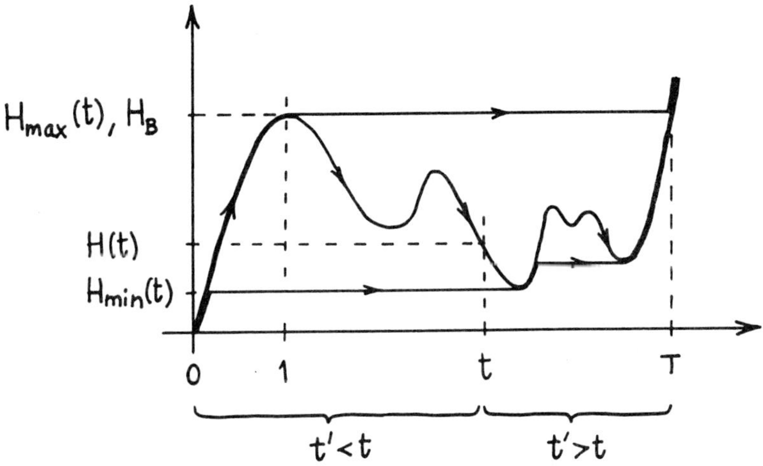

Figure 29: Temporal evolution of the applied field H and the corresponding monotonic evolution of the fields $H_{min}(t)$ and $H_{max}(t)$.

We have mentioned previously that the RFIM at $T = 0$ captures many essential features of the athermal first-order phase transitions in real systems. Figure 30 is an example. The two hysteresis cycles in the figure correspond to the martensitic transformation of the same copper-based alloy, and were obtained in an ordinary strain-controlled tensile test at room temperature. The first one was recorded after subjecting the crystal to a direct quench (a thermal treatment that is acknowledged to retain a large number of defects in this kind of materials) followed by a large number of stress-strain cycles. The second one was recorded after a slow quench (air cooling) of the crystal. The strikingly different behaviour can be attributed to the different amount of disorder quenched in the system after the two thermal treatments. A high defect concentration gives rise to transformation trajectories made up of a large number of minute avalanches, and a hysteresis cycle that extends on a wide range of the driving field. Internal trajectories display an amazingly perfect return-point memory. For much lower defect concentrations, on the other extreme, the transformation trajectories are nearly horizontal, reflecting that all the transformation takes place nearly at a single value of the stress field. Unfortunately, the test is carried out under conditions of strain-control (the most usual procedure), which prevent the transformation from taking place in a few giant transformation events. Instead, the load fluctuates wildly (in response to the strain changes of the crystal that cannot be accommodated by the tensile machine) and the memory properties of internal trajectories become difficult to analyze.

References

[1] Zemansky, M.W. and Dittmann, R.H.: Heat and Thermodynamics, McGraw-Hill, London 1981.

[2] Callen, H.B.: Thermodynamics, John Wiley and Sons, New York 1960.

[3] Stanley, H.E.: Introduction to Phase Transitions and Critical Phenomena, Oxford University Press, Oxford 1971.

[4] Huang, K.: Statistical Mechanics, John Wiley and Sons, New York 1987.

[5] Yeomans, J.M.: Statistical Mechanics of Phase Transitions, Clarendon Press, Oxford 1993.

[6] Binney, J.J., Dowrick, N.J., Fisher, A.J. and Newman, M.E.J.: The Modern Theory of Critical Phenomena, Clarendon Press, Oxford 1992.

[7] Langer, J.S.: An Introduction to the kinetics of First-Order Phase Transitions, in: Solids far from Equilibrium (Ed. C.Godrèche), Cambridge University Press, Cambridge 1992, 297-363.

[8] Sethna J.P., Dahmen K., Kartha S., Krumhansl J.A., Roberts B.W. and Shore J.D.: Hysteresis and hierarchies: dynamics of disorder-driven first-order phase transformations, Phys. Rev. Lett. 70 (1993), 3347-3350.

[9] Falk, F.: Model free energy, mechanics and thermodynamics of shape-memory alloys, Acta Metall. 28 (1980), 1773-1780.

[10] Toledano, J.C. and Toledano, P.: The Landau Theory of Phase Transitions, World Scientific Publ. Co., Singapore 1987.

[11] Nakanishi, N., Mori, T., Miura, S., Murakami, Y. and Kachi, S.: Pseudoelasticity in Au-Cd thermoelastic martensite, Phil. Mag. 28 (1973), 277-292.

[12] Wilson, K.G.: The renormalization group and critical phenomena, Rev. Mod. Phys. 55 (1983), 583-600.

[13] Guggenheim, E.A.: The principle of corresponding states, J. Chem. Phys. 13 (1945), 253-261.

[14] Heller, P. and Benedek, G.B.: Nuclear magnetic resonance in MnF_2 near the critical point, Phys. Rev. Lett. 8 (1962), 428-432.

[15] Thompson, D.R. and Rice, O.K.: Shape of the coexistence curve in the perfluoromethylcyclohexane-carbon tetrachloride system. II. Measurements accurate to 0.0001°, J. of the Am. Chem. Soc. 86 (1964), 3547-3553.

[16] Onsager, L.: Crystal statistics. I. A two-dimensional model with an order-disorder transition, Phys. Rev. 65 (1944), 117-149.

[17] Porter, D.A. and Easterling, K.E.: Phase Transformations in Metals and Alloys, Van Nostrand Reinhold, Wokingham 1981.

[18] Allen, S.M. and Cahn, J.W.: A microscopic theory for antiphase boundary motion and its application to antiphase domain coarsening, Acta Metall. 27 (1979), 1085-1095.

[19] Vives, E., Ortín, J., Mañosa, L., Ràfols, I., Pérez-Magrané, R. and Planes, A.: Distribution of avalanches in martensitic transformations, Phys. Rev. Lett. 72 (1994), 1694-1697; Meisel, L.V. and Cote, P.J.: Power laws, flicker noise and the Barkhausen effect, Phys. Rev. B, 46 (1992), 10822-10828; Field, S., Witt, J., Nori, F. and Ling X.: Superconducting vortex avalanches, Phys. Rev. Lett. 74 (1995), 1206-1209.

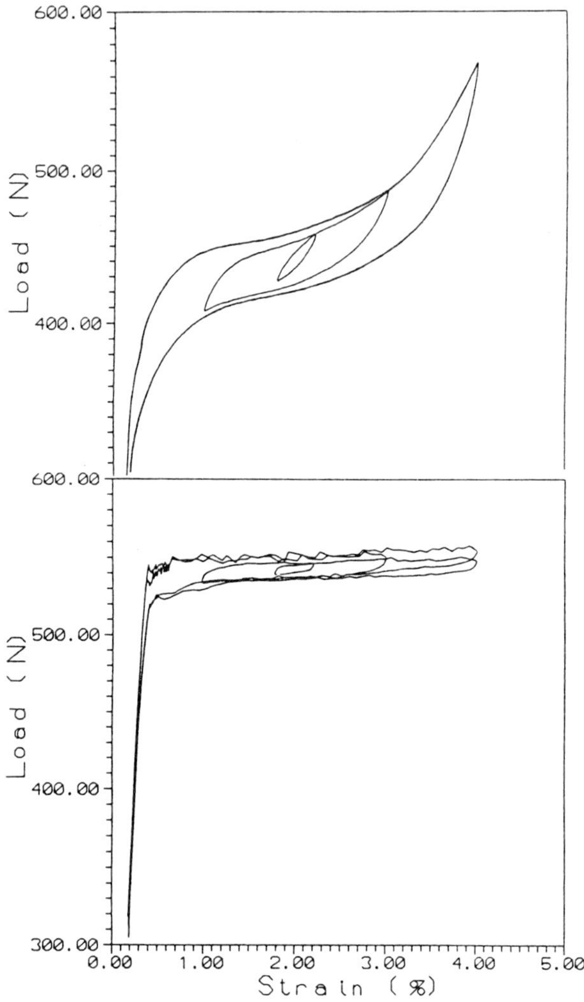

Figure 30: Stress-strain hysteresis cycles of a Cu-19.4Zn-13.1Al (at%) single crystal, recorded in a strain-controlled tensile machine operated at $\dot{\epsilon} = 0.05mm/min$. The behaviour is very different, depending on the thermal treatment used to retain the austenitic phase. The result of a direct quench is shown on top, and that of a slow quench on bottom.

INTERACTION OF STRESSES AND STRAINS WITH PHASE CHANGES IN METALS

PHYSICAL ASPECTS

H.P. Stüwe

University of Mining and Metallurgy, Leoben, Austria

Abstract

Phase changes in metals may cause stresses and strains, stresses and strains may cause phase changes. The first part of this chapter discusses such interactions for phase changes controlled by diffusion, the second part for diffusionless ("martensitic") transformations.

1 Diffusion Controlled Phase Changes

1.1 Thermodynamics[1]

Fig. 1 shows schematically the free enthalpy of two phases of equal chemical composition as function of the temperature.

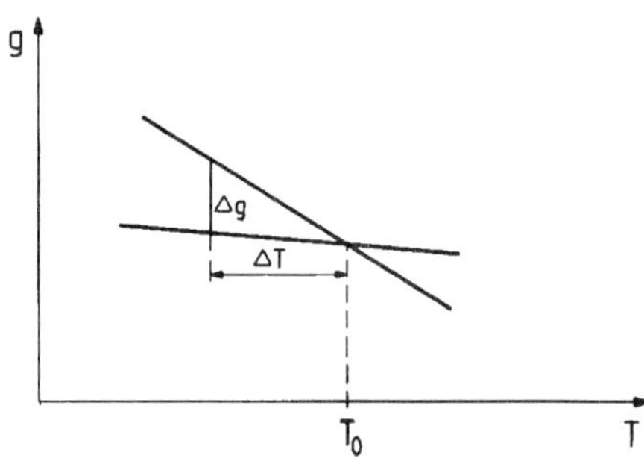

Figure 1: Specific free enthalpy of the two phases of equal chemical composition in the vicinity of equilibrium

At T_0 both phases are in equilibrium so that, by definition, no phase change happens. For a phase change with a finite reaction rate a "transgression" of the equilibrium temperature by ΔT is necessary; it provides a driving force Δg which can be approximated as

$$\Delta g \approx \Delta T \Delta S \tag{1}$$

where the transformation entropy ΔS is the heat of transformation divided by the equilibrium temperature. For the solidification of metals ΔS has values around 8 J/g-atom K. This uniformity reflects the change in configuration entropy between a close packed crystal and a melt. It is therefore independent of chemistry. Values for water are given in table 1.

Table 1:

Water		
	Q[J/mole]	ΔS[J/mole K]
melting	5760	21
evaporation	38800	104

[1] (see also chapter by Ortin)

1.2 Nucleation

Consider a spherical droplet of radius r formed in a water vapour undercooled by ΔT. Its free enthalpy is

$$\Delta G = -\frac{4\pi r^3}{3}\frac{\Delta g}{V} + 4\pi r^2 \gamma \tag{2}$$

where $\gamma \approx 2\,J/m^2$ is the specific surface energy and V the molar volume. This curve is shown in fig. 2.

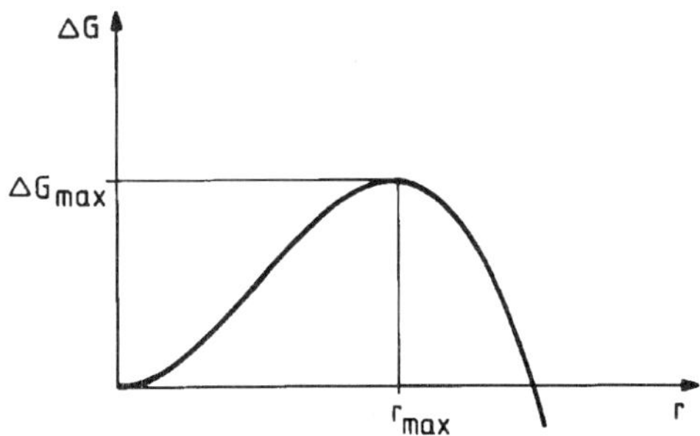

Figure 2: Free enthalpy of a droplet condensed from the vapor as a function of radius

It has a maximum at

$$r_{\mathrm{max}} = \frac{2\gamma V}{\Delta S \Delta T} = \frac{0.7\,\mu\mathrm{m\,K}}{\Delta T} \tag{3}$$

If N is the number of such droplets to make 1 mole, the molar activation energy for the formation is

$$N\Delta G_{\mathrm{max}} = -\Delta g + \frac{3\gamma V}{r_{\mathrm{max}}} = \frac{1}{2}\Delta g = \frac{\Delta S\,T}{2} \tag{4}$$

Even though the change in specific volume is large in condensation a term of the type $p \cdot V$ can be neglected in eq. (2). This is not so for the nucleation of a solid/solid phase change. Even though the specific changes in volume and shape are comparatively small they may lead to very high stresses. This may cause nuclei and growing phases to assume shapes other than spheres (e.g. to form rods or plates) to minimize strain energy rather than their surface. Eq. (3) shows that water vapour undercooled by 1 K is in equilibrium with droplets of $r = 0.7\,\mu$m. Although the equilibrium is unstable (smaller droplets should evaporate, larger ones should grow into rain drops) we know from experience that such a "fog" may be quite persistent. Similarly, particles of critical size formed in solid state reactions may be quite persistent. An example are oxide particles in molybdenum. They are quite persistent at room temperature with radii of 12Å which is very far from equilibrium [1].

1.3 Kinetics of dissipative reactions: the "nose" in ttt-diagrams

Figure 3 shows a schematic isothermal ttt-diagram. It shows two reactions: The formation of ferrite below A_3 and the formation of pearlite below A_1.

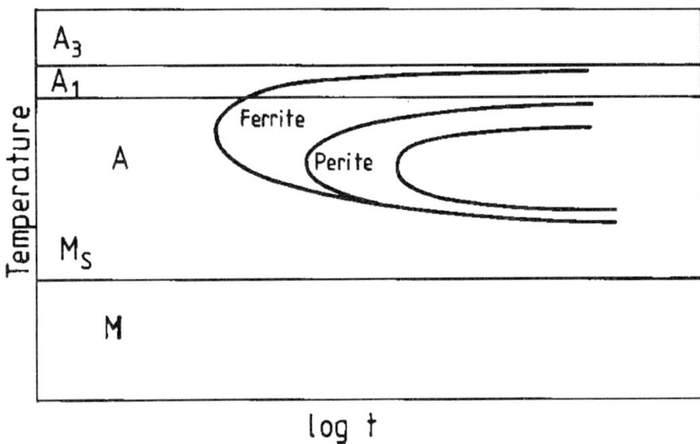

Figure 3: Isothermal ttt diagram, schematic (the bainitic reaction is suppressed, as e.g. in steel X40Cr13)

Both reactions have a "nose" (i.e. a maximum of reaction rate v) at $T_\mathrm{n} = T_0 - \Delta T_\mathrm{n}$ where T_0 is A_3 or A_1.
The reaction rate is proportional to a driving force which usually can be assumed in good approximation to be proportional to transgression (in our case to undercooling). It is also proportional to some diffusion coefficient. Thus, we can write

$$v \sim (T_0 - T)e^{-Q/RT} \quad . \tag{5}$$

Differentiating with respect to T we find the maximum for v at

$$T_\mathrm{n} = \frac{Q}{2R}\left[-1 + \sqrt{1 + \frac{4RT_0}{Q}}\right] \quad . \tag{6}$$

Developing the square root into a Taylor series and breaking off after the quadratic term we obtain

$$T_\mathrm{n} = T_0 - \Delta T_\mathrm{n} \approx T_0 - \frac{T_0^2 R}{Q} \quad . \tag{7}$$

We now assume that the rate controlling diffusion constant for the formation of both ferrite and pearlite is that for the diffusion of carbon atoms in ferrite with $Q = 18600\,\frac{\mathrm{cal}}{\mathrm{mole}}$. (In alloyed steels, Q may depend on the alloy content). We then obtain from eq. (7) the figures shown in table 2.

Table 2:

		from eq.(3.7)		from eq.(3.6)
	$T_0\,[K]$	$\Delta T_n\,[K]$	$T_n\,[°C]$	$T_n\,[°C]$
ferrite	$A_3 = 1070$	123	674	696
pearlite	$A_1 = 1020$	112	635	654

With this activation energy the term $4RT_0/Q \approx 0.44$ which is not really very small compared with 1. It is therefore better to use the exact eq. (6) which gives the values in the last column of the table.

1.4 Coarsening of second phase particles

Many technical alloys owe part of their strength to incoherent second phase particles. These cannot be cut by dislocations which leads to strengthening according to the Orowan-mechanism shown schematically in fig. 4: A dislocation is held back by two inclusions of diameter d at a distance Λ and bows out between them under the additional stress $\Delta\tau$.

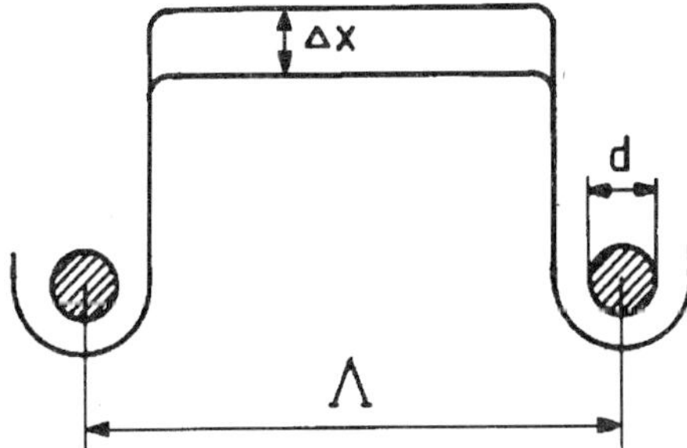

Figure 4: Dislocation bowing out between two inclusions (Orowan mechanism)

When the dislocation advances by a distance Δx the mechanical work done by $\Delta\tau$ must be sufficient to create two new pieces of dislocation of length Δx and specific line energy Gb^2. Thus,

$$\Delta\tau \approx \frac{2Gb}{\Lambda - d} \tag{8}$$

Since usually $d \ll \Lambda$ a steel, for instance, ($G \approx 84\,\text{GPa}$) will be strengthened by about 170 MPa if Λ is of the order of 300 nm. If $f \approx d^3/\Lambda^3$ is the volume fraction of the second phase, then

$$\Lambda \approx \frac{d}{\sqrt[3]{f}} \tag{9}$$

This equation shows that it is much more efficient to diminish the size d for a given volume fraction f than to increase f.
On the other hand very fine particles are quite unstable thermodynamically. They persist only at temperatures well below the nose of the ttt-diagram. At higher temperature they will experience Ostwald ripening which will increase the average value of d and, hence, of Λ without changing f. For cementite in steel fig. 5 gives a crude estimate of the times and temperatures involved in this type of softening.
A similar effect is observed in fine lamellar structures such as pearlite in steel which may "spheroidize" in the course of service [2,3]).

2 Reactions without diffusion ("military" reactions)

2.1 Mechanical twinning.

Fig. 6 shows a crystal of calcite. Its upper right hand part has been deformed plastically by external forces. Closer inspection reveals that not only the shape, but also the crystal orientation of this volume has changed: both are now mirror images of the matrix. We say that the crystal has formed a "twin". If a twin is formed in the interior of a crystal its shape is not so easily interpreted as in fig. 6. It is usually lenticular to minimize the elastic strain energy caused by its formation. An example is shown in fig. 7.

Even so, the orientation of the twin will be a mirror image of that of the matrix. One can define a "twinning system" by a twinning plane and a twinning direction, both of which are determined by the type of the crystal lattice. Examples are given in table 3.

Table 3:

Structure	twinning plane	twinning direction	number of variants
f.c.c.	$\{111\}$	$< 112 >$	12
b.c.c.	$\{112\}$	$< 111 >$	12
h.c.p.	$\{10\bar{1}2\}$	$< 10\bar{1}1 >$	6

By permutation of the signs of indices (keeping in mind the condition of mirror symmetry) one can see that there are more than one possible twinning systems in a crystal. External and internal stresses determine which of those "variants" will be operating. Two variants can be seen in fig. 7. Deformation twins form "instantaneously", i.e. the velocities involved are comparable to that of sound. They are not related to exernally imposed strain rates nor to any diffusion coefficient. This means that the twin must form by the cooperative movement of many atoms. Several models have been proposed

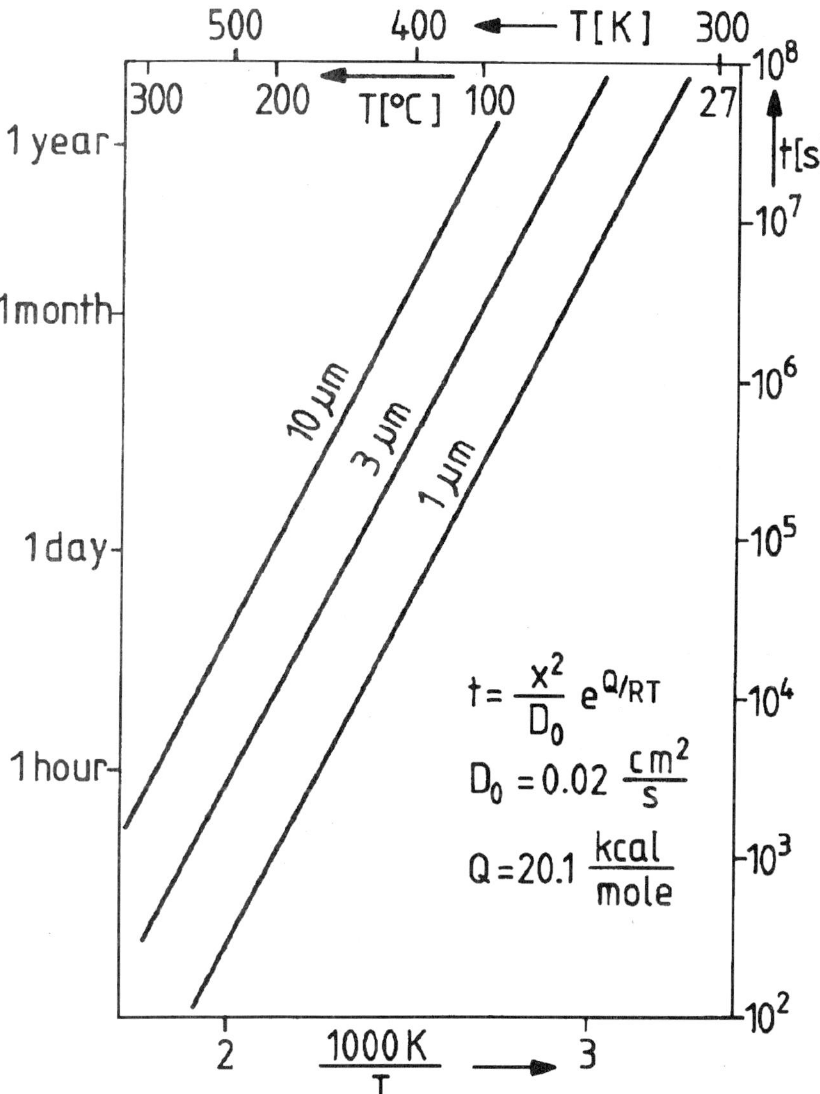

Figure 5: Estimate of diffusion times for coarsening of Fe_3C in steel (X is the distance between particles)

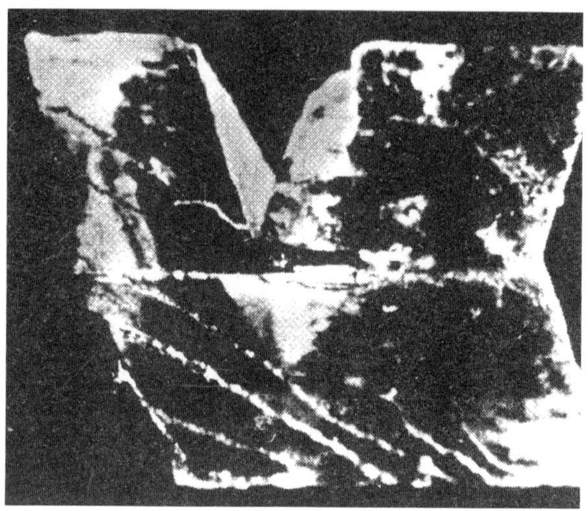

Figure 6: Mechanical twin in Calcite

Figure 7: Mechanical twins in a zinc single crystal

to show how such cooperative movement can be affected by the movement of suitable dislocations (see, e.g. [4–8]).

2.2 Martensite formation in steel

Looking again at fig. 3 one might plan to produce austenite supersaturated in carbon by a quench rapid enough to pass in front of the "noses". This is not possible. At the temperature M_s a diffusionless reaction starts to form martensite, a kind of supersaturated ferrite with a small tetragonal distortion due to the carbon atoms. The reaction ends at a lower temperature M_f i.e. after a certain time interval, the length of which is determined by the cooling rate only. Still, the transformation of each volume element is "instanteneous" in the same sense as in twinning. (At intermediate quenching rates one produces the bainite reaction which is not shown in fig. 3 and will not be discussed in this paper.)

The typical volume element of martensite can be approximated by an ellipsoid with axes of relative lengths 1:10:100. The two larger axes define a "habit plane". The relation of crystal orientation between the two phases depends on the alloy content. For carbon steels, Kurdjumow and Sachs found $(111)||(110)\alpha'$; $(1\bar{1}0)\gamma||[1\bar{1}1]\alpha'$. For a steel containing 30% Ni Nishiyama and Wassermann found $(111)\gamma||(110)\alpha'$; $[\bar{2}11]\gamma||(1\bar{1}0)\alpha'$. Indices of the habit plane are high and possibly irrational.

A very simple model for diffusionless transformation was proposed by Bain. It is illustrated in fig. 8 which shows two cells of a f.c.c. lattice.

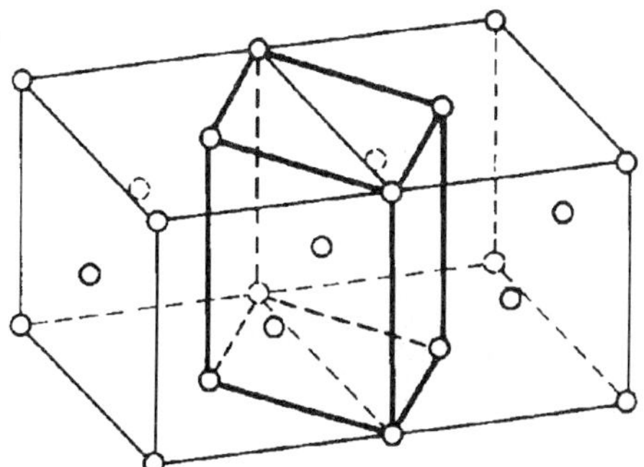

Figure 8: Bain model for martensite formation

The bold lines show that this lattice can also be seen as a bodycentered tetragonal lattice. The martensite lattice can therefore be formed simply by a compression of the c-axis and a corresponding dilatation of the other two axes. The orientation relationship between the phases should then be $(100)\gamma||(100)\alpha'$; $[100]\gamma||[110]\alpha'$ with a possible habit plane (100). Obviously there are 3 variants. This kind of martensite formation has indeed been observed in the transmission electron microscope i.e. in a thin foil that cannot support mechanical stress. The more complicated relations given above must therefore be enforced by the necessity to minimize internal stresses in a solid specimen. This theory has been worked out in detail by Liebermann, Wechsler and Read [9]. The theory makes use of the possibility to relieve shear stresses by internal twinning of the martensite.

2.3 Transformation induced plasticity [2]

Martensite formation causes stresses and strains. This of course also means that externally imposed stresses and strains will enhance the formation of martensite. One utilizes this effect in steels which contain a certain volume fraction of metastable austenite. During plastic deformation this austenite will transform to martensite which gives some contribution both to plastic strain and to work hardening. A classical example for this transformation induced plasticity (TRIP) are the Mn-alloyed Hadfield steels, but various low alloyed steels using the same principle are currently being developed. The effect can be studied very nicely in a model substance: iron can form a solid solution in copper at high temperatures and remain there after quenching. Upon annealing at moderate temperature small iron particles segregate which are austenitic because of epitaxy to the surrounding copper matrix. They persist even when cooled to room temperature. During plastic deformation of this material the inclusions deform martensitically to ferrite as can be easily detected by magnetic measurements or by the Mößbauer effect. Fig. 9 shows an electron micrograph of such particles. On close inspection one can discover in one particle striations which are evidence for the internal twinning [11].

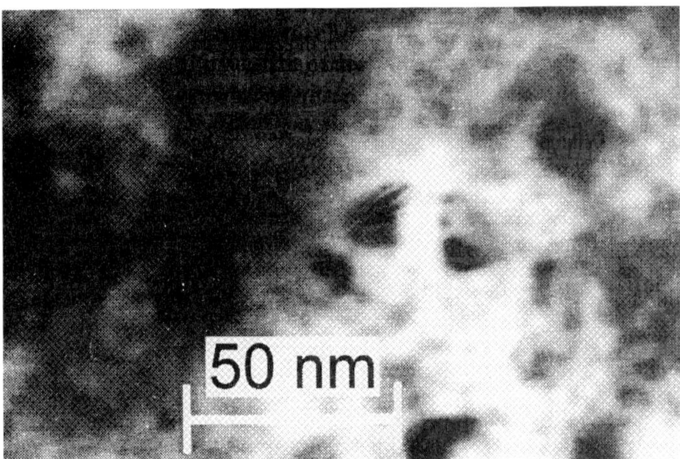

Figure 9: Ferrite formed in an inclusion in a deformed copper matrix

Usually there are several variants available for martensite formation. If the transformation is caused by external stresses, some of these variants will be favoured while others are not – the martensite formed will be textured rather than random.

[2]For a comprehensive review see [10]

2.4 Martensite in non-ferrous alloys

There are numerous non-ferrous alloys which also show diffusionless phase transformations. These new phases form plates that grow "instantaneously". Their orientation to the matrix crystal follows well defined orientation relationships, usually in several variants leading to different deformation tensors. Their formation can be induced by cooling the high temperature phase below a certain temperature but also by externally imposed stresses. On cooling, martensite formation starts at a temperature M_s and ends at a temperature M_f. Because of all these similarities the terminology of steel is applied to these alloys. The high temperature phase is often called "austenite", the low temperature phase "martensite" and the transformation "martensitic". There are, however, at least two important differences between ferrous and non-ferrous martensite:

a) In allotropic phase changes (i.e. without change in chemical composition) one expects a closer packing of the atoms in the low temperature phase than in the high temperature phase. This is because of the entropy of lattice vibrations. Qualitatively the effect can be easily visualized: In a close packed lattice there is less room for the thermal oscillations of the atoms of the crystal. All non-ferrous martensite reactions seem to follow this rule with the exemption of steel and related ferrous alloys. On closer inspection one sees that also iron follows this rule, namely in the transformation from δ-ferrite to γ-austenite at very high temperatures. That α-ferrite (and martensite) is (are) restabilized by cooling to lower temperatures is caused by magnetic ordering and is, therefore, an anomaly.

b) Ferrous martensite is formed irreversibly. Martensite formed by quenching can be austenitized only by an anneal at much higher temperatures. This reverse reaction is not martensitic but diffusion controlled. Similarly, stress induced ferrous martensite cannot be austenitized by simply reversing the stress. In this it resembles mechanical twins which cannot be "untwinned" by reversing the stress. The reason for this irreversibility seems obvious. If the formation of ferrous martensite or mechanical twins is caused by the motion of suitable dislocations then these dislocations will form their own network which cannot be dissolved easily. This is analogous to plastic deformation where upon reversal of strain most of the dislocations will not return to their sources but other, additional dislocations are created. The analogy is supported by the fact that the stresses necessary for martensite formation and for twinning are similar to the yield stress. For a more detailed discussion see [12].

c) Non-ferrous martensite, on the other hand, is reversible. Martensite formed by cooling will reconvert to austenite on heating. This reaction, which is not diffusion controlled and could therefore also be called martensitic, starts at a temperature A_s and ends at a higher temperature A_f. M_s and A_s are usually only a few K apart. Non-ferrous martensite produced by external stress may reconvert to austenite by heating, by reverting the stress or simply by releasing the stress (see section 3.2.5.). Stresses involved are lower than the true yield stress. The reason for this reversibility is not so obvious. There is a number of observations gained by inelastic neutron scattering which reveal "phonon softening" in certain lattice directions of the parent phase as it approaches the martensite temperature (see, e.g. [13]). This suggests that the matrix lattice may become unstable against shear in certain directions and transform as a whole at comparatively low stresses or at a certain temperature. Plastic deformation beyond the true yield point will introduce dislocations and destroy the reversibility [14].

2.5 Memory effect and pseudoelasticity [3]

In alloys forming reversible martensite, the relative positions of the temperatures A_s, A_f, M_s, M_f to the working temperature T will cause several effects which are quite different phenomenologically and, therefore, have different names. We want to add a temperature $M_d > M_s$ below which martensite is formed under external stress.

a) If $T < M_d$ and $< A_f$ then martensite will form under external stress. Only those variants will form that will lead to strains favoured by the stresses. Thus, the resulting structure is textured [16]. This deformation is persistent at T and might be called "plastic" in the meaning of TRIP. However, at moderately higher temperatures above A_f the martensitic reaction is reversed leading also to a reversal of strain. The effect is therefore called "pseudoplasticity". The reversible strain that can be supported by such materials which are called "memory alloys" is of the order of 10%.

b) If a reaustenitized memory alloy is cooled below M_f it will retain its shape because all variants of martensite are formed with equal probability. One can, however, "condition" a memory alloy by suitable mechanical treatment to contain internal stresses favouring certain variants. Such alloys have a "two way memory" i.e. they can change repeatedly between two shapes during repeated temperature cycles. The strains in this case are only between 1 and 2%.

c) If $A_f < M_d$ then martensite of suitable variants will form under stress and reverse to austenite (and the original shape) when the stress is released. The deformation is elastic (i.e. reversible) but the strains are much larger than in normal elasticity (of the order of 10%). Therefore, the term "pseudoelasticity" should be used. The term "rubber elasticity" has also been used; but as rightly has been pointed out in [17] it should be avoided because the physical background is different: In general elasticity is caused by an increase of free energy due an

[3]For a comprehensive (if early) review on this topic see [15].

imposed strain

$$\frac{dF}{d\epsilon} = \frac{dU}{d\epsilon} - \left(\frac{dS}{d\epsilon}\right) T \quad . \tag{10}$$

In metals (including alloys with reversible martensite) the first term on the right hand side of the equation is much more important than the second term. In rubber, as in some other polymers, the second term is much more important than the first.

d) Pseudoelasticity can also be found in materials that have already undergone martensitic transformation. It is then caused by the reversible growth of some variants at the expense of others.

2.6 The nickel-titanium system

The critical temperatures M_s etc. depend very sensitively on chemical composition. As an example, we shall discuss the alloy NiTi which was among the first to be investigated and which belongs to the few non-ferrous alloys with martensitic transformations that are now used commercially. The alloy has the approximate composition NiTi but with a certain solubility range. The structure of the high temperature phase is that of CsCl (i.e. ordered b.c.c.). Fig. 10 shows schematically the transformation temperatures as a function of concentration.

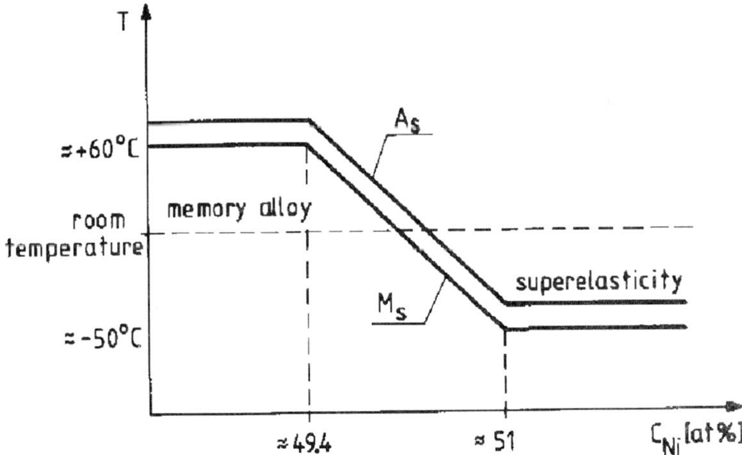

Figure 10: Semischematic plot of M_s and A_s in NiTi as a function of alloy content

One can see that alloys on the nickel-rich side are pseudoelastic at room temperature while alloys on the Ti-rich side are memory alloys. The figure is only schematic because the values of the transition temperatures given in the literature scatter very widely. There are at least two reasons for this:

a) The transition points depend very sensitively on concentration (maybe by up to 10 K/0.1% Ni).

b) the transition temperatures are sensitively influenced by the thermal and mechanical history of the alloy.

Shape memory alloys – both one-way and two-way – can be used to build machines and devices. These will be treated in the chapter by E. Patoor.

References

1. Shiwa,Y., H.P. Stüwe and E. Pink: Anelastic Effects in Molybdenum Due to the Precipitation and Dissolution of Oxides at Low Temperatures. Acta metall. mater. 38 (1990), 819–824.

2. Werner, E.: Thermal Shape Instabilities of Lamellar Structures. Z. Metallkde. 81 (1990), 790–798.

3. Werner E.: The Growth of Holes in Plates of Cementite. Mater. Sci. Engng A132 (1991), 213–223.

4. Cottrell, A.H. and B.A. Bilby: A Mechanism for the Growth of Deformation Twins in Crystals. Phil. Mag. 42 (1951), 573.

5. Mahajan, S.: Interrelationship between Slip and Twinning in b.c.c. Crystals. Acta Met. 23 (1975), 671.

6. Sleeswyk, A.W.: Emissary Dislocations: Theory and Experiments onf the Propagation of Deformation Twins in α-Fe. Acta Met 10 (1962), 705.

7. Reed-Hill, Hirth and Rogers: Deformation Twinning. AIME-Conference, Florida, 1963.

8. Christian, J.W. and S. Mahajan: Deformation Twinning. Progress in Materials Science, vol.39, 1-157.

9. see e.g. Liebermann, D.S.: Phase Transformations. Metals Park, Ohio: Amer. Soc. for Metals, 1970.

10. Fischer, F.D., Q.P. Sun and K. Tanaka: Transformation Induced Plasticity. Applied Mechanics Reviews 49, nr.6, June 1996, 317–364.

11. Tiefenthaler, B., G. Reisner and E. Werner: Verformungsinduzierte Martensitbildung in einer Kupfer-Eisen-Legierung: Experimente und Modellierung. Z. Metallkde. 86, 12 (1995), 845–851.

12. Hornbogen, E.: On Martensitic Transformation Cycles. Z. Metallkde. 86, 12 (1995), 656–664.

13. Herper, H.C., E. Hofman, Entel P. and W. Weber: Structural Phase Transformation and Phonon Softening in Iron-Based Alloys. Journal de Physique IV, Colloque C8, Supplement an J. De Physique III, Vol.5 (1995), 293–298.

14. Hornbogen, E. and E. Kobus: A Metallographic Study of Plysic Deformation of Martensitic NiTi. Z. Metallkde. 87 (1996), 442–447.

15. Delaey, L., R.V. Krishnan, H. Tas and H. Warlimont: Thermoelasticity, Pseudoelasticity and the Memory Effects Associated with Martensitic Transformations. Parts 1 and 2, J. of Mat. Sci. 9 (1974), 1521–1555.

16. see, e.g. Zhu, Z.S., J.L. Gu and N.P. Chen: Variant Selection and Phase Transformation Texture in Titanium. J. of Mat. Sci. Letters 14 (1995), 1153–1154.

17. Hornbogen, E.: On the Term "Pseudo-Elasticity". Z. Metallkde. 86 (1995), 341–344.

EXPERIMENTAL OBSERVATIONS FOR SHAPE MEMORY ALLOYS AND TRANSFORMATION INDUCED PLASTICITY PHENOMENA

E. Gautier
CNRS URA 159, Nancy, France

E. Patoor
CNRS URA 1215, Metz, France

Abstract

This chapter deals with the general aspects related to shape memory alloys and the TRIP phenomena. These two kinds of behavior originate from a particular solid-solid phase transformation : the martensitic transformation.

First two parts of this chapter present the typical characteristics of this first order diffusionless and displacive transformation. Differences between thermoelastic and non thermoelastic martensitic transformation are underlined. In the third part the different behaviors observed in shape memory alloys are detailed (superelasticity, one way shape memory, two way shape memory, rubberlike effect and damping capacity). Physical strain mechanisms at the origin of these behaviors are defined. The last section is devoted to TRIP phenomena. Respective importance of nucleation and growth of martensite plates are discussed in this case. Plate morphology modifications related to evolution in the Bain strain accommodation mechanism in presence of an applied stress are pointed out.

1. General aspects of the martensitic transformation

The martensitic transformation is a first order displacive transformation: the high temperature phase is called austenite, the product of the transformation is called martensite. Martensite designates originally the quenched-product in steels [1].
This transformation is mainly involved in:

a/ Thermal treatment in different alloys (steels, titanium alloys...). In steels applications come from the very strong hardness exhibited by the product phase.

b/ Shape memory alloys. In these materials a very large reversible strain is obtained that allows to define high and low temperature shapes.

This great diversity of behavior originates from differences in nucleation and growing processes in the different materials. In any cases, the martensitic transformation possesses well-defined characteristics that distinguish it among all the solid state transformations:

1. This solid state phase transition is associated with an inelastic deformation of the crystal lattice. No diffusive process is involved, transformation only results from a cooperative and collective motion of atoms on distance smaller than the lattice parameters. The lack of diffusion makes this transformation almost instantaneous [2].
2. Due to its first order character, parent and product phases coexist during the transformation. This is responsible for the existence of an invariant plane. Lattice systems of the two phases possess well defined mutual orientation relationships (the Bain correspondances) [3]. These relations depend on the nature of the alloy.
3. Transformation of an elementary volume element produces a volume change and a shearing along well-defined planes. This shearing can be twenty times larger than elastic ones. This transformation is crystallographically reversible [4].
4. Since the martensite crystal lattice has a lower symmetry than the parent phase one, several variants of martensite can be formed from the same parent phase crystal [5].
5. Stress and temperature have a large influence on the martensitic transformation. Transformation takes place when the free energy difference between the two phases reaches a critical value [6].

Crystallographic theories of the martensitic transformation are based on these characteristics [7] [8]. In these theories, existence of a particular lattice invariant strain (LIS) is assumed (by shearing, twinning or with stacking faults). The amount of shear (or the relative twin thickness) is determined when postulates that the invariant plane such obtained is the habit plane for the transformation. These theories were successfull for transformation in which the lattice distorsion is small. However, the predictions shows deviations for materials whose transformation strains are larger [9] [10]. Several corrections have been made to initial theories [11] [12].

2. Characteristics of the martensitic transformation

Martensitic transformation is characterized by the existence of a macroscopic shape change associated with the modification of the crystalline structure. Many phenomena find their origin in this fundamental mechanism.

2.1. Plates formation and morphology of the martensite

To minimize the interaction energy between the martensite and the austenite, martensite plates adopted a very flat shape along the habit plane [2]. Orientation of this plane is function of the alloy composition. In most cases it is not a crystallographical plane. The habit plane orientation constitutes a major characteristic of the transformation. In shape memory alloys, the transformation direction is almost parallel to this plane and the volume change is very low. For ferrous alloys, the volume change is larger (difference of one order of magnitude).

Existence of several variants of the same martensitic phase is another important point. According that crystal lattice symmetry is lower in martensite than in austenite, each crystal of the high temperature phase can transform into several variants of martensite [5]. A variant is characterized by its habit plane normal and its transformation direction. These variants are crystallographically equivalent but with different orientations, depending on the alloy composition. Due to the existence of these variants, a self-accommodated growing of the martensitic phase is possible. In this case, no macroscopic shape change is observed, only a volume variation. Variants arranged in that way formed a self-accommodated group. In copper-based shape memory alloys, self-accommodated groups are constituted by the four variants grouped around the <011> poles of the parent phase (figure 1) [6].

Microstructural aspects related to the martensitic transformation lead to distinguish three kinds of interfaces: the austenite/martensite one (first order phase transformation), the martensite/martensite that are produced between two different variants and the twin boundaries inside the martensite plate itself (due to the LIS). Such a distinction is very important to define the physical strain mechanisms in these alloys [14].

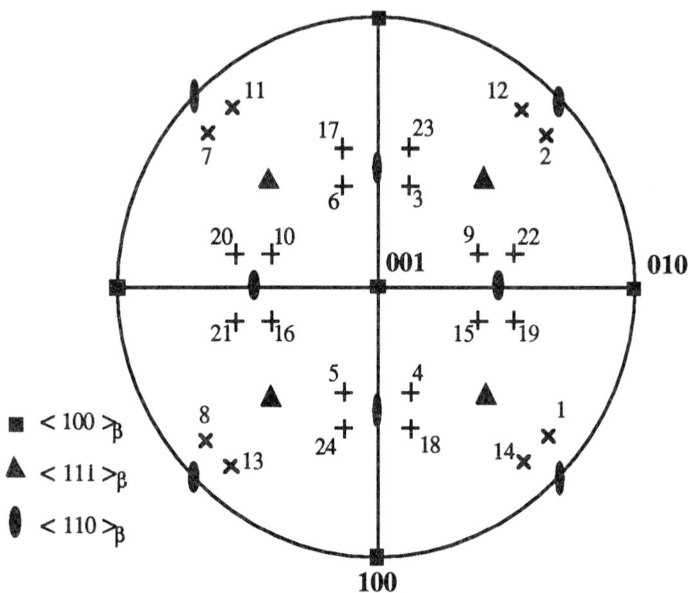

Figure 1 : Position of the twenty-four habit plane normals in a Cu-Zn-Al shape memory alloys [13].

2.2. *Transformation temperatures in a stress-free state*

Martensitic transformation appears when the free energy of the martensite becomes smaller than that of the austenite. This happens at temperatures below a critical one, denoted by T_0, where the free energies of the two phases are equal. However, when cooling, one observes that the transformation does not begin exactly at T_0 but at a temperature called M_s slightly below T_0. Free energy necessary for nucleation and growth is responsible to this shift [6]. The transformation continuously goes on as the temperature is lowered until a temperature denoted by M_f. For shape memory alloys, the M_s - M_f temperature difference is low compared to ferrous alloys (≈ 40 °C for SMA and ≈ 200°C for ferrous alloys). This M_s - M_f temperature difference is an important factor to characterize the shape memory behavior. Martensitic transformation is generally an anisothermal transition. When the temperature is kept constant inside the transformation domain the amount of martensite remains constant too. However some isothermal character of the transformation can be observed for ferrous alloys [15].

The M_s temperature depends on the alloy composition. Empirical relationships have been established either for thermoelastic alloys or for ferrous alloys.

For Cu-Zn-Al alloys for exemple, Ahlers gives :

$$M_S(°C) = 2485 - 66{,}9\,(1{,}355\,Al + 1\,Zn) \quad (\text{atomic } \%)\ [12]$$

For ferrous alloys, Andrews proposed [17]:

$$M_S(°C) = 539 - 423\,(\%C) - 30.4\,(\%Mn) - 17.7\,(\%Ni) - 12.1\,(\%Cr) - 11\,(\%Si) - 7.5\,(\%Mo) \quad (\text{for wt } \% \text{ and } \%C < 0.6 \text{ wt}\%)$$

Such equations constitute a useful approximation to the material elaboration. One notices a very great sensitivity of the M_s temperature with the alloying elements. Moreover, microstructural defects, degree of order in the parent phase, grain size of the parent phase can also modify the transformation temperature of several degrees [18].

2.3. *Stress-induced transformation*

Due to the displacive character of martensitic transformation, applied stress plays a very important role. Application of a macroscopic stress σ on a volume of austenite provides a mechanical energy contribution:

$$dW = \sigma_{ij}\, dE_{ij}$$

where E is the macroscopic strain of the crystal. Denoting by S the entropy, change in the free energy dU of the system is then expressed as:

$$dU = T\,dS + \sigma_{ij}\, dE_{ij}$$

and variation in Gibbs free energy turns as:

$$dG = -S\,dT - E_{ij}\, d\sigma_{ij}$$

Thermodynamical equilibrium condition between the two phases ($G^M = G^A$) gives:

$$-[\Delta S]^{A \to M}\, \Delta T - [\Delta E_{ij}]^{A \to M}\, \Delta\sigma_{ij} = 0$$

where $[\Delta E]^{A \dashrightarrow M}$ denotes the jump in the total strain between the two phases. This equilibrium condition gives a Clausius-Clapeyron type equation that relates the applied stress and the temperature T_0 [19]. The M_s variations with stress are then supposed to be equivalent to the variations in T_0. This implies that the resistive force for nucleation and growth are taken independant on the temperature.

In general cases $[\Delta S]^{A \dashrightarrow M}$ and $[\Delta E]^{A \dashrightarrow M}$ are supposed to be temperature independent and a linear relationship is obtained (figure 2). A stress-induced transformation temperature M_s^σ and a critical transformation stress σ_s, that is a function of the temperature, are defined. More accurate analysis consider the change in $[\Delta E]^{A \dashrightarrow M}$ under the applied stress [11]. Moreover, they are inhomogeneities in the stress and strain fields inside the material and thus a transformation stress range is often observed. These parameters are very sensitive to the internal state of the material and to the loading mode (tension, compression, shear...).

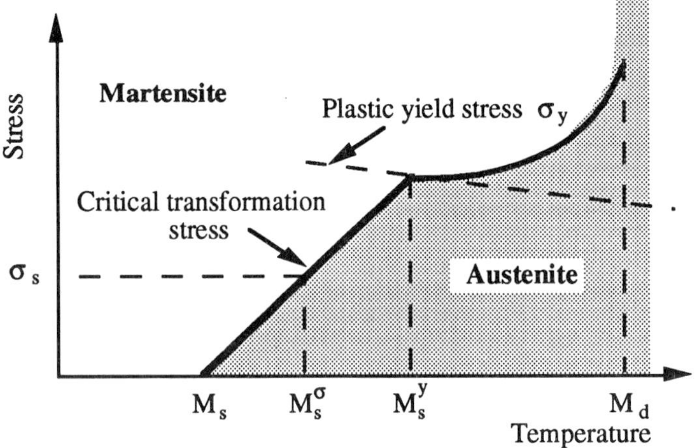

Figure 2 : Relationship between applied stress and temperature to induce the martensitic transformation. No transformation occurs above M_d.

Diagram 2 has to be completed by the definition of an upper temperature limit M_d. Above this temperature plasticity by motion of dislocations become the dominant strain mechanism in the parent phase. Inside interval M_s-M_d two distinct mechanisms are observed to induce the martensite. From M_s to a temperature called M_s^y the martensite can be stress-induced and the critical transformation stress σ_s linearly increases with the temperature. Temperature

M_s^y defines the temperature at which σ_S and the austenitic plastic yield stress σ_y are equal [20]. At temperature between M_s^y and M_d plastic strain occurs in austenite prior to the phase change. One distinguishes two types of martensite, the stress-induced ($M_S < T < M_s^y$) and the strain-induced ($M_s^y < T < M_d$).

2.4. Thermoelastic and non-thermoelastic martensitic transformation

Transformation strains are associated with the transformation and are composed by a shearing, inducing the shape change and an expansion component giving the volume change. Accommodation of these transformation strains leads to strain energy which plays a very important role. The growth of a martensitic plate within the parent phase matrix produces an important stress field. If this stress reaches the yield stress of the parent phase (or the martensite one) plastic accommodation occurs by motion of dislocations. In such case there is no thermoelastic balance for the transformation. This is observed in steels [21]. In this case, despite the reversible character of the phase change, the macroscopic transformation strain is not reversible. Reverse transformation must occur by nucleation of the high temperature phase in martensite [22]. If the plastic yield stress is large enough in the two phases (like in ceramics) or if the strain energy is weak the transformation strain is accommodated in a fully elastic way [21]. In this case, plate growth proceeds by a succession of thermoelastic balances. The martensite plates grow or shrink according with the stress or the temperature change. This distinction between thermoelastic martensitic transformation (in shape memory alloys) and non-thermoelastic (transformation in steel) is very important. Macroscopic properties associated to the martensitic transformation are entirely different in these two classes of alloys [23].

2.5. Hysteresis of the transformation

Reversion of martensite is characterized by a hysteresis loop. According to thermoelastic or non thermoelastic character of the transformation, different physical mechanisms are responsible to this phenomenon. Hysteresis size is strongly different in these two cases, around 5 to 15 °C in copper-based alloys, and between 200 and 400 °C in steel [2].

In thermoelastic alloys, real process of transformation implies large motion of interfaces. Many factors are involved in these displacements [11] [24]. Among them, friction stresses exerted by the crystal lattice on a moving interface play a large role [21]. Lovey has shown [25] that austenite/martensite interface interacts with the parent phase dislocations.

Chrysochoos had established that in addition to this intrinsic dissipation, an additionnal contribution is related to thermomechanical coupling [26].

These mechanisms act in different way for forward and reverse transformations. Heating from a temperature below to M_f, the reverse transformation only begins at a temperature, noted A_s, larger than M_f. Completion to a fully austenitic state is reached for a temperature A_f larger than M_s (figure 3). During this reversion, all the macroscopic transformation deformation produced by the forward transformation is reverted and initial orientation of the parent phase crystal lattice is recovered. The width of the hysteresis cycle is in relation with the energy dissipated during the transformation. This size of the hysteresis cycle is an important factor to describe the shape memory behavior. Its width depends on the alloy composition and on the thermomechanical loading history (in relation with the amount of defects in the material).

In non thermoelastic alloys, irreversible plastic deformation often occurs with the transformation, thus the crystallographical orientation of the parent phase lattice can not be recovered during the reversion of the transformation. Moreover, in these alloys the T_0-M_s temperature difference is very large due to the large strain energy and also to friction forces. Moreover the mobility of the γ/α' interfaces is very low, even zero for the existing plates. The transformation during heating will thus be controlled by the nucleation conditions in the two phases. In addition, in steels very often the martensite which is a metastable phase should decompose into the more stable phases alpha and carbides during the reheating. For these different reasons hysteresis size is different for TRIP steels and SMA [2].

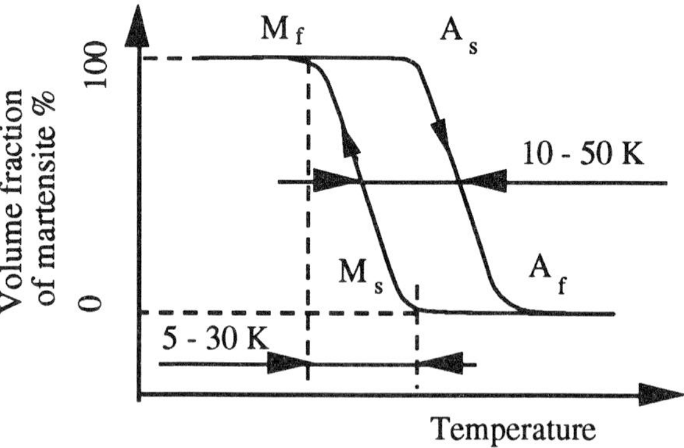

Figure 3 : Amount of martensite versus the temperature for forward and reverse transformation in thermoelastic alloys. Definition of the characteristic temperatures (usual values given for copper-based and Ni-Ti shape memory alloys).

3. Shape memory behavior

Behavior of shape memory alloys is more complex than that of usual materials, but this complexity is at the origin of their utilization in many applications. In these materials, the stress-strain curve is strongly non-linear and one obtains a very large reversible strain. This behavior is strongly temperature dependent and very sensitive to the nature, to the number and to the sequence of thermomechanical loading cycles. A hysteresis phenomenon more or less important is observed on unloading.

Different behaviors observed in shape memory alloys find their physical origin in the thermoelastic martensitic transformation previously described. The so-called shape memory effect covers only a part of the different phenomena observed in this type of alloy. Very different behaviors are observed, according to the loading path and according to the thermomechanical history of the material. Five classes of behavior are usually distinguished: super-thermoelasticity, one-way shape memory, two-way shape memory, rubberlike effect and a large damping capacity. In this section, main characteristics associated with these classes of behavior are presented and the different strain mechanisms at the origin of these effects are described.

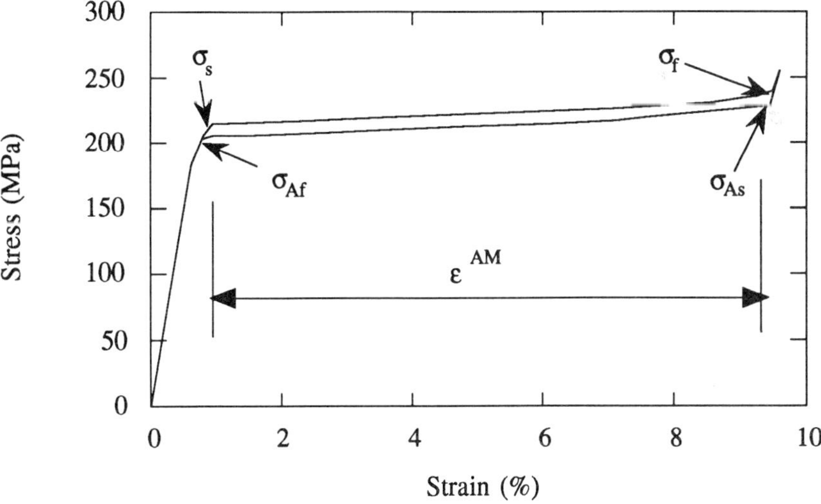

Figure 4 : Superelastic behavior observed on a Cu-16 at.% Zn-15 at.% Al single crystal ; M_S = -60°C ; isothermal loading (T = 18°C).

3.1. *Super-thermoelasticity*

This behavior, also called super-elasticity or pseudo-elasticity originates in the stress-induced martensitic transformation from an initial austenitic state [27]. In this condition a very large reversible strain is observed when the material is stressed. One distinguishes isothermal loadings where the transformation results from an increase of the applied stress at constant temperature (figure 4) and anisothermal loadings where the transformation is thermally induced under a constant applied stress (figure 7).

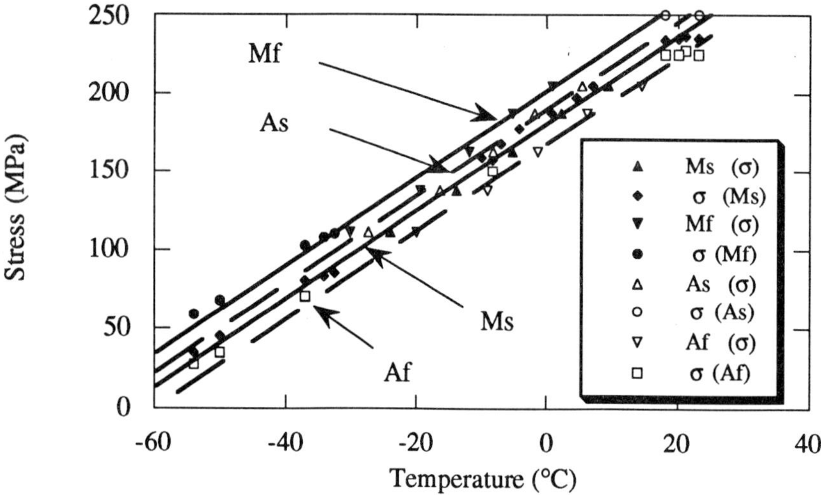

Figure 5 : State diagram experimentally defined on a Cu-16 at.% Zn-15 at.% Al single crystal from tensile tests using isothermal and anisothermal loadings.

From several isothermal tests performed at different temperatures and/ or anisothermal ones with different applied stress, a state diagram defining the different stability domains for parent phase and martensite can be drawn in a stress-temperature space (figure 5). A linear relationships is experimentally obtained between the transformation stress and temperature. As martensitic transformation is a first order transition, diagram of figure 5 presents a two-phases domain. A similar diagram can be defined for the reverse transformation martensite —> austenite. The difference $A_f - M_s$ gives the hysteresis of the transformation. Diagram 5 can be compared to this defined in figure 2. From this comparison it can be conclude that only stress-induce martensite is responsible to the superthermoelastic behavior. Thus in order to obtain good superelastic property a large $M_s - M_s^y$ range is required. This is obtained increasing the yield stress of the parent phase, using appropriate thermomechanical treatments (strengthening, structural hardening...) or using affining elements [28].

3.1.a. Superelasticity

Observed for the time first on a Cu-Zn alloy by Reynolds and Bever [29] in 1952, this behavior is obtained increasing the stress level at a constant temperature higher than M_S. It is characterized by a critical stress σ_S and a transformation stress range σ_f - σ_S. The critical stress looks like a plastic yield limit, but it corresponds to the beginning of the stress-induced phase transformation. Uniaxial tensile test is well-adapted to characterize this type of behavior. Three distinct stages are observed on the stress-strain curve (figure 4): at stress above σ_S, the material behaves in a purely elastic way. As soon as the critical stress is reached the transformation strain play the major role. At stress larger than σ_f the elastic behavior of the martensite is observed. On unloading, the same general shape is observed but an hysteresis loop is obtained. Characteristic stresses σ_{As} for the beginning of the reverse transformation and σ_{Af} for the end of reverse transformation are defined.

Four essential characteristics can be extracted from these curves:

- The critical stress σ_S above that austenite no longer presents the classic linear elastic behavior.

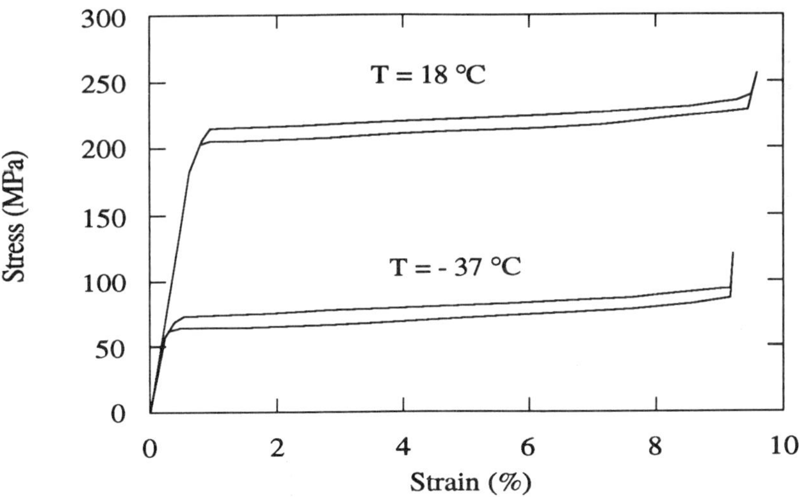

Figure 6 : Influence of a temperature change on the superelastic behavior (T = 18°C and T = -37°C) on a Cu-16 at.% Zn-15 at.% Al single crystal (M_S = -60°C).

- The strain ε^{AM} which is the maximal reversible strain obtained by the phase transformation.
- The transformation stress range $E_\sigma = \sigma_f - \sigma_s$
- The hysteresis stress size $H_\sigma = \sigma_s - \sigma_{Af}$

Metallographical observations show that in single crystal, this behavior is associated with the growing of one unique variant of martensite [27]. This variant possesses the most favorable orientation according to the applied stress. An increase of the temperature test is mainly responsible for an increase of the critical transformation stress (figure 6), the three other characteristics remaining almost constant.

3.1.b. Super-thermal behavior

When cooling under a constant applied stress from a fully austenitic state, one observes that the transformation is now characterized by a martensitic start temperature M_s^σ and a martensite finish temperature M_f^σ, that are functions of the applied stress. Macroscopic transformation strain obtained in that way (figure 7) is of several orders of magnitude greater than usual thermal strain (that explains the denomination of super-thermoelasticity).

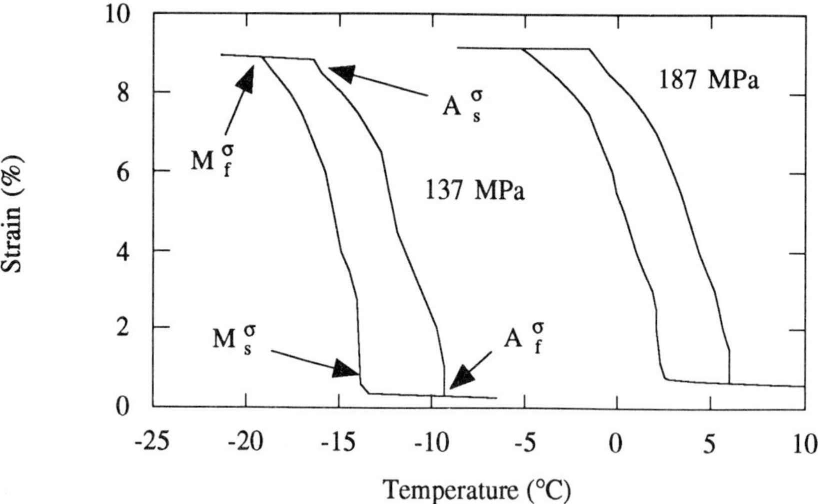

Figure 7 : Superthermal behavior of a Cu-16 at.% Zn-15 at.% Al single crystal under different constant stress level (σ =137 MPa and σ =162 MPa)

An hysteresis loop is observed; the reverse transformation begins and achieves at different temperatures to the forward transformation. One defines a temperature transformation range $E_T = M_s^\sigma - M_f^\sigma$ and a hysteresis width $H_T = (A_s^\sigma - M_f^\sigma)$. In single crystal, the maximal transformation strain ε^{AM} such obtained has the same amplitude than in superelasticity. Metallographical observations reveal the same kind of microstructure that for isothermal tests [30]. Under a variation of the applied stress, temperatures A_s^σ, M_f^σ, M_s^σ, A_f^σ evolve, but characteristics E_T and H_T remain little affected.

3.1.c. Polycrystalline structure effects

Microstructural aspects exert a strong influence on the stress-strain curve and on the temperature-strain curve. In polycrystals, the differences in crystallographical orientation between grains produce different transformation conditions in each grain. Thus on loading from an initial austenitic state the martensitic transformation is progressively induced in the different grains and, at the difference of the single crystal case, no well defined critical transformation stress is observed.

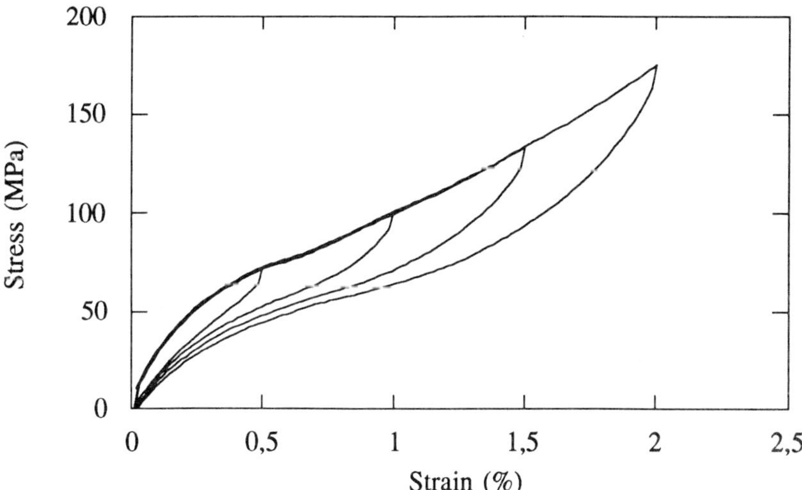

Figure 8 : Superelastic behavior at room temperature on a Cu-Zn-Al-Ni polycrystalline wire (Ms = -18°C)

Polycrystalline structure also imposes to respect compatibility conditions for the total strain at the grain boundaries. Consequently in polycrystals, the phase transition occurs less sharply, the hysteresis size increased and the macroscopic transformation strain obtained decreases (figure 8). In situ optical microscopy observations reveal that several variants of martensite are stress-induced inside each grain. The number of these variants increases with the progress of the transformation. The macroscopic transformation strain so obtained is smaller than for single crystalline material.This deformation becomes very sensitive to the applied stress level on cooling. At large stress level, variants of martensite induced are strongly oriented by the applied stress. At low stress level, growing of self accommodated groups becomes preponderant.

In addition, large strain incompatibilities occurring at grain boundaries are responsible of crack nucleation and reduce the fatigue life for polycrystalline materials in comparison with the single crystal one (figure 9) [31].

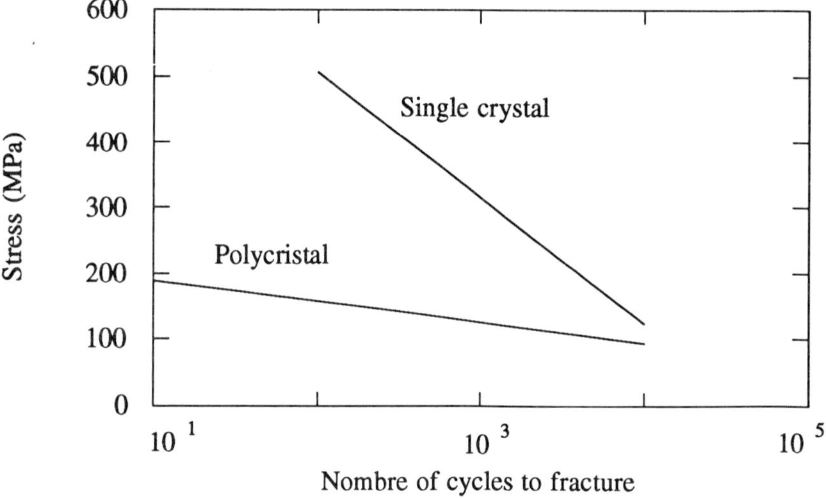

Figure 9 : Comparaison between fatigue life of a polycrystal and a single crystal for superelastic tensile tests on Cu-Al-Ni alloy [31].

3.2. *Shape memory effect*

In this paragraph, characteristics of the so-called shape memory effect are described. In this process, the microstructure of the martensitic phase is very different to that observed during stress-induced transformations. The stress-free cooling of a single crystal of austenite produces a complex arrangement of several variants of martensite. A self-accommodated growing is obtained such that the average macroscopic transformation strain is equal to zero

[13]. The self-accommodated morphology is a characteristic of the system of alloy used. In copper-based alloys twenty-four variants of martensite constitute six self-accommodated groups scattered around the < 0 1 1> poles of austenite with a typical diamond morphology. The growing of such groups gives no macroscopic transformation strain, however the multiple interfaces present in these structures (boundaries between the martensite variants and twinning interfaces) are very mobile. This great mobility is at the origin of the shape memory effect. Motion of these interfaces is obtained at stress level far smaller than the plastic yield limit of martensite. This mode of deformation, called reorientation of variants, dominates at temperature lower to M_f.

The nature of this mechanism explains the shape memory effect is observed after a sequential loading including a stress-free cooling to produce a fully martensitic state, followed by application of a stress in the low temperature phase ($T < M_f$) and finally a return in the parent phase by heating (until $T > A_f$). Several different mechanisms successively occur during such a thermomechanical loading.

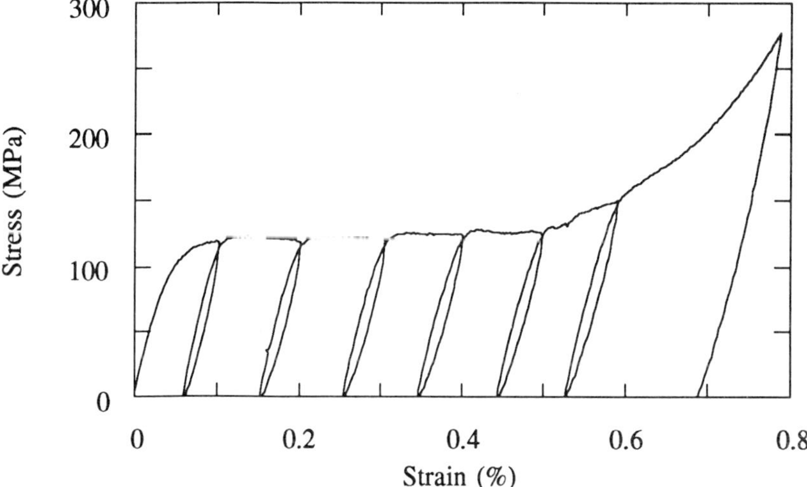

Figure 10 : Tensile behavior of a equiatomic NiTi wire (M_s = 60°C) tested at room temperature.

The first sequence induces the development of the self-accommodated martensitic structure, no macroscopic shape change is observed. During the second stage, the mechanical loading in the martensitic state produces a reorientation of the variants that gives a large inelastic strain. This behavior seems alike the superelastic one but the deformation obtained is not reversible on unloading (figure 10). During the last sequence, reverse transformation by heating remove this residual strain. Since martensite plates have been "stress-reoriented", their reversion in austenite gives a large transformation strain having the same amplitude but the opposite direction and the solid turns to its original austenitic shape. This phenomenon is called one-way shape memory because the shape reversion is only obtained on heating.

3.3. Two-way shape memory effect

In two-way shape memory, a shape change is obtained both during heating and cooling. The solid exhibits two stable shapes, a high temperature one (in the austenitic domain) and a low temperature one (in martensite). Transition from the high temperature shape to the low temperature one (and from low to high) is obtained without any applied stress assistance. In contrast to the preceding behaviors (superelasticity, one-way shape memory), that are intrinsic, the two-way shape memory is an acquired characteristic.

The superthermoelastic behavior described in section 3.1 only constitutes an approximation of the actual stress-induced behavior of shape memory alloys. In fact, one observes only a partial reversion of the maximal strain. A small residual strain remains after each unloading. After several loading-unloading cycles, one observes that this residual strain increases then saturates. Further cooling of the material, in absence of applied stress is now related to the occurrence of a macroscopic transformation strain contrarily to what is observed on a material before cycling. The material has been trained. This phenomenon has been observed since 1974 [32]. Many different training sequences can be used [33].

Microstructural observations [34] have shown that the fundamental mechanism involved in all these methods consists to the production of a microstructure of oriented defects (dislocations array, precipitates, residual martensite...). The training process induces a microstructural dissymmetry. Such a dissymmetry is responsible for a preferential growing of the martensitic variants. However the exact nature of this dissymmetry and the precise way that it participates to the two-way shape memory is still in discussion. It is clearly established that the first variants formed to the cooling insure the most part of the macroscopic transformation strain [35]. One observes equally that variants formed during the two-way memory effect are identical to those induce in the course of the training.

In fact, two-way shape memory effect corresponds to a superthermal effect in which applied stresses are replaced by internal stress induced by the training process. This explains that the super-thermal effect is sometimes called stress-assisted two-way memory.

If for various reasons (aging, overloading, ordering...), the internal stress field is modified, the two-way memory strain is reduced or can be vanished [36].
The training treatment also induces secondary effects: evolution of the transformation temperatures, change in the hysteresis size, evolution in the transformation stress range and decrease of the macroscopic transformation strain. These effects relate to these observed during fatigue tests performed on this type of alloy. This imposes to optimize the training treatment. A low number of training cycles produces an unstable two-way effect but an overtraining generates undesirable effects that reduce the efficiency of the treatment [35].

3.4. Rubberlike effect

This effect, first observed by Ölander in 1932 on an Au-Cd alloy [37], constitutes the first studied manifestation of shape memory effect. This type of behavior is a characteristic of the martensitic phase ($T < M_f$). It gives a similar stress-strain curve as the superelastic behavior but without any phase change. The elementary mechanism involved is a reversible motion of interface between martensitic domains having different crystallographic orientation. These domains are generally twin related. In shape memory alloys, to assure the existence of the invariant plane, the invariant lattice strain to be added to the Bain strain, occurs generally by twinning. Under an applied stress, these twin boundaries are moving and thus induce a macroscopic strain. Twins boundaries being very mobile, the critical stress to move them is very weak (some MPa). When this displacement is not reversible, loading-unloading path gives a residual strain that is alike a plastic strain but with a limited amplitude.
When this displacement is reversible on unloading (like in Au-Cd alloys), the reversible macroscopic strain obtained is composed with the usual elastic strain and a reversible component associated to the interface motion. For a given stress, a reversible total strain larger than the usual elastic one is then obtained. The stiffness modulus is therefore largely smaller to the elastic one. This phenomenon is called rubberlike effect or in some cases (by analogy with super-thermoelasticity) as pseudoelasticity by reorientation. Temperature plays only a secondary role in this behavior since there is no phase change.

3.5. Damping capacity

This last aspect is not a characteristic behavior proper to shape memory alloys. It consists in the degradation of mechanical energy in heat. All materials present this property. However shape memory alloys exhibit a damping capacity far larger than that of usual materials. This is linked to the existence of the numerous interfaces related to the martensitic transformation: between austenite and martensite, between the different variants of martensite and due to twin boundaries inside the martensit itself.

This property is linked to dissipative aspects linked to all physical phenomenon. In spite of the thermoelastic character of the transformation many irreversible events occur (production of defects, motion of dislocations...). Hysteresis observed in superelasticity is one of this energy dissipation manifestation [38].

During a transformation cycle (or a reorientation one), one characterizes the damping capacity of an isotropic material by the ratio of the dissipated energy over the total energy developed for one cycle. This ratio classically depends on the frequency of the solicitation, the amplitude of vibrations and the temperature. In shape memory alloys it also depends on the difference T-M_S to the transformation temperature. One generally distinguishes three damping regimes in these materials (figure 11).

a) For temperatures larger than M_S and under weak mechanical solicitations, the material remains in the austenitic state. The damping capacity is weak, and comparable to that of metallic materials.

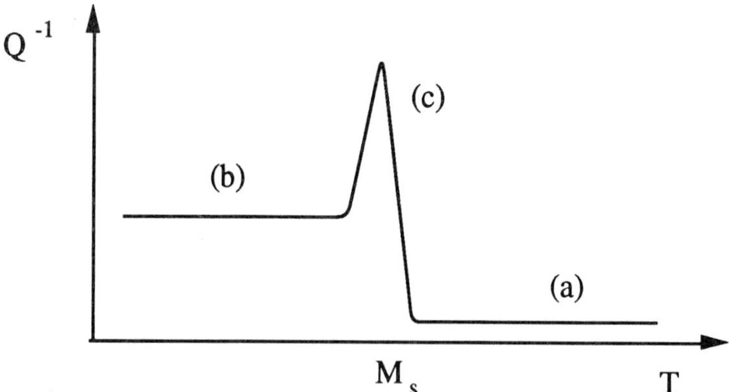

Figure 11 : Evolution of the damping capacity with the temperature for different material states (a) pure austenite, (b) martensite, (c) inside the transformation domain.

b) For temperatures below M_f, the damping capacity becomes more important. That is linked to the large amount of interfaces present in the material in the low temperature phase.

c) The maximum is reached for temperatures closer to the transformation temperature. Creation and displacement of austenite-martensite interfaces being accompanied by a very strong level of defect production and large thermomechanical coupling.

The control of vibrations being a problem of first importance in many industrial sectors (aeronautic, space, high level athletic equipment ...) damping properties linked to the martensitic transformation constitute a source of potential development for shape memory alloys.

3.6. Strain mechanisms

The different behaviors presented in the preceding paragraphs are related to the elementary strain mechanisms observed in shape memory alloys. All these mechanisms are linked to particular properties presented by the martensitic transformation and by the martensite. The polyvariant structure of this phase and the existence of self-accommodated patterns are essential aspects. First of all, the possibility to select the growing of preferential variants using a stress field (applied or internal) is at the origin of superelasticity and two-way memory effect. Then, the great mobility of interfaces is responsible to the large inelastic strain observed in the martensitic state and therefore at the origin of the so-called shape memory effect. These mechanisms are also at the origin of the rubberlike effect and of the large damping capacity presented by these materials.

To summarize, it is possible to classify these mechanisms in four categories:

(a) production of stress-induced oriented martensite (due to applied or internal stress field) and reversion into the austenitic state (observed in superelasticity, two-way shape memory effect and for damping capacity),

(b) growing of non oriented martensite plates (self accommodated groups of variants) during a stress-free cooling (one-way shape memory effect),

(c) irreversible reorientation of the variants in the martensitic state applying an external stress (one-way shape memory effect, damping capacity),

(d) partial but reversible reorientation of variants under the action of an applied stress in the martensitic state (rubberlike effect, damping capacity).

To model the shape memory behaviors implies to adopt a scale of description adapted to the different strain mechanisms involved. That is presented in chapter V including a kinematical analysis of the transformation in a crystal of parent phase and a thermodynamical approach to define constitutive equations. This framework of resolution is applied on stress-induced transformation behavior.

4. Ferrous alloys. TRIP behavior

One very important difference between thermoelastic and non thermoelastic alloys is the way the transformation strains are accommodated during the martensitic growth, respectively by fully elastic deformation or by elastic and plastic deformations. In TRIP phenomena, plastic deformation will play an important role, either for the nucleation, for the growth of the martensite plate and thus the resulting plate morphology, and finally for the associated mechanical behavior.

As presented in § 2.3. martensitic transformation can be promoted by the applied stress. Considering experiments, applying a stress during martensitic transformation leads to three basic modifications :

- Variations in the M_s temperature and in the transformation kinetics
- Variations in the mechanical behavior
- Variations in the morphology of the martensite plates

These variations have been studied for different kinds of experiments, either during cooling under a constant stress, or at a constant temperature during deformation.

4.1. Stress-Assisted and Strain-Induced Transformation

In the case of non thermoelastic ferrous alloys, the critical stress to induce martensitic transformation is considered to follow the scheme defined in figure 2. The critical stresses at which martensite is formed at different test temperatures are given figure 12 for two ferrous alloys. For the Fe-Ni-C alloy, in the temperature range M_s - M_s^Y, the first observed inelastic strain is due to the martensitic transformation and it is observed for a critical stress σ_C. The martensite formed in this temperature range is referred as to stress-induced martensite. In the range M_s^y - M_d, the plastic strain occurs first and transformation is then simultaneous to the plastic deformation in the parent phase. This martensite is called the strain-induced martensite. For temperatures above M_d, only slip deformation is observed in the parent phase.

In the stress-assisted temperature range, the relation between the critical stress to assist martensitic transformation and the temperature test is obtained considering the additional driving force related to the applied stress as given in § 2.3. In this way the M_s temperature variations with the applied stresses were very early predicted by Patel and Cohen [39].The thermodynamical criterion was further developed in order to take into account the different "twins or polydomains" arrangement when transformation occurs under stress, and which results in a change of the macroscopic shear strain γ_0 [40, 11].

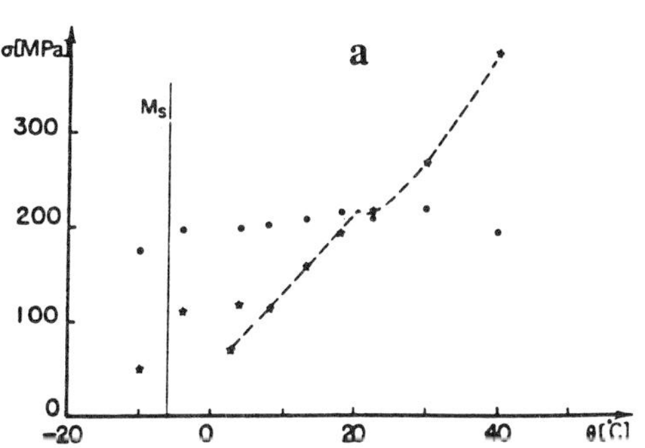

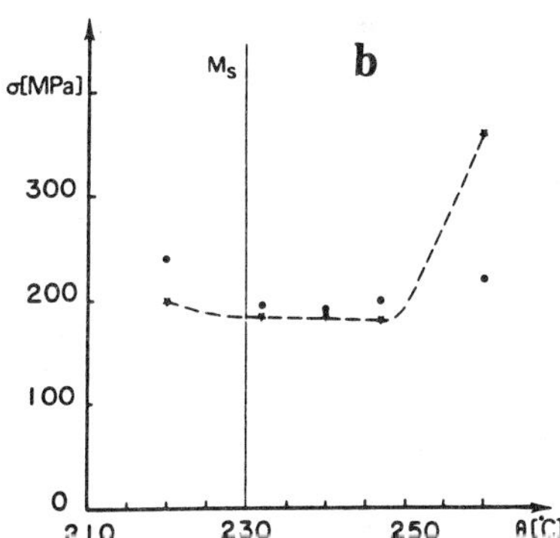

Figure 12 : Variations in the critical stress to induce martensite
a) Fe-Ni-C alloy b) 60NCD11 steel
* Critical stress to induce martensite measured by thermomagnetic measurements
• 0.2% Yield stress

However, the linear relation between the critical stress and the test temperature is not always observed as shown in figure 12 (60NCD11 steel) [20, 41]. Olson et al. [20] attributed such variations to the isothermal character of the martensitic transformation in some alloys. Such behavior can also be understood considering that the transformation strain accommodation which is an important parameter in the martensitic transformation varies from a self accommodating process to an elasto plastic process [42].

In the strain-induced transformation range, the lowering of the critical stress in regard to the predicted one considering the thermodynamic approach has been related to different origins :
• formation of new nucleation sites as proposed by Olson and Cohen [20, 43]. Lecroisey and Pineau [44] clearly evidenced such origins by observation of new martensite plates at intersection of deformation bands. The formation of new nucleation sites is a function of the stacking fault energy (SFE). If SFE is low, new nucleation sites are favoured. The M_s-M_d temperature range is thus enlarged as shown in table 1.

Table 1 : Variations in M_s-M_d temperature range versus stacking fault energy [45].

COMPOSITION	M_s-M_d (°C)	γ ergs/cm^2	M_s (°C)
Fe -30Ni	28	>60	- 25
Fe-7Cr-19Ni	100	50	- 25
Fe-16Cr-11Ni	170	20 - 23	- 20

• Stress concentration occurs at obstacles (grain boundaries, twin boundaries, ...). The local stress which assists the transformation is thus larger than the macroscopic applied stress [46].
• If transformation occurs in the plastic deformation range of austenite the transformation strains will be accommodated by plastic deformation in the austenite and thus reducing the resistive force for the transformation [47].
A last case can be considered, if plastic deformation occurred before the transformation. A decrease in the M_s temperature can then be observed for low amounts of plastic deformation [48]. This decrease can be linked to the work hardening of the austenite and to the friction forces associated with the plate growth which are increased due to the higher dislocations density in the matrix.

4.2. Kinetics of the transformation

For transformation during tensile test, the amount of martensite progresses as the stress or the strain increases. Onodera and Tamura [46] proposed a model based on thermodynamical considerations where the progress of the transformation (f amount of martensite formed) is related to the external applied stress σ :

$$f(\sigma) = \frac{1}{\alpha}\int_0^{\alpha}\frac{2}{\pi}\int_0^{\frac{\pi}{2}} K\left(\frac{A}{B}\right)\left\{\frac{\sigma}{2}\left[\gamma_0 \sin 2\theta \cos 2\alpha + (1+\cos 2\theta)\right] - W\right\} d\theta\, d\alpha$$

K is a constant ;
A is a constant for transformation during cooling without stress such as $f = A\Delta T$;
$B = \Delta S$;
α and θ were defined previously (α_0 is a maximal)

More usually, the kinetics of the transformation has been treated as a function of the plastic strain ε.

Different relations were proposed [49]. Olson and Cohen [43] proposed a model considering that new nucleation sites were produced when plastic deformation occurs. The formation of the new nucleation sites was dependent on the formation rate of shear bands, this deformation being function of the SFE, and the deformation rate. The fraction of martensite is :

$$f = (1 - \exp[-\beta(1 - \exp(-\alpha\,\varepsilon)^n)]$$

α is related to the production of shear bands intersection = f (SFE)
β is the probability that a shear band intersection acts as a nucleation site
n is a constant

This relation was further developed by Stringfellow et al. on the same physical basis considering an additional effect of the stress [50]. They considered on one hand the plastic strain effect which affects the formation of new nucleation sites through the parameter α, and on the other hand the stress effect which affects the propability distribution.

Each of this model can be used depending on the temperature of the tests and the range of the applied stress. Indeed, Zhang [47] showed that the kinetics can be described by Onodera and Tamura model at temperatures near the M_s^y temperature, and for stresses larger than the yield stress of the parent phase when the transformation strains are not self-accommodated. For the larger temperature tests, Olson and Cohen relationship fits better. Its is clear that the effect of strain and stress have both to be considered.

For transformation during cooling under constant stress the M_s variations is considered essentially. However the kinetics of the transformation can be slightly modified [47].

4.3. Transformation plasticity deformation

Transformation plasticity deformation occurs when a specimen transforms under stress, for stresses even lower than the yield stress of each constituent of the material. The occurrence of this phenomenon in martensitic transformation can lead in some cases to the observation of very large deformations associated with the strain induced martensitic transformation as shown in figure 13.

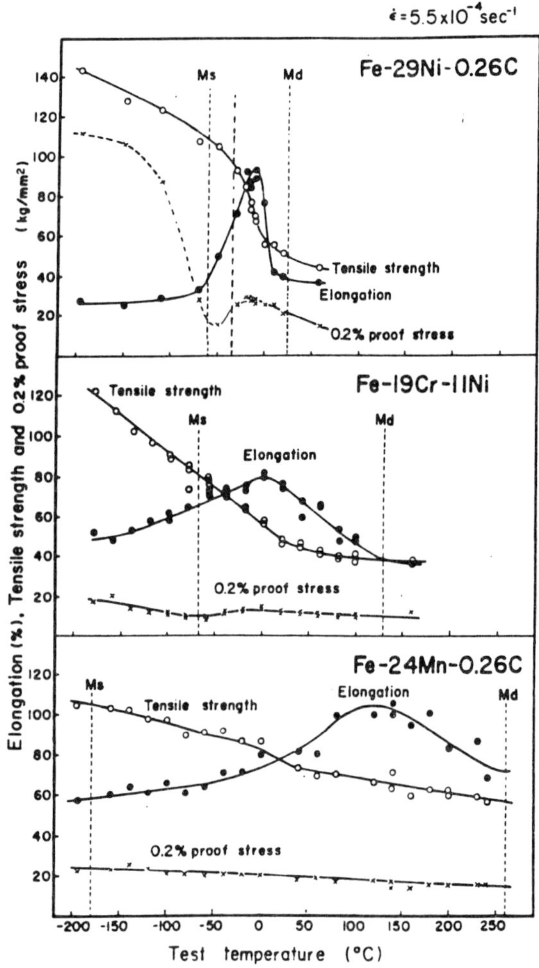

Figure 13 : Effect of test temperature on tensile properties for three metastable austenitic alloys [49].

For martensitic transformation, transformation plasticity deformation is generally attributed to two basic mechanisms :

1 - Orientation of the martensite plates by the applied stress [51]. This effect is due to the shear component of the phase transformation deformation. It is widely illustrated by superelasticity or shape memory effect (§ 3). The deformation associated with the transformation of a single crystal is not strongly dependent on the level of the applied stress during transformation. It mainly depends on the relative orientation of the single crystal in regard to the applied stress. The deformation should be the same if the specimen (single crystal) transforms during a tensile test at constant temperature, or during cooling in the transformation range under a constant applied stress (§ 3). The deformation may be reversible when martensite transforms in austenite in a reversible way .
The contribution of this mechanism is then dependent on the orientation distribution of the martensite plates. For the polycrystal, two points have to be considered : i) The polycrystal presents a grain orientation distribution. The most favourable orientation of the martensite plates in regard of the applied stress will be different in each grain, and so the deformation in the direction of the applied stress as modelled by Magee [51]. ii) When plates are formed, deformation incompatibilities (near grain boundaries for example) will lead to internal stresses. These internal stresses will locally modify the stress tensor. The plate orientation will no longer be determined by the external applied stress but by the local stress state existing in the material [52].
Moreover, for this mechanism one generally considers that the shear component is not modified when the transformation occurs under stress. However Kosenko et al. [53] and Pankova and Roytburd [40] have shown that for stress - induced martensite in ferrous alloys, the distribution of internal twins is modified and in consequence the value of the macroscopic shear deformation is modified too. The modification of the internal structure of the plate (modification of domains distribution) due to a different stress state in the presence of an applied stress, is an additionnal source of transformation plasticity deformation [54] to the one defined by Magee.

2 - Orientation of the plastic yielding around the transforming particles when transformation deformations are accommodated by plastic deformation. The contribution of this mechanism is dependent on the transformation strain, the applied

stress, the mechanical properties of the phases and the transformation progress [55-58].

For martensitic transformation of ferrous alloys, it is necessary to consider the accommodation process of both the volumic variations and the macroscopic shear deformation. The volumic variations can be accommodated by elastic or plastic deformation. For the macroscopic shear deformation different process have to be considered. The shear deformation can be accommodated by elastic deformation in the parent phase, by formation of self accommodating plates, by elasto plastic accommodation mainly in the parent phase or by an inelastic deformation at the plate tips in the product phase [59]. The effect of the external stress is then to modify the process of the shear strain accommodation.

Experimental studies of transformation plasticity are usually performed for transformation during cooling under a fixed stress. However some experiments were conducted during straining at temperatures in the M_s-M_d temperature range showing large elongations which is often refered as the TRIP effect (figure 13). This phenomenon is also refered as a "dynamical" softening of the material [60]. The main difficulty in the experimental study of transformation plasticity deformation is to obtain the variations of the transformation plasticity deformation associated with the amount of martensite formed under the given stress.

In figure 14 are reported transformation plasticity deformation variations measured versus the content of martensite during cooling under different constant applied stresses for two Fe-Ni-C alloys. Transformation plasticity deformation ε_{pt} is defined as : $\varepsilon_{pt}(f) = \varepsilon_{\sigma}(f) - \varepsilon_0(f)$ where $\varepsilon_{\sigma}(f)$ and $\varepsilon_0(f)$ are the deformations due to the formation of a content of martensite f, under a stress equal to σ and to 0 respectively. Figure 15 shows the variations of ε_{pt} normalised to martensite content ($\Delta\varepsilon_{pt}/\Delta f$) versus σ for three given martensite contents.

Three stress ranges can be defined where $\Delta\varepsilon_{pt}/\Delta f$ increases slowly with σ (50 MPa for Fe-20Ni-0.5C alloy), and then increases rapidly (until 200 MPa). For higher stresses (> 200 MPa), $\Delta\varepsilon_{pt}/\Delta f$ increases again slowly. These variations are similar for all given values of martensite content.

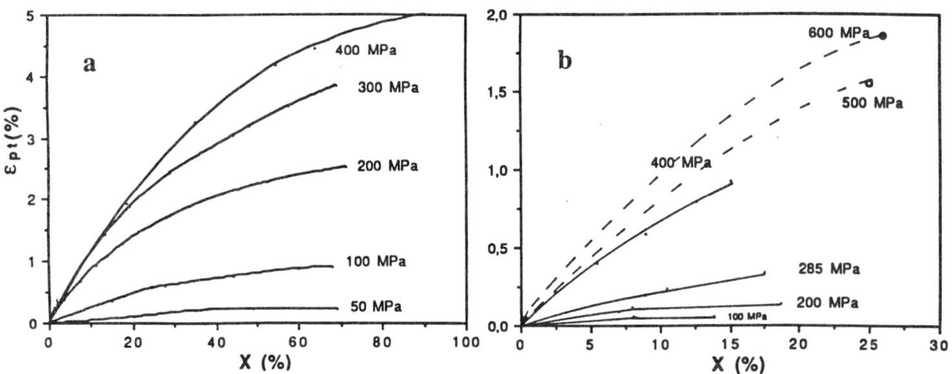

Figure 14 : Variations of transformation plasticity deformation versus the content of martensite obtained during cooling under different applied stresses.
a) Fe-20Ni-0.5C alloy M_s = -35°C ;
b) Fe-25Ni-0.66C alloy M_s = -153°C [47, 61]

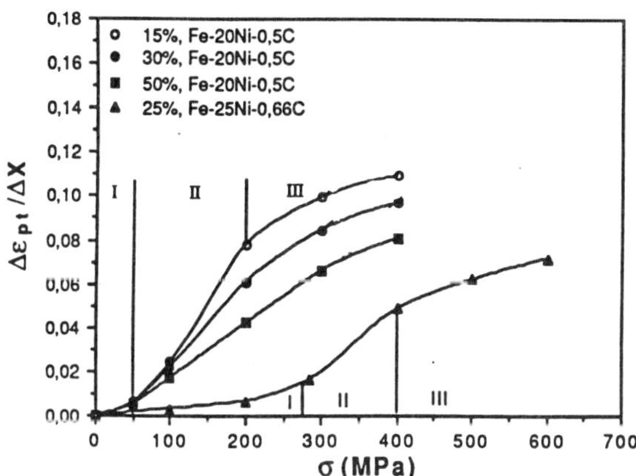

Figure 15 : Variations of transformation plasticity deformations normalized to the amount of phase formed versus the applied stress for different content of martensite. [61]

The transformation plasticity deformation varies as a function of the amount of martensite, the applied stress and the nature of the alloy (mechanical properties of the phases, transformation strains). The detailed study of theses alloys [61, 62] showed that in the first stress range I, the transformation plasticity deformation is due to the first mechanism, orientation of the plates or the domains, and the deformation is reversible as the reverse transformation proceeds. Moreover, the results for the two alloys show that the rapid increase of transformation plasticity in stress range II is linked to a non reversible mechanical behavior of the alloys. The microstructures observed for specimens transformed under different stresses (figure 16) reveal that a favourable orientation of the plates is observed in the grain when the stress is increased. The orientation is clearly observed when the stress reaches a value of 400MPa. It is interesting to note that these values are near the yield strength of the austenite [47].

Figure 16 : Optical micrographs of specimen transformed during cooling under various constant applied stresses for the Fe-25Ni-0.66C alloy.
a) 0 MPa ; b) 285 MPa ; c) 500 MPa ; d) 600 MPa

The occurrence of a clear orientation of the martensite plates when stress is increased is significant of a large influence of the applied stress to the orientation of the plate. This is reached if the internal stresses due to deformation incompatibilities are relaxed. As this orientation effect becomes more sensitive when the applied stress reaches a value near the yield stress of the austenite, one can conclude that the relaxation of the internal stresses occurs by plastic deformation of the austenite. Also as the shear component is relaxed by plastic deformation of the austenite, less self accommodating plates will be formed and an increasing contribution of the first mechanism is observed. When stress increases and its value is near the yield stress of the austenite, the two mechanisms are acting, and the second mechanism favors the first one.

For stresses in range III, the decrease in transformation plasticity as stress increases is related to the saturation effect of the plate orientation. Most of the plates present the favorable orientation, and the increase in transformation plasticity deformation is related to the second mechanism.

These different mechanisms and their contribution versus the applied stress for a low martensite content are schematically drawn in figure 17. As martensite content increases, the limits of the domains at which these mechanisms intervene are modified. The degree of plate orientation decreases as the martensite content increases. Experimentaly this could be observed considering transformation plasticity variations versus the martensite content for an applied stress of 50MPa and for the Fe-20Ni-0.5C alloy. For martensite contents larger than 25%, the transformation plasticity deformation does not increase further. That means that only plate orientation occurs at that stress level, and the orientation distribution of the plates is random at $f > 0.25$. For larger stresses, this orientation distribution becomes random at much larger martensite contents.

Recent experiments were performed studying the effect of multiaxial stress on the martensitic transformation [63]. These results showed that the transformation plasticity expressed as an equivalent plastic strain cannot simply be related to the equivalent stress. This can be easily understood considering the above analysis about the role of the internal stress on the different transformation plasticity contributions which will be different for different thermomechanical path.

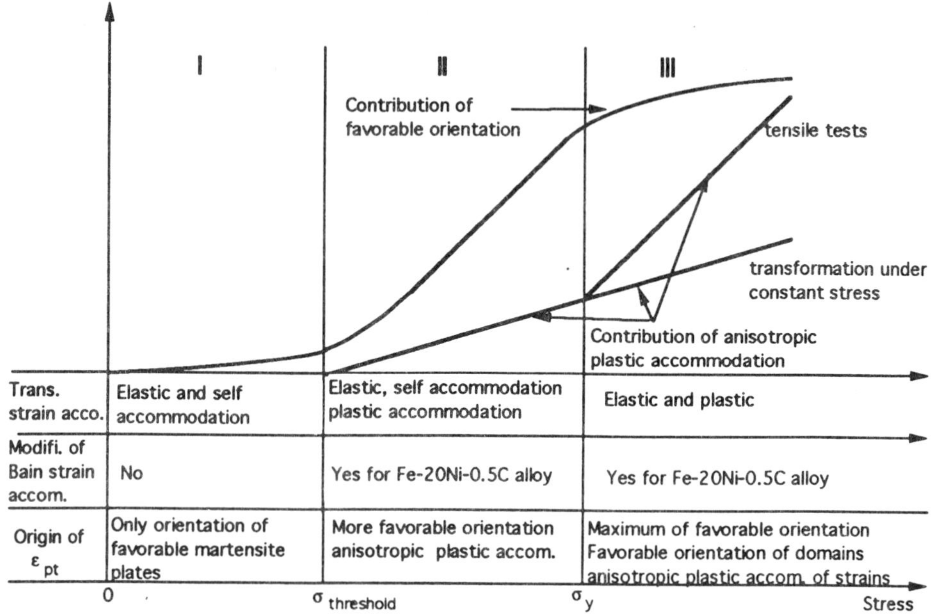

Figure 17 : Schematic contribution of the different mechanisms to the transformation plasticity deformation.

The study of the transformation plasticity during a tensile test revealed the variations of the transformation plasticity deformation is larger than the one obtained during transformation under constant stress for similar conditions of stresses and temperatures as shown figure.18 [47]. Moreover, the large deformations are only observed in the plastic deformation range. These large deformations are simultaneous to changes in the plate morphology [64 - 66]. The habit plane of the plates are modified either from $(225)_f$ to $(111)_f$ or from $(259)_f$ to $(111)_f$ when the applied stress or the test temperature increase.

The plate morphology modifications (habit plane, internal structure) are related to modifications in the Bain strain accommodation mechanism [64-66]. In analogy to the transformation plasticity deformation proposed by Roytburd [54] where the behavior of the phases are elastic and the interfaces stay coherent, domains for a given Bain strain with preferred orientation will be formed. Moreover, the internal stresses due to Bain strain will be more and more relaxed by plastic deformation in the parent phase. A limiting case can be a complete relaxation by plastic deformation in the parent phase. As a consequence, the transformation plasticity deformation increases. Again both "mechanisms" of transformation plasticity are acting : orientation of domains (the strain tensor of the domains being the Bain strain tensor) and plastic accommodation of the Bain strain as is schematized in figure 17, leading to larger transformation plasticity deformations

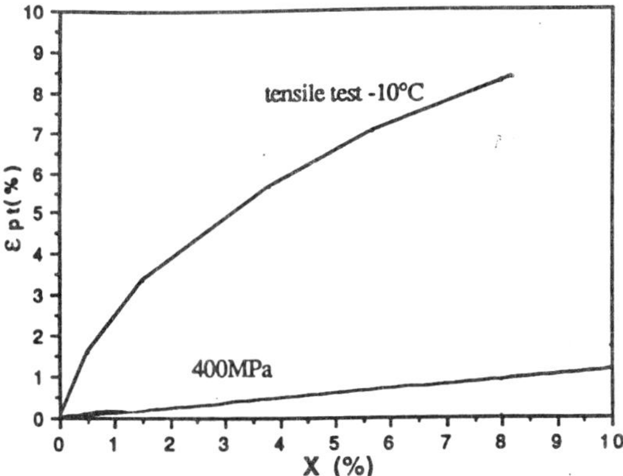

Figure 18 : Comparison of the two tests. Variations of ε_{pt} versus martensite content during a tensile test and a cooling test under constant stress.

At last it must be noted that the large transformation plasticity deformations are only observed for transformation during plastic deformation (not during cooling). This is related to the dynamical character of both the plastic deformation and the transformation. A further analysis of the transformation during plastic deformation would be necessary to better understand these phenomena.

References

1. R. MADDIN, "*A History of Martensite : Some Thoughts on the Early Hardening of Iron*", Chapter 2, "Martensite", (Eds. G.B. OLSON, W.S. OWEN) ASM International, pp. 11-19 (1992).
2. Z. NISHYAMA, "Martensitic Transformation", Academic Press (1978).
3. J.S. BOWLES, C.M. WAYMAN, "*The Bain Strain, Lattice Correspondances and Deformations Related to Martensitic Transformations*", Metall. Trans. Vol. 3, pp. 1113-1121 (1972).
4. L. KAUFMAN, M. COHEN, "*Martensitic Transformations*", in Progress in Metal Physics 7, (Eds. B. CHALMERS and R. KING), Pergamon Press, pp. 165-246 (1958).
5. J. DE VOS, E. AERNOUDT, L. DELAEY," *The Crystallography of the Martensitic Transformation of B.C.C into 9R a Generalized Mathematical Model*", Z. Metallkde, Bd. 69, H7, pp. 438-444 (1978).
6. L. DELAEY, "*Diffusionless Transformations*", chapter 6 in Materials Science and Technologies, Vol. 5 : Phase Transformations in Materials, (Eds. R.W. CAHN, P. HAASEN, E.J. KRAMEN) Ed. VCH Publishers, ISBN 3-527-26818-9, pp. 339-404 (1991).
7. M. S. WECHSLER, D. S. LIEBERMAN, T. A. READ," *On the Theory of the Formation of Martensite*", Trans AIME, Vol. 197 , pp. 1503-1515 (1953).
8. J.S BOWLES, J.K. MACKENZIE, "*The crystallography of martensitic transformations I and II*", Acta Metall., Vol. 2, pp. 129-137 and 138-147 (1954).
9. K. OTSUKA, "*Crystallography of Martensitic Transformations and Lattice Invariant Shears*", Materials Science Forum, Vols. 56-58, pp. 393-404 (1990).
10. K. BHATTACHARYA, "*Comparison of the geometrically nonlinear and linear theories of martensitic transformation*", Continuum Mech. Thermodyn, Vol. 5, pp. 205-242 (1993)
11. A. L. ROYTBURD, M.N. PANKOVA "*Effect of external stresses on habitus orientation and substructure of stress-induced martensite plates in ferrous alloys*", Phys. Met. Metall., Vol. 59, pp. 131-140 (1985).
12. J.M. BALL, R.D. JAMES, "*Fine Phase Mixture as Minimizers of Energy*", Arch. for Rational Mechanics and Analysis, Vol. 100, pp. 13-52 (1987).
13. T. SABURI, C.M. WAYMAN, K. TAKATA , S. NENNO, " *The Shape Memory Mechanism in 18R Martensitic Alloys*", Acta metall., Vol. 28, pp. 15-32 (1980).
14. J.W. CHRISTIAN, "*Deformation by Moving Interfaces*", Metall. Trans. A, Vol. 13A, pp. 509-538 (1982).
15. V. RAGHAVAN, "*Kinetics of Martensite Transformation*", Chapter 11, "Martensite", (Eds. G.B. OLSON, W.S. OWEN) ASM International, pp. 197-225 (1992).
16. M. AHLERS, "*On the Stability of the Martensite in β-Cu-Zn Alloys*", Scripta Met., Vol. 8, pp. 213-216 (1974).
17. K.W. ANDREWS, "*Empirical Formulae for the Calculation of Transformation Temperatures*", Journal of the Iron and Steel Institute, Vol. 203, No. 7, pp. 721-727 (1965).
18. J.L. MACQUERON, M. MORIN, G. GUENIN, A. PLANES, J. ELGUETA, T. CASTAN, "*Atomic Ordering and Martensitic Transition in a Cu-Zn-Al Shape Memory Alloy*", Journal de Physique IV, Coll. C4, Vol. 1, No. 9, pp. C4-259-263 (1991).
19. P. WOLLANTS, M. DE BONTE, J.R. ROSS, "*A Thermodynamic Analysis of the Stress-Induced Martensitic Transformation in a Single Crystal*", Z. Metallkde, Bd. 70, H.2, pp. 113-117 (1979).

20. G.B. OLSON, M. COHEN, "*Stress Assisted Isothermal Martensitic Transformation : Application to TRIP Steels*", Metall. Trans. A, Vol. 13A, pp. 1907-1914 (1982).
21. M. GRUJICIC, H.C. LING, D.M. HAEZEBROUCK, W.S. OWEN, "*The Growth of Martensite*", Chapter 10, "Martensite", (Eds. G.B. OLSON, W.S. OWEN) ASM International, pp. 175-196 (1992).
22. G.B. OLSON, M. COHEN, "*Thermoelastic Behavior in Martensitic Transformations*", Scripta Met., Vol. 9, pp. 1247-1254 (1975).
23. M. COHEN, "*Martensitic Transformations in Material Science and Engineering*", Transactions of the Japan Institute of Metals, Vol. 39, No. 8, pp. 609-624 (1988).
24. M. AHLERS, P. PASCAL, R. RAPACIOLI, "*Transformation Hardening and Energy Dissipation in Martensitic β-Brass*", Materials Science and Engineering, Vol. 27, pp. 49-55 (1977).
25. F.C. LOVEY, A. AMENGUAL, V. TORRA, M. AHLERS, "*On the Origin of the Intrinsic Thermoelasticity Associated with a Single-Interface Transformation in Cu-Zn-Al Shape Memory Alloys*", Phil. Mag. A, Vol. 61, No. 1, pp. 159-165 (1990).
26. A. CHRYSOCHOOS, H. PHAM, O. MAISONNEUVE, "*Une analyse expérimentale du comportement thermomécanique d'un alliage à mémoire de forme de type Cu-Zn-Al*", C.R. Acad. Sci. Paris, t. 316, Série II, pp. 1031-1036 (1993).
27. K. OTSUKA, C.M. WAYMAN, K. NAKAI, H. SAKAMOTO, K. SHIMIZU, "*Superelasticity Effects and Stress-Induced Martensitic Transformations in Cu-Al-Ni Alloys*", Acta metall., Vol. 24, pp. 207-226 (1976).
28. K. SHIMIZU, "*Ageing and Thermal Cycling Effects in Shape Memory Alloys*", J. Electron Microsc., Vol. 34, No. 4, pp. 277-288 (1985).
29. J.E. REYNOLDS, M.B. BEVER, "*On the Reversal of the Strain Induced Martensitic Transformation in the Copper-Zinc System*", Journal of Metals, pp. 1065 (1952).
30. E. PATOOR, A. EBERHARDT, M. BERVEILLER, "*Potentiel pseudoélastique et plasticité de transformation martensitique dans les mono et polycristaux métalliques*", Acta metall., Vol. 35, pp. 2779-2789 (1987).
31. H. SAKAMOTO, "*Fatigue Behavior of Monocrystalline Cu-Al-Ni Shape Memory Alloys under Various Deformation Modes*", Trans. of the Japan inst. of Metals, Vol. 24, No. 10, pp. 665-673 (1983).
32. J. PERKINS, "*Residual Stresses and the Origin of Reversible (Two-Way) Shape Memory Effect*", Scripta Met., Vol. 8, pp. 1469-1476 (1974).
33. L. CONTARDO, G. GUÉNIN, "*Training and Two Way Memory Effect in Cu-Zn-Al Alloy*", Acta metal. mater., Vol. 38, No. 7, pp. 1267-1272 (1990).
34. D. RIOS JARA, G. GUENIN," *On the Characterization and Origin of the Dislocations Associated with the Two-Way Memory Effect in Cu-Zn-Al Thermoelastic Alloys*", Acta metall., Vol. 35, pp. 109-119 et pp. 121-126 (1987).
35. R. STALMANS, J. VAN HUMBEECK, L. DELAEY, " *The Two Way Memory Effect in Copper-Based Shape Memory Alloys - Thermodynamics and Mechanisms*", Acta metall. mater., Vol. 40, No. 11, pp. 2921-2931 (1992).
36. P. RODRIGUEZ, G. GUÉNIN, "*Stability of the Two Way Memory Effect During Thermal Cycling of High M_s Temperature Cu-Al-Ni Alloy*", Material Science Forum, Vol. 56-58, No. 7, pp. 541-546 (1990).
37. A. ÖLANDER, "*An Electrochemical Investigation of Solid Cadnium-Gold Alloys*", Journal of the American Chemical Society, Vol. 54, pp. 3819-3833 (1932).
38. J. STOIBER, J. VAN HUMBEECK, R. GOTTHARDT, "*Hysteresis Effects During Martensitic Transformation in a Cu-Zn-Al Studied by Internal Friction Measurements*", Material Science Forum., Vol. 56-58, No. 7, pp. 505-510 (1990).

39. J.R. PATEL, M. COHEN, "*Criterion for the Action of Applied Stress in the Martensitic Transformation*" Acta Metall. Vol 1, pp. 531-538 (1953)
40. M. PANKOVA, A. L. ROYTBURD, "*Orienting Influence of Applied Stress on the Martensitic Transformation in Alloys Based on Iron*" Phys. Met. Metall., Vol. 58, pp. 81-90 (1984).
41. E. GAUTIER, A. SIMON, G. COLLETTE, G. BECK, "*Effect of Stress and Strain on Martensitic Transformation in a Fe-Ni-Mo-C Alloy with a High M_s Temperature*" Journal de Physique Colloque C4, Vol.43, pp. C4 473-477 (1982).
42. E. AEBY-GAUTIER, "*Transformation perlitique et martensitique sous contrainte dans les aciers*", Thèse de Doctorat d'État ès Sciences Physiques, Institut National Polytechnique de Lorraine, Nancy, France (1985).
43. G.B. OLSON, M. COHEN "*Kinetics of Strain Induced Martensitic Nucleation*" Metall. Trans. A, Vol. 6A, pp. 791-795 (1975).
44. F. LECROISEY, A. PINEAU, "*Martensitic Transformations Induced by Plastic Deformation in the Fe-Ni-Cr-C System*", Met. Trans., Vol 3, pp. 387-396 (1972).
45. F. LECROISEY, "*Transformations martensitiques induites par déformation plastique dans le système Fe-Ni-Cr-C*", Thèse de Doctorat ès Sciences Physiques, Université de Nancy I (1971).
46. H. ONODERA, I. TAMURA, "*Effect of Stress and Strain on Deformation Induced Martensitic Transformation in Austenitic Steels*", Proc. U.S. Japan Seminar on Mechanical Behavior of Metals and Alloys Associated with Displacive Tansformation Try, NY, p. 24 (1979).
47. J. ZHANG,"*Influence de la contrainte sur la transformation martensitique d'alliages Fe-Ni-C*", Thèse de Doctorat de l'Institut National Polytechnique de Lorraine, Nancy (1993).
48. E. GAUTIER, A. SIMON, G. BECK, "*Martensitic Transformation Kinetics during Anisothermal Creep and Tensile Tests*", Proc. ICOMAT 1986, The Japon Institute of Metals, pp. 503 - 508 (1986).
49. I. TAMURA, C. M. WAYMAN, "*Martensitic Transformations and Mechanical Effects in Martensite*", Chapter 12, "Martensite", (Eds. G.B. OLSON, W.S. OWEN) ASM International, pp. 227-242 (1992).
50. R.G. STRINGFELLOW, D.M. PARKS, G.B. OLSON, "*A Constitutive Model for Transformation Plasticity Accompaying Strain-Induced Martensitic Transformations in Metastable Austenitic Steels*", Acta metall. mater., Vol. 40, pp. 1703-1716 (1992).
51. C.L. MAGEE, Ph D Thesis, Carnegie Mellon University (1966).
52. E. GAUTIER, X.M. ZHANG, A. SIMON, "*Role of Internal Stress State on Transformation Induced Plasticity and Transformation Mechanisms during the Progress of Stress Induced Martensitic Transformation*", Proc. ICRS2, Eds. G. BECK, S. DENIS, A. SIMON (Elsevier Applied Science) pp. 777-782 (1988).
53. N.S. KOSENKO, A. L. ROYTBURD, L.G. KHANDROS, "*Thermodynamics and Morphology of Martensitic Transformations under External Stress*", Phys. Met. Metall., Vol. 44, pp. 48-55 (1979).
54. A. L. ROYTBURD, "*Deformation through Transformations*", Journal de Physique IV, Colloque C1, sup. Journal de Physique III, Vol 6, pp. C1 10-25 (1996).
55. M. DE JONG, G.W. RATHENAU, "*Mechanical Properties of Iron and some Iron Alloys while undergoing Allotropic Transformation*", Acta Metall., Vol. 7, pp. 246-253 (1959)
56. M. DE JONG, G.W. RATHENAU, "Mechanical Properties of an Iron Carbon Alloy during Allotropic Transformation", Acta Metall., Vol. 9, pp. 714-720 (1961).

57. G.W. GREENWOOD, R.H. JOHNSON, "*The Deformation of Metals under Small Stresses during Phase Transformations*", Proc. R. Soc., Vol. 283A, p. 403 (1965).
58. E. GAUTIER, A. SIMON, G. BECK, "*Plasticité de transformation durant la transformation perlitique d'un acier eutectoïde*", Acta Metall., Vol. 35, p. 1367 (1987).
59. A. L. ROYTBURD, G.V. KURDJUMOV, "*The Nature of Martensitic Transformation*", Materials Science and Engineering, Vol. 39, pp. 141-167 (1979).
60. T. NARUTANI, G.B. OLSON, M. COHEN, "*Constitutive Flow Relations for Austenitic Steels During Strain-Induced Martensitic Transformation*", J. de Physique, Vol. 43, pp. C4 429-434 (1982).
61. J.S. ZHANG, E. GAUTIER, A. SIMON "*Reversible and Irreversible Transformation Plasticity Deformations in Fe-Ni-c Alloys*", Proc. ICOMAT 1992, Monterey Institute of Advanced Studies, pp. 503 -508 (1992).
62. E. GAUTIER, J.S. ZHANG, X.M. ZHANG, "*Martensitic Transformation under Stress in Ferrous Alloys. Mechanical Behavior and Resulting Morphologies*", Journal de Physique IV, Colloque C8, sup. Journal de Physique III, Vol 5, pp. C8 41-50 (1995).
63. J.C.VIDEAU, G. CAILLETAUD, A. PINEAU, "*Experimental Study of the Tansformation Plasticity in a Cr-Ni-Mo-Al-Ti Steel*", Journal de Physique IV, Colloque C1, sup. Journal de Physique III, Vol 6, pp. C1 465-474 (1996).
64. X.M. ZHANG, E. GAUTIER, A. SIMON, "*Martensite Morphology and Habit Plane Transitions during Tensile Tests for Fe-Ni-C Alloys*", Acta metall., Vol. 37, pp 477-487 (1989).
65. X.M. ZHANG, D.F. LI, Z.S. XING, E. GAUTIER, J.S. ZHANG, A. SIMON, "*Morphological Transitions of Deformation Induced Lenticular Martensite in Fe-Ni-C Alloys*", Acta metall. mater., Vol. 41, pp. 1683 - 1689 (1993).
66. X.M. ZHANG, T.R. HU, X.M. MENG, Y.Y. LI, E. GAUTIER, J.S. ZHANG, "*Transformation Mechanisms of Deformation-Induced Compact Martensite in Fe-Ni-C Alloys*", Journal de Physique IV, Colloque C8, sup. au Journal de Physique III, Vol 5, pp. C8 41-50 (1995).

INTERACTIONS BETWEEN STRESSES AND DIFFUSIVE PHASE TRANSFORMATIONS WITH PLASTICITY

E. Gautier
CNRS URA 159, Nancy, France

ABSTRACT

When diffusive phase transformations occur under applied stresses different modifications appear depending on the nature of the applied stress. The effect of the hydrostatic stresses is essentially to modify the transformation kinetics and the composition of the phases formed. These results are briefly reviewed. For non hydrostatic stresses, different modifications are observed, changes in the transformations kinetics, in the mechanical behaviour i.e. Transformation Plasticity Deformation, and in some cases in the morphology of the product phases. The effect of monoaxial stresses will be shown considering both kinetics changes and modifications of the mechanical behaviour. The paper is focused on cases were both transformation kinetics modifications and transformation plasticity are observed. Most examples are given for ferrous alloys and namely the pearlitic transformation.

1. INTRODUCTION

The effect of stresses on solid solid phase transformations has to be considered because the mechanisms involved, the composition of the precipitates, the morphology of the plates ... can be altered due to these stresses. Solid solid phase transformations are often concerned with coherency stresses. Their effect is more and more considered in the analysis of the phase transformation mechanism [1]. In this paper we will report about the experimental effect of "macroscopic" stressses for which changes in the microstructure and in the mechanical properties can be observed. These stresses can either be applied during the transformation or exist in the parts when transformation proceeds

The influence of stresses on phase transformations has been studied by different authors. Applying a stress during the transformation will modify the thermodynamics of the transformation and thus will lead to changes in the phase diagrams, the kinetics of the transformation and the morphology of the transformation. Different studies consider the influence of hydrostatic stresses on phase diagrams.

Another group of studies is dealing with homogeneous precipitation under stress. Drastic changes in the morphology can be observed. Also, favorable orientation of precipitates transformed under stress has been observed in aluminium alloys for example [2]. In the case of Nickel base superalloys, applying a stress at a constant temperature (and a constant amount of precipitates) can lead to precipitates raftening [3]. The direction of the precipitates will depend on the direction of the applied stress, the misfit between the precipitate and the matrix, and the differences in elastic properties. Different models were developed to analyse such variations [4], to predict the change in morphology in the case of isolated precipitate [5,6], or the thermodynamic driving force for the raftening [3,7]. These results will not be reported in this paper.

The role of the applied stress on the transformation kinetics and the mechanical behaviour during the phase transformation has been experimentally studied for different metallic alloys. Authors studied either the kinetics of the transformation under stress or the mechanical behaviour. Very few studies report on both effect studied simultaneously. The paper will thus essentially report about experimental results that we get considering the effect of stresses on both the kinetics and the mechanical behaviour in the case of the pearlitic transformation of an Fe-0.8%wtC steel. Some other references will be recalled presenting either similar results, or interesting experimental features of transformation under stress governed by diffusion.

2. INFLUENCE OF HYDROSTATIC STRESSES

The effect of hydrostatic stress on phase transformation is essentially to modify the phase diagrams and the kinetics of the transformation. It appears from different studies that transformation is inhibited by an hydrostatic pressure [8-13]. Results of Hilliard for an Fe-C diagram [11] showed that the A_3 and A_1 temperatures drop with increasing pressure and that the eutectoïd composition is lowered. During the decomposition of austenite, the continuous cooling transformation (CCT) and isothermal transformation (IT) curves are displaced to longer times and lower temperatures [9, 11, 13-15]. Some authors [13, 14] have reported a greater modification of the nucleation rate than of the growth rate.

Most of these variations can be explained by considering the changes in the thermodynamic phase diagram due to pressure variations. In the case of different alloys Kaufman [10] predicted the equilibrium temperature variations as the equilibrium concentrations variations between the phases (limits of solubility). For low hydrostatic pressures, the variations in the equilibrium temperature can be approached by the Clausius-Clapeyron relationship :

$$\frac{dT_e}{dP} = \frac{\Delta V . T_e}{\Delta H}$$

where T_e is the equilibrium temperature.
ΔV the molar volume variation.
ΔH the enthalpy of the transformation
P the pressure.

ΔV and ΔH are dependent on pressure and more importantly, on temperature. With ΔV and ΔH values at 1 atm. the initial slope of the T_e -P curves can be obtained.
Also for an eutectoïd steel, one obtains $dT_e/dP = 0.011 \text{ K MN}^{-1} \text{ m}^2$
with $\Delta H = 607 \text{ MJm}^{-3}$ [16], $T_e = 1000$ K and $\Delta V/V = 6.6\ 10^{-3}$

These variations concerns the equilibrium diagram. Considering the kinetics of the transformation, Nilan [15] introduced the effect of pressure into the classical nucleation theory to explain the time displacement of the the IT curves.

3. INFLUENCE OF UNIAXIAL STRESSES

When a stress is applied during the phase transformation different modifications are observed. The dilatometric recording during an isothermal pearlitic transformation under tensile stress is shown figure 1. One observes that the kinetics of the transformation is accelerated and that the length variations associated with the transformation are modified. These two modifications will be considered.

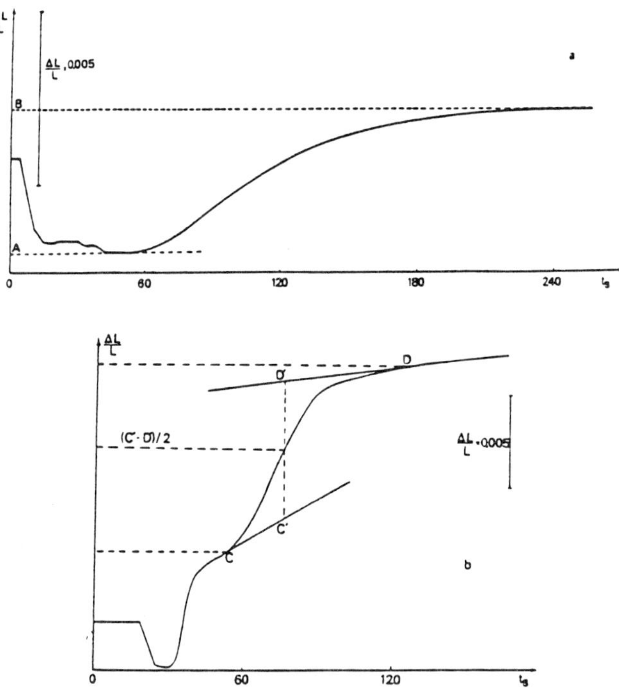

Figure 1 : Dilatometric recordings during isothermal phase transformation at 673°C (Fe-0.8 wt%C) under two constant stresses.

a) 30Mpa without creep

b) 60Mpa with creep

3.1 Kinetics modifications

In order to study the kinetics of the transformation most of the studies have been performed under isothermal conditions with constant applied stresses during the transformation. Phase transformations are studied considering the dilatometric variations associated with the volumic variations during the phase transformations or for exemple electric resistivity variations. At least the microstructure of the samples transformed under stress are considered.

The diffusional decomposition of austenite in pearlite is accelerated as a result of applied tensile or compressive stresses, so displacing the IT curve towards shorter times [17-22]. Our results on the influence of stresses on the kinetics of pearlitic transformation of an eutectoïd carbon steel are shown in figure 1. Both the beginning and ending times of transformation are shortened under stress as illustrated in figure 2. Moreover, the magnitude of this effect may be different for different transformation temperatures. In figure 2 we observe that for transformation at 663°C (fig. 2 (a)) the effect of stress on the beginning and ending times is less than for transformation at 673°C (fig. 2 (b)).

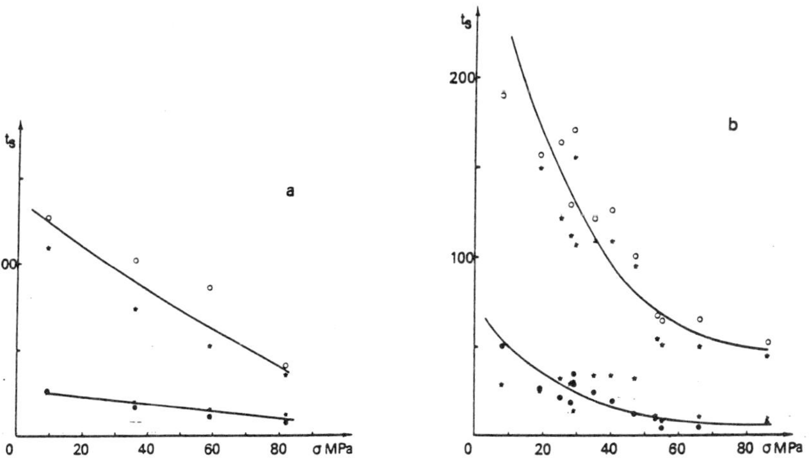

Figure 2: Influence of constant applied tensile stresses on the beginning and ending time of the isothermal pearlitic transformation (at 673 and 663°C) of an eutectoid carbon steel [22, 23] a) 663°C b) 673°C

★● beginning times ☆O ending times

This effect of stress on pearlitic transformation which is dependent on the temperature of transformation is also reported by Nocke [20]. In the same study, Nocke made measurements for tensile and compressive applied stresses. He shows that the same variations of the beginning and ending times of transformation are obtained for stresses of the same range (independent of their direction).

The transformation kinetics have also been analysed according to the laws developed by Johnson Mehl Avrami : $y = 1 - \exp(-bt^n)$ or by Zener $y = 1 - \exp - (t/\tau)^n$ (y is the amount of new phase formed, t the time, b, τ, n are constants). From experimental results on the pearlitic transformation under applied tensile stresses an increase in the reaction order n is observed [20, 22].

Studies in terms of nucleation and growth of the transformed specimens have revealed that it is the nucleation rate which is the most affected by the applied stress [18, 22]. This is clearly illustrated by figure 3 where the microstructure for specimens transformed under different stresses are shown. The pearlitic transformation has been stopped by rapid quenching of the specimen at different times. For theses specimens, the growth time (time between the quenching and the beginning of the transformation) is nearly the same for the different specimens. We observe that the size of the pearlite nodules is the same. However the number of nodules is quite larger when the specimen are transformed under stress (800 mm^{-3} for σ = 6Mpa and 1900 mm^{-3} for σ = 66MPa). Detailed analysis [24] showed that the growth is not modified by the applied stress (within the stress range investigated) but that the nucleation rate increases strongly. Moreover if the nucleation rate is described according to Russel ($N = N_S \exp(-\tau_N/t)$ N nucleation rate, N_S stationary nucleation rate, t time , τ_N incubation time for nucleation) we have shown that the effect of stress results essentially in a decrease of τ_N. The nucleation sites (grain boundaries) remain the same ; the stress leads to a more rapid site saturation. This can explain the lower effect of the stress for lower temperatures.

The microscopic phenomena responsible for this increase in nucleation rate are complex. Factors, such as a decrease in the interfacial energy (modification of grain boundary coherency due to microplasticity) or an increase in grain boundary diffusion can occur. The increase in driving force is generally considered as negligible.

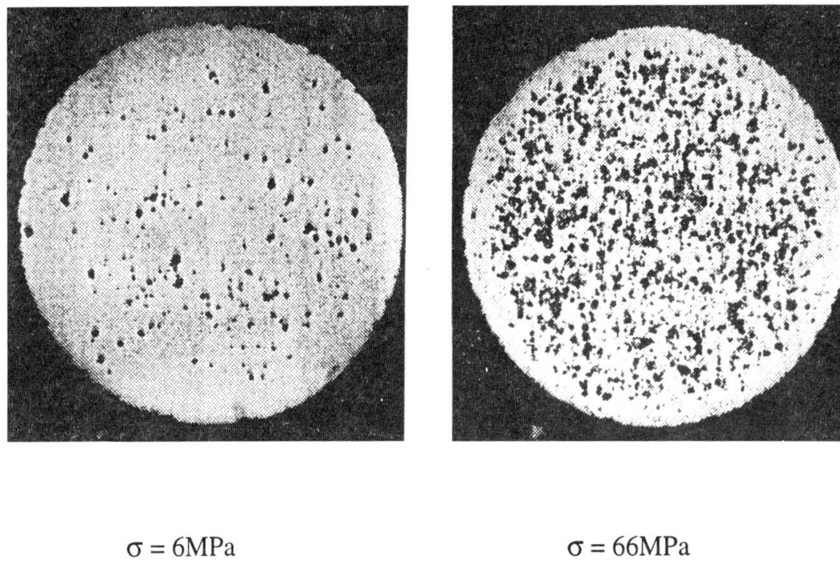

$\sigma = 6MPa$	$\sigma = 66MPa$
f = 0.04	f = 0.40
Growth time : 27s	Growth time : 29s

Figure 3 : Microstructure of specimens partially transformed under various stresses.

In our study we did not observe any noticable influence of the stress on the growth rate. However Sulonen [25, 26] observed that for discontinuous precipitation in Cu-Cd alloys the growth can be modified by the applied stress (Figure 4). The growth was favoured in the transverse direction of the applied stress and inhibited in the direction of the applied stress. The volumic variation is negative. The author attributed this effect to the coherency stresses. Hillert [27] modified the theory considering that the elastic strain energy modifies the solute concentration ; these variations are responsible for the oriented growth. Dryden and Purdy discussed further these results [28]. They calculated a low effect of the coherency stresses on the growth orientation by the modification of the solute concentration. They attributed the oriented growth to the effect of the volume change associated with the transformation. However, this last interpretation is still discussed by Sulonen [29].

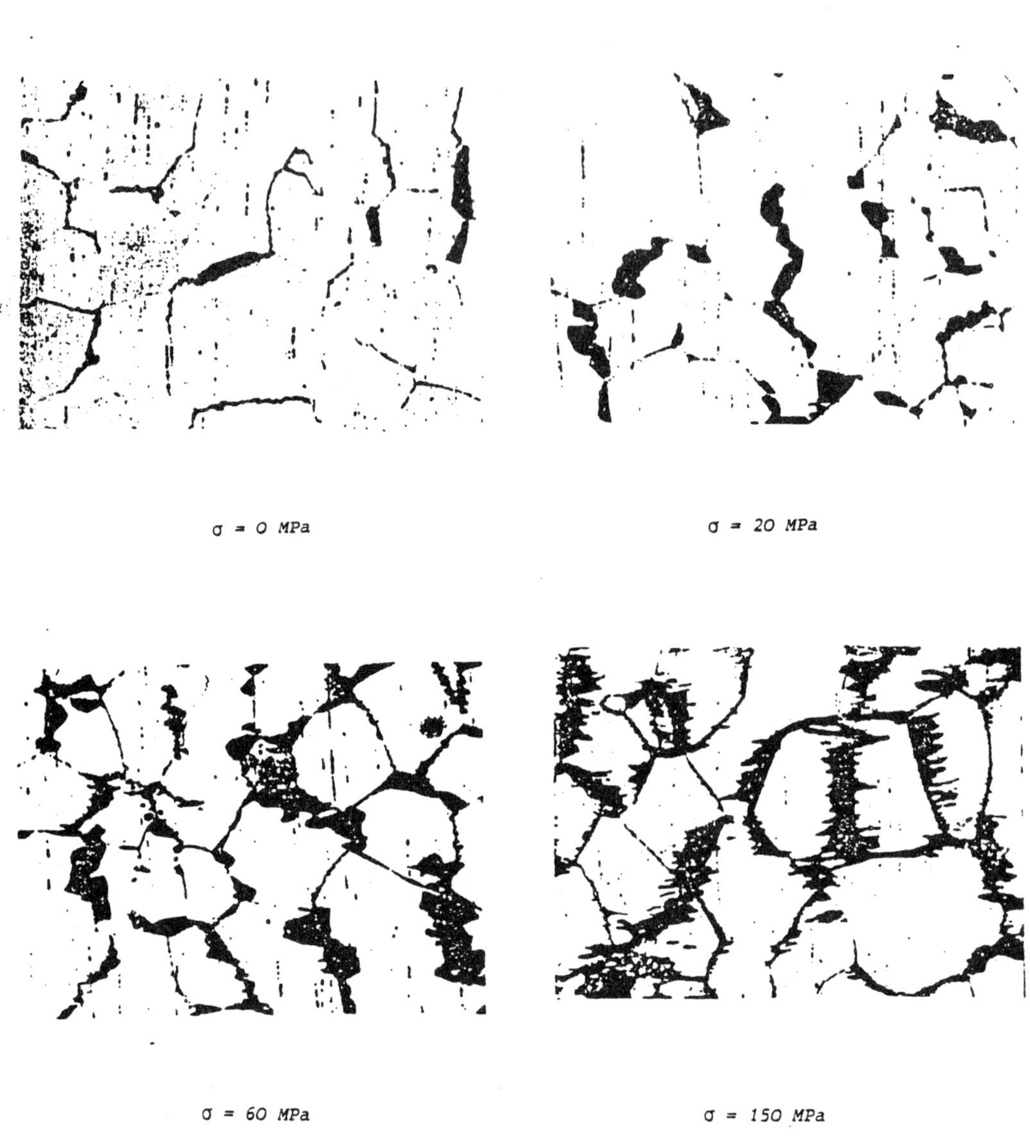

Figure 4 : Morphology variations due to transformation under stress for Cu-Cd alloy [25, 26]

3.2. Transformation plasticity deformation

In figure 1 we showed that the length variations measured during the transformation under stress are increased. This phenomenon is observed even for very low stresses, lower than the yield stress of each phase. The transformation plasticity is mainly studied during transformation under constant stress. However, it can also be observed during tensile testing and thus lead to unusual mechanical behaviour as shown in § 4.

Measurements of the length variations during isothermal transformation under constant applied stress have shown that the deformation associated with the transformation varies with the applied stress (Fig.5). Such additionnal deformation is observed in the direction of the applied stress and for stresses even lower then the yield stress of the softer phase. It is refered as to transformation plasticity deformation. For an eutectoïd transformation, the transformation plasticity deformation, for a complete tranformation, was observed to vary nearly linearly with the applied stress (Fig. 5). Similar results were observed by different authors and for other alloys [31-37]. For larger stresses a deviation from linearity is observed [38].

De Jong and Rathenau, Greenwood and Johnson... evidenced that the slope is dependent on the volumic variations $\Delta V/V$ associated with the transformation and the yield stress of the weaker phase (σ_e). The tranformation plasticity deformation (ε_{pt}) is then expressed as :

$$\varepsilon_{pt} = K \frac{\Delta V}{V} \frac{\sigma}{\sigma_e}$$

where σ is the applied stress and K a constant.

Less results concern the variations of transformation plasticity with the amount of phase formed. The very few results obtained showed that for the pearlitic transformation [30], as for the ferritic transformation [32] a linear relation between the transformation plasticity deformation and the amount of phase transformed is observed too. Results for the pearlitic tranformation are shown figure 6 . The transformation plasticity deformation is then expressed as :

$$\varepsilon_{pt} = K\, f \frac{\Delta V}{V} \frac{\sigma}{\sigma_e}$$

with f the amount of phase formed

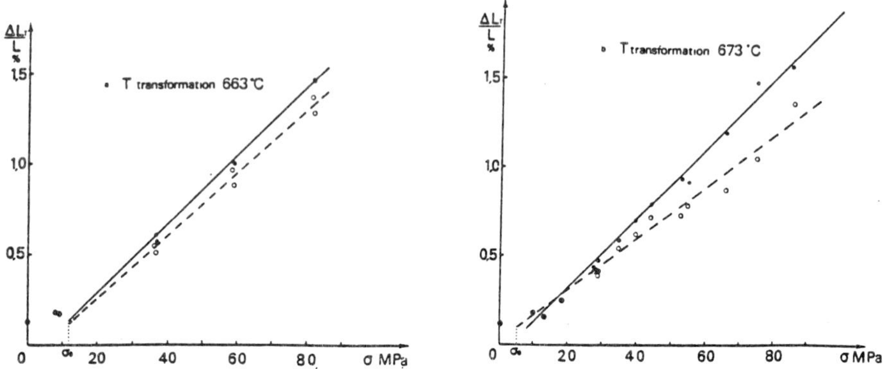

Figure 5 : Isothermal length variations for a complete transformation as a function of the applied stress (pearlitic transformation at different temperatures) [30]
● without creep correction ○ with creep correction

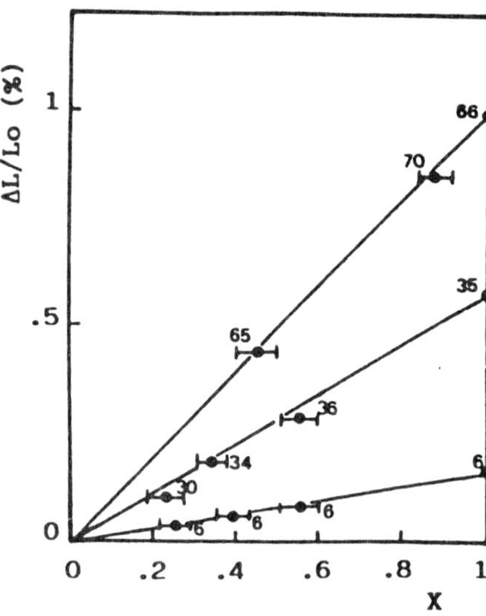

Figure 6 : Length variations versus the amount of pearlite formed (X) under constant stress [30]. The brackets represent the error for the measurement of pearlite. The applied stress is given for each test.

3.3 Origin of transformation plasticity deformation

One of the first interpretations of tranformation plasticity takes account of the specific volume variations during tranformation, as a result of which internal stresses are created. If an external stress is applied during tranformation the material - subjected to both internal and external stresses - yields plastically for an applied stress lower than the yield stress. Plastic yielding will occur in the direction of the applied stress [31,32].

Greenwood and Johnson [34] have established a quantitative relation of transformation plasticity deformation versus the applied stress based on this concept :

$$\varepsilon_{pt} = \frac{5}{6} \frac{\Delta V}{V} \frac{\sigma}{\sigma_e}$$

where ε_{pt} is the transformation plasticity deformation
σ the applied stress.
σ_e the yield stress of the "weaker" phase
$\Delta V/V$ is the specific volume variation.

This macroscopic model is based on the Levy von Mises yielding criterion. Different assumptions - open to criticism - have been made an are well reviewed by Abrassart. [38]. This relation describes the variation of observed transformation plasticity with yield stress, with $\Delta V/V$, and its linear dependence on stress. The authors found good agreement with measured values.

Analytical models of transformation plasticity deformation were further developed by Leblond [40] or by Fischer [41 and this book]. These models lead to similar relationships, the main differences being the correlation with the amount of phase formed and the coefficient K.

At last, for pearlitic transformation, numerical simulations were performed by Ganghofffer et al [42] calculating the transformation plasticity deformation for monoaxial stress state or multi axial stress states [43] by 3D finite element modelling. The order of magnitude of the transformation plasticity deformation was calculated, however a linear variation of the transformation plasticity deformation was not obtained. The calculated values were larger than the experimental ones at the beginning of the transformation and lower at the end. These calculations showed that the plastic deformation was located very near to the transforming front. Moreover simulating a complex stress state it was shown that the transformation plasticity deformation was proportional to the deviatoric part of the stress tensor.

4. MECHANICAL BEHAVIOUR DURING PHASE TRANSFORMATION

Applying a deformation rate during the transformation leads also to transformation plasticity deformation. However, the behaviour is quite different. Figure 7 shows the stress strain curves obtained during tensile tests performed during the austenite→pearlite transformation [44,45] for two transformation temperatures (665 and 675°C) and a deformation rate of $5 \cdot 10^{-5}\ s^{-1}$. For 675°C, we obtain a continously increasing stress with the increase of deformation however the apparent yield stress is lower than that of austenite at higher temperature. For 665°C, a softening is observed at the low deformation values followed by an increase of the stress. Simultaneous electrical resistivity measurements allow to link the softening to the occurence of the phase transformation.

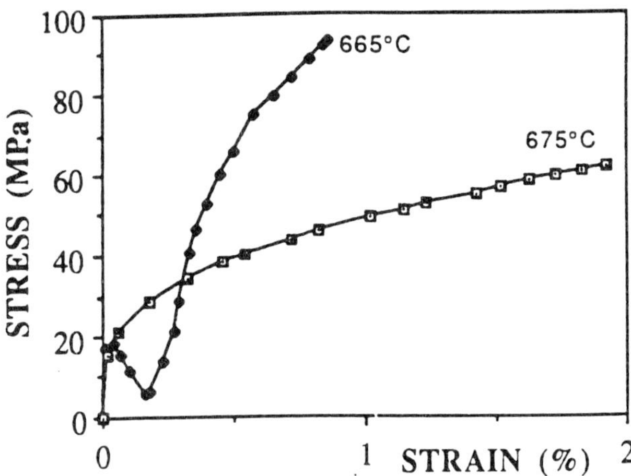

Figure 7 : Tensile stress strain curves at 665 and 675°C.

The amount of the softening is dependent on the deformation rate relative to the transformation rate. The influence of the deformation rate is shown for tensile tests during ferrito pearlitic transformation at constant cooling rate (Fe-0.2C steel). In Figure 8 the stress is plotted versus the temperature (the deformation is proportional to the decrease in temperature). It is clearly shown that the larger softening is observed for the lower deformation rate.

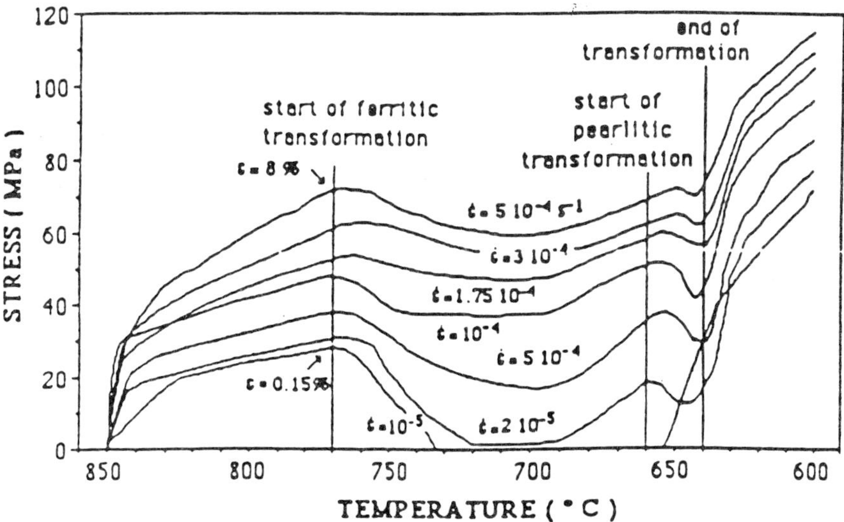

Figure 8 : Tensile stress - strain curves observed at different deformation rates during cooling an Fe-0.2C steel..

In these tests, a constant deformation rate is imposed to the specimen during the transformation. The observed softening can be easily understood if one considers that the volumic variation as the transformation plasticity deformation associated with the amount of phase formed will totaly or partly contribute to the imposed deformation.

In order to put out the contribution of transformation plasticity deformation, we calculated the stress strain curve during the transformation [44,45]. The results are presented in this book by S. Denis for different transformations, ferritic pearlitic and bainitic transformations. Thses calculations show that transformation plasticity deformation is a large contribution to the total deformation and cannot be neglected. The softening observed is dependent on the ratio deformation rate/transformation rate.

5. CONCLUSIONS

The effect of macroscopic stresses on phase transformation kinetics and phase transformation plasticity has been presented and illustrated. It is shown that applying a stress during phase transformation will modify the transformation kinetics, the mechanical behaviour and the morphology of thestructure. The transformation plasticity deformation is due to the anisotropic plastic yielding occuring when the internal stresses associated with the volume change and the applied stress coexist. Considering the kinetics of the transformation, nucleation is enhanced and growth is less modified. However, in some cases the interface migration can be modified.

REFERENCES :

1 A. G. KHACHATURYAN Theory of Structural Tranformations in Solids, John Wiley Interscience New York1983.

2 T. ETO, A.SATO and T. MORI, Acta Met **26** (1978) pp 499-508.

3 F.R.N. NABARRO Metall. Mater. Trans. A 27A (1996) pp 513-530.

4 A. HAZOTTE, A. RACINE and S. DENIS Journal de Physique IV, Colloque C1, supplémént au Journal de Physique III, (1996), C1 119 - C1 128.

5 A. PINEAU Acta Metall. **24** (1976) pp 559-564.

6 I.M. KAGANOVA and A. L. ROYTBURD Sov Phys. JETP **67** (1988) 1173-1183.

7 F.R.N. NABARRO, C. M. CRESS and P. KOTSCHY Acta Mater **44** (1996) 3189-3198.

8 L. KAUFMANN, in Materials under Pressure Honda Memorial Series in Materials Science N°2, **65** (1974) Tokyo, Maruzen Company.

9 E SCHMIDTMANN, H. GRAVE and F.S. CHEN, Trait. Therm. **115** (1977) p 57.

10 L. KAUFMANN, "Solids under pressure" p 303, 1963, New York, Mac Graw Hill.

11 J.E. HILLIARD Trans AIME **227** (1963) p 429.

12 S. V. RADCLIFFE and M. SCHATZ Acta Metall **10** (1962) p 201.

13 S. V. RADCLIFFE, M. SCHATZ and A. S. KULIN J. Iron and Steel Inst. **201** (1963) p 143.

14 M. FUJITA and M. SUZUKI, Trans Iron Steel Inst. Jpn **14** (1974) p44.

15 T.G. NILAN, Trans AIME **239** (1967) p 899.

16 J.J. KRAMER, G.M. POUND, and R.F. MEHL, Acta Metall **6** (1958) p 763.

17 S. BHATTACHARYYA and L. KEHL Trans ASM **47** (1955) p 351.

18 L. KEHL and S. BHATTACHARYYA Trans Asm **48** (1956), p 234.

19 L. F. PORTER and P.C. ROSENTHAL Acta Metall **7** (1959) p 504.

20 G. NOCKE, E. JÄNSCH and P. LENK Neue Hütte **21** (1976) p 468.

21 M.D. JEPSON and F.C. THOMPON J. Iron and Steel Intitute **187** (1949) p 49

22 E. AEBY-GAUTIER "Transformations perlitique et martensitique sous contrainte de traction dans les aciers" Thèse de Docteur es Sciences, Institut National Polytechnique de Lorraine, Nancy-F 1985.

23 S. DENIS, E. GAUTIER, S. SJÖSTRÖM and A. SIMON Acta Metall. **35** (1987) pp 1621-1632.

24 E. GAUTIER and A. SIMON Proc " Phase Transformation 1987" Edts G.W. Lorimer, The Institute of Metals, (1988), pp 451-454.

25 M.S. SULONEN Acta Met. **12** (1964) pp 749.

26 M.S. SULONEN Acta Polytechnica Scandin. **28** (1964) 5-20.

27 M. HILLERT Metall Trans. **3** (1972) 2729.

28 J. R. DRYDEN and G. R. PURDY Acta Metall Mater. **38** (1990) pp 1255-1261.

29 M.S. SULONEN Proceedings "Solid Solid Phase transformations" Edts W. C. Johnson; J. M. Howe, D.E. Laughlin and W. A. Soffa The Mi,erals, Metals and Materials Society 1994 pp 517-520.

30 E. GAUTIER, A. SIMON and G. BECK, Acta Metall. **35** (1987) pp 1367-1375.

31 M. De JONG and G.W. RATHENAU Acta Metall. **7** (1959) p 246.

32 M. De JONG and G.W. RATHENAU Acta Metall. **9** (1961) p 714.

33 F. N. CLINARD and O. D. SHERBY Acta Metall. **12** (1964) p 911.

34 G. W. GREENWOOD and R. M. JOHNSON Proc. Roy. Soc. **283A** (1965) p 403.

35 D. OELSCHLÄGEL and V. WEISS Trans ASM **59** (1966) p143.

36 R. A. KOT and V. WEISS Metall Trans. **1** (1970) p 2685.

37 G. YODER and V. WEISS Metall Trans. **3** (1972) p 675.

38 D.C. DUNAND and C.M. BEDELL Acta mater **44** (1996) pp 1063-1076.

39 F. ABRASSART Thèse de Doctorat ès Sciences Physiques Nancy (1972).

40 J. B. LEBLOND J. DEVAUX and J. C. DEVAUX, Int. Journal of Plasticity, **5** (1989) 551-572.

41 F.D. FISCHER Acta metall. mater. **38** (1990), 1535-1546.

42 J. F. GANGHOFFER, S. DENIS, E. GAUTIER, A. SIMON and S. SJÖSTRÖM, Eur. J. Mech, A/Solids, **12** (1993) pp 21-32.

43 S. SJÖSTRÖM, J.F. GANGHOFFER, S. DENIS, E. GAUTIER and A. SIMON, Eur. J. Mech, A/Solids, **13** (1994) pp 803-817.

44 Ch. LIEBAUT Thèse de Doctorat de l'Institut National Polytechnique de Lorraine Nancy 1988.

45 E. GAUTIER, S. DENIS, Ch. LIEBAUT, S. SJÖSTRÖM and A. SIMON, Journal de Physique IV, Colloque C3, supplémént au Journal de Physique III, (1994), C3 279 - C3 284.

MICROMECHANICAL MODELLING OF THE THERMOMECHANICAL BEHAVIOR OF SHAPE MEMORY ALLOYS

E. Patoor and M. Berveiller
CNRS URA 1215, Metz, France

Abstract

In this work we developed a model for the behavior of shape memory alloys based simultaneously on thermodynamical and micromechanical concepts. The basic field equations including moving boundary concepts are recalled and applied to the description of the transformation by discrete internal variables. To point out the different characteristic length scales appearing in shape memory alloys, two models are developed. The first one concerns the behavior of a grain in polycrystalline materials and the second one uses the self-consistent approximation for the intergranular interaction.

The results obtained from this model well agree with the experimental observations. In particular, the model is able to predict the dissymmetry observed during a tensile-compression test as well as the behavior during multiaxial loading. The parameters of the model are identified from experiments only.

For structure calculation applications, we develop also a simplified analytical model using only two internal variables for which some aspects may be identified from the crystallographic model.

1. Introduction

The aim of this work is to establish the thermomechanical behavior of materials during martensitic phase transformation using modern scale transition methods.

Compared with other classes of materials, phase transition problems present more complex microstructure at different length scale and coupling between different strain mechanisms (elasticity, plasticity by dislocation motions, thermal dilatation, phase transition).

Fortunately, the case of Shape Memory Alloys (SMA) appears more simple since thermo-elasticity may be assumed homogeneous. Plastic flow is usually negligible but residual stresses related with plastic strain incompatibilities have some strong effects on phenomena like the two way shape memory effect (TWSME) and the behavior during thermomechanical cycling.

In first approximation, the fundamental mechanism appearing in SMA may be only related with the transformation strain of a volume element when a martensitic transformation occurs.

Nevertheless, such problem remains complex since several length scales have to be considered :

- in the case of polycrystalline materials, due to the limited value of the transformation strain, the grain to grain interactions are increased with respect to classical plasticity or elasticity.
- the formation of a multivariant pattern inside the grain induces high intragranular stresses.

In addition, the progress of the transformation with respect to the thermomechanical load occurs by nucleation of new domains, growing of existing martensitic domains by interface motion or (and) exchange between variants.

Since the problem has to be considered from multiscale point of view taking into account the microstructure evolution, only partial solutions have been developed during the past years.

Some models ([1][2]) represent the transformation using only a single macroscopic parameter (the overall volume fraction of martensite) neglecting the orientation of the transformation strain with respect to the applied stress.

On the other hand, several works deal with the microscopic self organization of the martensitic domains neglecting the grain to grain interactions and their influence on the overall behavior [3] [4] [5].

More global approaches were proposed using scale transition methods either by analytical description [6] [7] [8] or by finite element computations [9] [10].

In the present work we derive the overall behavior of SMA from micromechanical considerations starting from a thermo-micromechanical model for the behavior of grains and using the classical self-consistent model for the grain to grain interactions.

This paper is divided in three parts.

In the first one, kinematics, kinetics and thermodynamical aspects are described from a continuum point of view using moving boundary concepts.

The second part deals with a macroscopic description in which the transformation is represented using two internal variables : the volume fraction of martensite and the mean transformation strain.

In the last part, the transformation strain is assumed to be piecewise uniform and known inside the grain so that each variant is represented by its volume fraction.

By this way, a large number of internal variable is introduced but the associated transformation strain is now a well defined quantity.

This last approach leads to excellent agreement with experimental measurements without any fitting parameter. The model is able to predict the dissymmetry observed during a tensile-compression test and the behavior during multiaxial loading.

2. Continuous description of martensitic phase transformation

When phase transitions are involved, the change of some mechanical characteristics of a "particule" (i.e. the volume element in continuum physics) must be specified. Since for SMA, elastic constants may be approximated as identical in austenite and martensite and since plasticity is neglected, only the stress free strain of the "particule" has to be specified for the special martensitic transformation occuring in a given material (see § 2.1.1).

Since we are interested in the behavior of a macroscopic volume element, the macroscopic strain and stress for a Representative Volume Element (RVE) are defined from standard definitions introduced in the mechanics of inelastic or inhomogeneous media (see § 2.1.2).

During evolution of the transformation, special mechanisms involving moving boundaries must be taken into account (see § 2.1.3).

Derivation of the overall behavior of the austenite-martensite Representative Volume Element requires the calculation of the free energy of the RVE in which the elastic and the chemical parts have to be taken into account (see § 2.2.1). The dissipation which is deduced from the comparison betwen external power and the change in free energy (see § 2.2.2) defines the driving force for the evolution of the transformation described in this section using a moving boundaries framework.

2.1. Kinematics and kinetics of martensitic phase transition

2.1.1. Bain and Transformation strains for a volume element. Microstructure (habit plane, variants) of an Austenite-Martensite system

Martensitic phase transformation are considered [11] [12] [13] as first order phase transition occuring without diffusion which means that the chemical composition remains constant but the atoms are organized in a new crystallographical lattice.
An inelastic lattice strain called Bain strain $\left(\varepsilon^{B}\right)$ describes the geometrical transformation from the parent lattice (Austenite) to the product lattice (Martensite). Since first order phase transition involves the coexistence of the two phases, the strain fields $\varepsilon^{B}(r)$ (which is ε^{B} inside the martensite, and zero inside the austenite) is usually strongly incompatible. These incompatbilities lead to high internal stresses which are relaxed (at least partially) by a lattice invariant strain ε^{Lis} appearing simultaneously with and inside the martensite.
Thus, the transformation strain ε^{T}, which may also be considered as a stress free or inelastic strain in the sense of Eshelby [14] or Kröner [15], is obtained by :

$$\varepsilon^{T} = \varepsilon^{B} + \varepsilon^{Lis} \tag{2.1}$$

Usually, ε^{B} is given from crystallographic and atomic arguments and ε^{Lis} is deduced from compatibility conditions taking into account the observed morphology of the martensitic domains which have laths or plates shape. Two equivalent theories may be used to derive these compatibility conditions.

a) The Wechsler-Lieberman and Read Theory (WLR) [16]

In this theory, the Bain strain is assumed to be given and the lattice invariant strain is assumed to be realized from plastic slip having an amplitude γ, slip direction $\vec{m}$ and acting on plane with unit normal $\vec{n}$.
Thus ε^{Lis} is given by :

$$\varepsilon_{ij}^{Lis} = \frac{1}{2}\left(m_i n_j + m_j n_i\right)\gamma \qquad \text{or} \qquad \varepsilon_{ij}^{Lis} = R_{ij}\gamma \tag{2.2}$$

The transformation strain ε^T follows from ε^B and ε^{Lis}

$$\varepsilon_{ij}^T = \varepsilon_{ij}^B + R_{ij}\gamma$$

Since ε^T (without elastic strain) is imposed to be compatible across the interface between austenite and martensite the classical jump condition

$$\left[\varepsilon_{ij}\right] = \varepsilon_{ij}^{\ +} - \varepsilon_{ij}^{\ -} = \frac{1}{2}\left(M_i N_j + M_j N_i\right) \tag{2.3}$$

is here

$$\varepsilon_{ij}^T = \varepsilon_{ij}^B + R_{ij}\gamma = \frac{1}{2}\left(M_i N_j + M_j N_i\right) \tag{2.4}$$

where N denotes the normal to the habit plane, i.e. the interface between austenite and martensite. A non trivial solution $\left(\varepsilon^B \neq 0,\ \gamma \neq 0\right)$ of (2.4) may be obtained if one eigen value of ε^T vanishes so that we get the condition :

$$Det\left(\varepsilon_{ij}^B + R_{ij}\gamma\right) = 0 \tag{2.5}$$

from which the unknown γ can be determined and consequently vectors $\vec{N}$ and $\vec{M}$.
The structure of (2.4), given explicitely in Figure (2.1), indicates that generally ε^T contains a spherical part $\varepsilon_{kk}^B \delta_{ij}$ and a deviatoric part corresponding to a shear parallel to the habit plane. For shape memory alloys the volume change is generally very weak in comparison

with the deviatoric strain (~0.1). Due to the high symmetry of the austenite lattice, several Bain strain (differently oriented) are possible so that different variant may be formed in a grain or a single crystal.

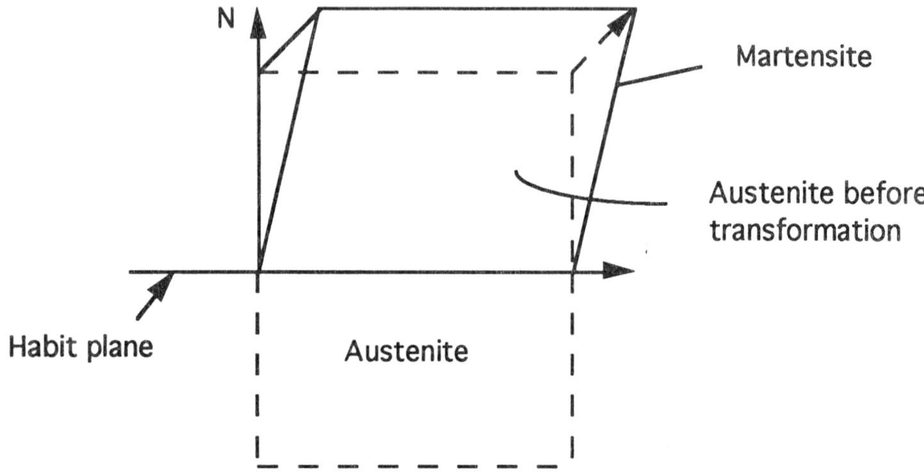

Figure. 2.1: Transformation strain of a volume element

b) The Eshelby inelastic inclusion problem [14]

Here it is assumed that the stress free strain ε^{T} inside an (unknown) ellipsoidal inclusion with volume V_{I} and shape A is composed of the Bain strain ε^{B} and the lattice invariant strain ε^{Lis}. The Eshelby model indicates that the elastic interactions between V_{I} and the surrounding matrix introduce uniform elastic strain ε^{e} inside V_{I} so that

$$\left(\varepsilon^{\mathrm{T}}+\varepsilon^{e}\right)=S^{\mathrm{E}}\varepsilon^{\mathrm{T}} \tag{2.6}$$

where again $\varepsilon^{\mathrm{T}}=\varepsilon^{\mathrm{B}}+\varepsilon^{Lis}$

Free energy ϕ per unit volume of an infinite matrix containing the inclusion V_{I} is given by (see § 2.2.1) :

$$\phi = \frac{1}{2}\frac{V_I}{V}\varepsilon^T C\left(I - S^E(A)\right)\varepsilon^T \tag{2.7}$$

where $S^E(A)$ is the Eshelby tensor depending on the (uniform) elastic moduli C and the shape A of the inclusion. From (2.7) the "compatible" transformation microstructure (for which $\phi = 0$) is obtained using the condition Det ε^T=0 that defines a penny shape inclusion whose plane is parallel to the shear plane of the transformation strain.

c) Application to the Fe-31 wt%Ni system

Data for a Fe-31 wt%Ni system are taken from [17].
The lattice transformation corresponds to a fcc --> bcc transformation with a Bain strain given by

$$\varepsilon^B = \begin{pmatrix} a & 0 & 0 \\ 0 & a & 0 \\ 0 & 0 & c \end{pmatrix} \qquad \begin{aligned} a &= 0.132 \\ c &= -0.199 \end{aligned}$$

defined in the reference system constituted by the lattice of the austenite parent phase.
The lattice invariant strain is assumed to occur by slip on the (101) $\left[\bar{1}01\right]$ system, so that :

$$\varepsilon^{Lis} = \begin{pmatrix} \bar{1} & 0 & 0 \\ 0 & 0 & 0 \\ 0 & 0 & 1 \end{pmatrix}\gamma$$

According to these definitions, the transformation strain $\varepsilon^T = \varepsilon^B + \varepsilon^{Lis}$ is given by

$$\varepsilon^T = \begin{pmatrix} a - \frac{\gamma}{2} & 0 & 0 \\ 0 & a & 0 \\ 0 & 0 & c + \frac{\gamma}{2} \end{pmatrix}$$

In this case, the compatibility condition Det $\varepsilon^T = 0$ leads to

$$\left(a - \frac{\gamma}{2}\right)a\left(c + \frac{\gamma}{2}\right) = 0$$

from which the two roots $\gamma_1 = 0.264$ and $\gamma_2 = 0.398$ are obtained. Keeping only the lowest value $\gamma_1 = 0.264$, the normal $\vec{N}$ to the habit plane and the transformation direction $\vec{M}$ are deduced from the condition $\varepsilon_{ij}^T = \frac{1}{2}\left(M_i N_j + M_j N_i\right)$.

In the present case, one gets $\vec{N} = (0,\ 0.662,\ 0.580)$ $\vec{M} = (0,\ 0.199,\ -0.115)$

Due to the small strain hypothesis used, these values are sligthly different from those given by the WLR theory.

2. 1. 2. Transformation strain of a Representative Volume Element (RVE)

Let the reference configuration of the RVE be the austenite grain of a polycrystal in its natural state (stress free, $\sigma = 0$) at a temperature $T > A_f$. Due to the thermomechanical loading, some part V_M of this RVE of volume V is transformed into martensite (Fig. 2.2). Each part V_I of V_M undergoes a uniform transformation strain ε^{TI} and internal stresses are built due to the incompatibility of the $\varepsilon^T(r)$ field. Here the $\varepsilon^T(r)$ field presents incompatibilities since the extension of the martensitic variants is limited by the interaction between the domains. The total deformation field $\varepsilon(r)$ such obtained results from several contributions :

- an elastic part denoted $\varepsilon^e(r)$
- a thermal part denoted $\varepsilon^{th}(r)$
- the transformation part $\varepsilon^T(r)$

so that for the displacement field $u_i(r)$, one obtains :

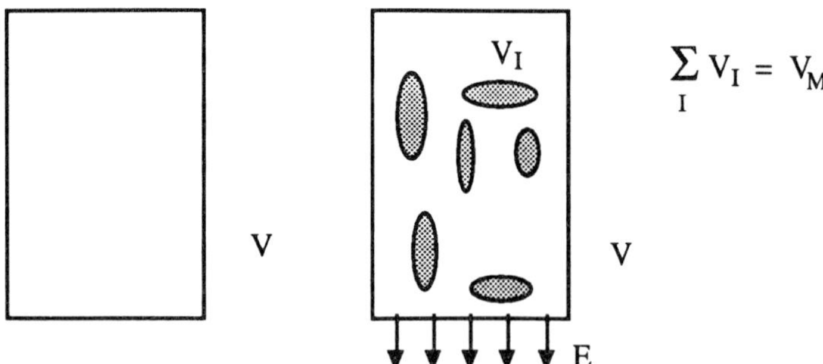

Figure 2. 2: Definition of the transformation strain field in the Representative Volume Element of volume V.

$$u_{(i,j)} = \varepsilon_{ij}(r) = \varepsilon_{ij}^{e}(r) + \varepsilon_{ij}^{th}(r) + \varepsilon_{ij}^{T}(r) \tag{2.8}$$

where (i,j) means symetrisation of $u_{i,j}$. By integration over the volume V and the assumption that, at ∂V, $u_i = E_{ij}x_j$, the overall mean deformation E_{ij} is given by :

$$E_{ij} = \frac{1}{V}\int_V \varepsilon_{ij}(r)dV \tag{2.9}$$

or considering (2.8)

$$E_{ij} = \frac{1}{V}\int_V \left\{\varepsilon_{ij}^{e}(r) + \varepsilon_{ij}^{th}(r) + \varepsilon_{ij}^{T}(r)\right\}dV \tag{2.10}$$

The overall strain E may be also decomposed into elastic $\left(E^e\right)$ thermal $\left(E^{th}\right)$ and transformation $\left(E^T\right)$ parts so that :

$$E = E^e + E^{th} + E^T \tag{2.11}$$

Only if the elastic compliances $s(r)$ $\left(\varepsilon^e = s\sigma\right)$ and the thermal dilatations $\alpha(r)$ ($\varepsilon^{th} = \alpha\,\theta$) are assumed homogeneous, the overall strains E^e and E^{th} can be identified with the corresponding mean values of the local parts. From the decomposition :

$$E = \frac{1}{V}\int_V s(r)\sigma(r)dV + \frac{1}{V}\int_V \alpha(r)\theta dV + \frac{1}{V}\int_V \varepsilon^T(r)dV \tag{2.12}$$

the uniformity hypothesis of s and α leads to

$$E = s\frac{1}{V}\int_V \sigma(r)dV + \alpha\frac{1}{V}\int_V \theta dV + \frac{1}{V}\int_V \varepsilon^T(r)dV \tag{2.13}$$

From the property $\Sigma = \frac{1}{V}\int_V \sigma dV$ and the uniformity of θ, one gets :

$$E = E^e + E^{th} + E^T \tag{2.14}$$

with $E^e = s\Sigma$ and $E^{th} = \alpha\theta$

Considering (2.12) and (2.14), the global transformation strain E^T is defined by :

$$E^T = \frac{1}{V}\int_V \varepsilon^T(r)dV \tag{2.15}$$

Since the field $\varepsilon^T(r)$ is piecewise uniform, considering it keep a uniform value inside each variant of martensite, it can be represented by a set of indicator functions $\theta^I(r)$ so that

$$\varepsilon^T(r) = \sum_I \varepsilon^{TI}\theta^I(r) \tag{2.16}$$

where ε^{TI} is the (uniform) transformation strain inside V_I and $\theta^I(r) = \begin{cases} 1 \text{ if } r \in V_I \\ 0 \text{ if } r \notin V_I \end{cases}$

From (2.15) and (2.16), one gets

$$E^T = \frac{1}{V}\sum_I \int_{V_I} \varepsilon^{TI} dV \quad \text{or} \quad E^T = \sum_I f^I \varepsilon^{TI} \tag{2.17}$$

where parameter $\mathrm{f}^{\mathrm{I}} = \frac{\mathrm{V_I}}{\mathrm{V}}$ represents the volume fraction of variant I.

In that case, the overall transformation strain is described by (known) transformation strains ε^{TI} and the corresponding volume fractions f^I.

A more global description may be obtained introducing the whole transformed volume of martensite $V_M \left(= \sum_I V_I\right)$ and the mean transformation strain $\overline{\varepsilon}^{TM} = \frac{1}{V_M}\int_{V_M} \varepsilon^T(r)dV$ inside this global volume. Since $\varepsilon^T(r)$ is zero inside the austenite, E^T is also given by :

$$E^T = \frac{1}{V}\int_{V_M} \varepsilon^T(r)dV \tag{2.18}$$

$$E^T = \frac{V_M}{V}\frac{1}{V_M}\int_{V_M} \varepsilon^T(r)dV \tag{2.19}$$

or

$$E_{ij}^T = f\overline{\varepsilon}_{ij}^{T_M} \tag{2.20}$$

where $f = \frac{V_M}{V}$ is the total volume fraction of martensite and $\bar{\varepsilon}^{T_M} = \frac{1}{V_M}\int_{V_M} \varepsilon^T(r)dV$ represents the mean transformation strain over V_M.

In this two phase description, the overall transformation strain $E^T = f\bar{\varepsilon}^{T_M}$ depends on two unknowns, i.e. a scalar one, the volume fraction f and a tensorial one, the mean transformation strain $\bar{\varepsilon}^{T_M}$.

2. 1. 3. Evolution of the transformation inside the RVE

Progress of the transformation results both from the motion of the internal interphase boundaries between the austenitic and martensitic phases and from the motion of interfaces between the different variants of martensite. In the small strain approximation framework, the time derivative of equation (2.15) is given by (figure 2. 3) :

$$\frac{dE^T}{dt} = \frac{1}{V}\int_V \frac{\partial \varepsilon^T}{\partial t}(r,t)dV - \frac{1}{V}\int_\Sigma \left[\varepsilon^T\right] w_\alpha n_\alpha d\Sigma \tag{2.21}$$

where:

- $\frac{\partial \varepsilon^T}{\partial t}$ denotes the variation of the transformation strain with respect to time
- $\left[\varepsilon^T\right] = \left(\varepsilon^{T+} - \varepsilon^{T-}\right)$ is the jump of ε^T across the interface Σ with the unit normal $\vec{n}$ from minus side to plus side
- $w_\alpha n_\alpha$ is the normal velocity of the boundary Σ.

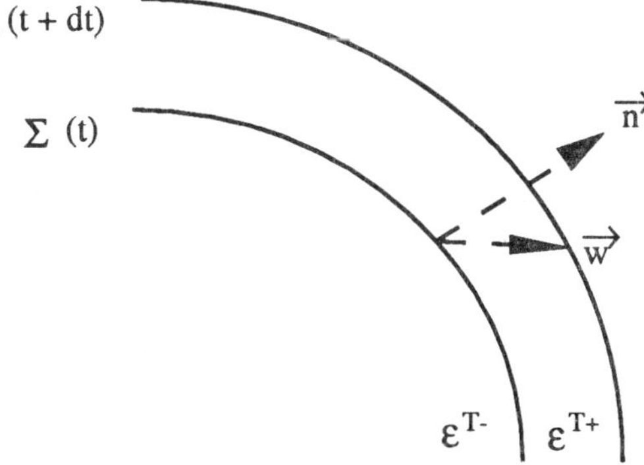

Figure 2. 3: Definition of jumps of strain and interface motion across a moving interface

Considering the internal interface Σ is composed with austenite-martensite interfaces Σ_{AM} and martensite-martensite interfaces Σ_{MM} and taking into account that $\frac{\partial \varepsilon^T}{\partial t} = 0$, the overall transformation strain rate (2.21) is given by :

$$\frac{dE^T}{dt} = \frac{1}{V}\int_{\Sigma_{AM}} \varepsilon^{T-} w_\alpha n_\alpha d\Sigma - \frac{1}{V}\int_{\Sigma_{MM}} \left[\varepsilon^T\right] w_\alpha n_\alpha d\Sigma \qquad (2.22)$$

where the two mechanisms involved in the behavior of the shape memory alloys appear :

- the first integral in relation (2.22) corresponds to transformation with deformation
- the second integral corresponds to deformation without phase transformation $(\dot{f} = 0)$ and describes the mechanism of reorientation of the martensite variants.

Similar conclusions may be obtained using a crystallographical approach or the macroscopic two phase description. In the first case, evolution of E^T with respect to time is related to changes in the f^I parameters. From (2.17) it comes :

$$\frac{dE^T}{dt} = \sum_I \varepsilon^{TI} \frac{df^I}{dt} = \sum_I \varepsilon^{TI} \dot{f}^I \qquad (2.23)$$

Since $\dot{f}^I$ may be positive or negative, equation (2.23) includes both the transformation and the reorientation mechanism. In the second case, using the two phase description, the transformation strain rate is given from (2.20) by :

$$\frac{dE^T}{dt} = \bar{\varepsilon}^{T_M} \frac{df}{dt} + f \frac{d\bar{\varepsilon}^{T_M}}{dt} \qquad (2.24)$$

Discussion of equation (2.24) makes explicit the various mechanisms involved in the behavior of shape memory alloys.

- if $\bar{\varepsilon}^{T_M} = 0$ and $\frac{d\varepsilon^{T_M}}{dt} = 0$, one gets a phase transformation $\left(\frac{df}{dt} \neq 0\right)$ without deformation $\left(\frac{dE^T}{dt} = 0\right)$;

• if $\bar{\varepsilon}^{T_M} \neq 0$ and $\frac{d\varepsilon^{T_M}}{dt} = 0$, the phase transformation $\left(\frac{df}{dt} \neq 0\right)$ produces a deformation $\frac{dE^T}{dt} = \bar{\varepsilon}^{T_M}\frac{df}{dt}$ without reorientation;

• if $\frac{df}{dt} = 0$ and $\frac{d\varepsilon^{T_M}}{dt} \neq 0$, one gets a deformation $\frac{dE^T}{dt} = f\frac{d\bar{\varepsilon}^{T_M}}{dt}$ without transformation (only reorientation);

• in the general case, the deformation $\frac{dE^T}{dt} = \bar{\varepsilon}^{T_M}\frac{df}{dt} + f\frac{d\bar{\varepsilon}^{T_M}}{dt}$ is produced by transformation $\left(\frac{df}{dt} \neq 0\right)$ and reorientation $\left(\frac{d\varepsilon^{T_M}}{dt} \neq 0\right)$.

2.2 Free energy and dissipation

2. 2. 1 - Free energy related to a martensitic transformation in the RVE

Let us consider a RVE in a two phases state (austenite and martensite). Total free energy is determined considering both the RVE and the loading system. The rate at which total free energy varies if if the configuration of the system changes by a (virtual) interface movement defines the thermodynamical driving force for this evolution. The total potential energy of the system is composed of :

+ the elastic strain energy with density $w(r) = \frac{1}{2}\sigma_{ij}\,\varepsilon^e_{ij}$

where local stresses σ are related to internal and applied stresses

+ The crystallographic (usually called chemical) free energy with density $\varphi(r)$

+ The interfacial energy between all the constituents. This last term is generally neglected in the case of thermoelastic shape memory alloys.

Other sources of energy (dislocations, points defects) are often neglected but may play some role for specific effects in the behavior of shape memory alloys (training effect or cyclic behavior). Here again the reference configuration of the RVE with volume V is the stress free austenitic state. We assume that the thermomechanical loading process is sufficiently slow so that the temperature remains uniform over the RVE. For a unit volume V, the free energy Φ is given by :

$$V\Phi = \int_V \varphi(r)dV + \int_V w(r)dV \tag{2.25}$$

Since $\varphi(r)$ is $\varphi_A(T)$ if $r \in V_A$ and $\varphi_M(T)$ if $r \in V_M$ the first term in (2.25) may be written as :

$$\int_V \varphi(r)dV = V_A\varphi_A(T) + V_M\varphi_M(T) \tag{2.26}$$

so that

$$\begin{aligned}\frac{1}{V}\int_V \varphi(r)dV &= f\varphi_M + (1-f)\varphi_A \\ &= \varphi_A + (\varphi_M - \varphi_A)f\end{aligned} \tag{2.27}$$

Since the austenite is stable at high temperature, a temperature $T=T_0$ for which $\varphi_M = \varphi_A$ does exist (figure 2. 4)

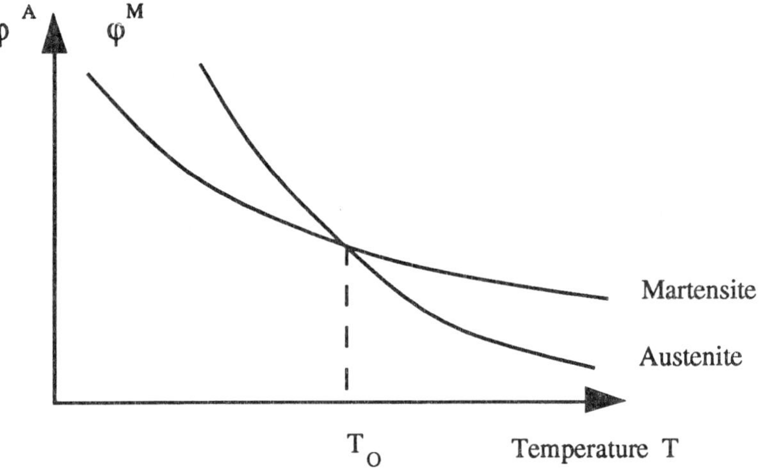

Figure 2.4 : Evolution of the chemical free energy with temperature T

In the vicinity of T_0, the chemical part of the free energy may be linearized by

$$\frac{1}{V}\int_V \varphi(r)dV = B(T - T_0)f \tag{2.28}$$

The elastic part of the free energy :

$$W = \frac{1}{V}\int_V w(r)dV = \frac{1}{2V}\int_V \sigma_{ij}\,\varepsilon^e_{ij}\,dV \tag{2.29}$$

may be transformed using results (2.8) from the kinematics :

$$\varepsilon^e_{ij} = u_{(i,j)} - \varepsilon^T_{ij} \quad \text{inside V} \qquad \text{and} \qquad u_i = E_{ij}\,x_j \quad \text{at } \partial V \tag{2.30}$$

so that

$$W = \frac{1}{2V}\int_V \sigma_{ij}\left(u_{i,j} - \varepsilon^T_{ij}\right)dV \tag{2.31}$$

By partial integration and taking into account boundary conditions, (2.31) is reduced to

$$W = \frac{1}{2}E_{ij}\,\Sigma_{ij} - \frac{1}{2}\int_V \sigma_{ij}(r)\varepsilon^T_{ij}(r)dV \tag{2.32}$$

where $\Sigma_{ij} = \frac{1}{V}\int_V \sigma_{ij}\,dV$ are the overall stresses.

Decomposing the stresses σ_{ij} into macroscopic stresses Σ_{ij} and internal stresses τ_{ij} (with properties $\tau_{ij,j} = 0$ and $\frac{1}{V}\int_V \tau_{ij}\,dV = 0$) and using the Hooke's law $\Sigma = C\left(E - E^T\right)$, (2.32) can be written as

$$W = \frac{1}{2}\left(E - E^T\right)C\left(E - E^T\right) - \frac{1}{2V}\int_V \tau_{ij}(r)\varepsilon^T_{ij}(r)dV \tag{2.33}$$

where the internal stresses τ_{ij} depend only on the incompatibilities of the transformation strain field $\varepsilon^T(r)$. Thus W depends on boundary conditions E, transformation strain field $\varepsilon^T(r)$ and material properties C. Calculation of the internal stress field τ may be performed using the Green function techniques [18] [19]. The classical equations of the problem

$$\sigma_{ij} = C_{ijkl}\left(\varepsilon_{k,l} - \varepsilon_{kl}^{T}\right)$$
$$\sigma_{ij,j} = 0$$
$$u_i = E_{ij}x_j \quad \text{on } \partial V$$

are solved by

$$\varepsilon_{ij}(r) = E_{ij} + \int_V \Gamma_{ijkl}(r - r')C_{klmn}\left(\varepsilon_{mn}^{T}(r') - E_{mn}^{T}\right)dV' \tag{2.34}$$

where the modified Green function Γ_{ijkl} is related to the displacement Green function G by

$$\Gamma_{ijkl} = G_{ik,jl'(ij)(kl)}$$

The application of (2.9) to elementary inclusion-matrix problems leads to the famous Eshelby [14] and Kröner [15] solutions. For a single martensitic ellipsoïdal domain, the transformation strain field $\varepsilon^{T}(r)$ is given by :

$$\varepsilon^{T}(r) = \varepsilon^{T_I}\theta^{I}(r) \tag{2.35}$$

where ε^{T_I} is the uniform stress free strain inside V_I and θ^{I} the indicator function of V_I. In that case, $\varepsilon\left(=\varepsilon^{I}\right)$ is uniform inside the inclusion and given by

$$\varepsilon^{I} = E + S^{E}\,\varepsilon^{T_I} \tag{2.36}$$

and

$$\sigma^{I} = \Sigma - C\left(I - S^{E}\right)\varepsilon^{T_I} \tag{2.37}$$

where S^{E} is the Eshelby tensor. Elastic energy W is given by :

$$W = \frac{1}{2}\left(E - E^{T}\right)C\left(E - E^{T}\right) + \frac{1}{2}\frac{V_I}{V}\varepsilon^{T_I}C\left(I - S^{E}\right)\varepsilon^{T_I} \tag{2.38}$$

Here, one may remark that W depends on E, ε^{T_I}, but also on the volume V_I and the morphology of the transformed region. For a non dilute concentration of transformed domains (I=1,N) with volume V_I and transformation strain ε^{T_I}, elastic interactions between domains contribute also to the elastic energy W. The method initially proposed by Kröner [15] and developed by Mori-Tanaka [20] is used in order to model these interactions. A single domain with Eshelby tensor S^{EI} and transformation strain ε^{T_I} (figure. 2. 5) can be considered as an Eshelby inclusion with the stresses σ_A inside the matrix (austenite) as long range stresses . Thus

$$\sigma^I = \sigma^A - C\left(I - S^{EI}\right)\varepsilon^{T_I} \tag{2.39}$$

From the equilibrium conditions

$$\Sigma = (1 - f)\, f\ \sigma^A + \sum_I f^I\ \sigma^I \tag{2.40}$$

one gets

$$\sigma^I = \Sigma - C\left(I - S^{EI}\right)\varepsilon^{T_I} + \sum_J f^J\left(I - S^{EJ}\right)\varepsilon^{T_J} \tag{2.41}$$

and for the internal stresses τ^I

$$\tau^I = \sigma^I - \Sigma = -C\left(I - S^{EI}\right)\varepsilon^{T_I} + \sum_J f^J C\left(I - S^{EJ}\right)\varepsilon^{T_J} \tag{2.42}$$

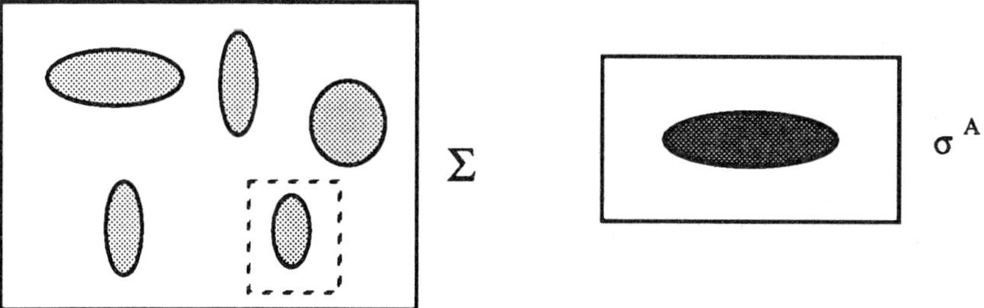

Figure 2.5 : The Mori-Tanaka [20] problem

The elastic energy is now

$$W=\frac{1}{2}\left(E-E^{T}\right)C\left(E-E^{T}\right)+\frac{1}{2}\sum_{I}\varepsilon^{T_I}C\left(I-S^{EI}\right)\varepsilon^{T_I}f^{I}$$
$$-\frac{1}{2}\sum_{I,J}\varepsilon^{T_I}C\left(I-S^{EJ}\right)\varepsilon^{T_J}f^{I}f^{J} \tag{2.43}$$

The first term proportional to f^I is usually small compared with the second one (proportional to $f^I f^J$) describing the interaction between the variants. One has to note that this description assumes some disorder for the spatial position in the variant which is not necessary observed in single crystals or in grains of a polycrystal. A different and more realistic approach is proposed in chapter 3.

A more precise but rather complex interaction matrix may be obtained using the discrete interactions tensor between two inclusions [21]. We write integral equation (2.9) for the internal stresses in a different way :

$$\tau(r)=\int_{V'}\Gamma^{\sigma}(r-r')\varepsilon^{T}(r')dV' \tag{2.44}$$

where Γ^{σ} represents a modified stress Green function which can be easily derived from Γ and C [22]. The internal stress part of the free energy (2.33)

$$W^{int}=-\frac{1}{2V}\int_{V}\tau(r)\varepsilon^{T}(r)dV' \tag{2.45}$$

is now

$$W^{int}=-\frac{1}{2V}\int_{V}\int_{V'}\varepsilon^{T}(r)\Gamma^{\sigma}(r-r')\varepsilon^{T}(r')dV'\,dV \tag{2.46}$$

Since ε^T is piecewise uniform, we get

$$W^{int}=-\frac{1}{2V}\sum_{I}\sum_{J}\int_{V'}\int_{V}\varepsilon^{TI}\theta^{I(r)}\Gamma^{\sigma}(r-r')\varepsilon^{TJ}\theta^{J(r)}dVdV' \tag{2.47}$$

$$=\frac{1}{2}\sum_{I}\sum_{J}\varepsilon^{TI}T^{IJ}\varepsilon^{TJ} \tag{2.48}$$

where the interaction tensors T^{TJ} given by

$$T^{IJ} = -\frac{1}{2V}\int_{VI}\int_{VJ}\Gamma^{\sigma}(r-r')dVdV' \tag{2.49}$$

depend on the volume, the shape and the orientation of V_I and V_J and on their relative position. This exact calculation of the interaction term W^{int} requires the spatial description of the transformation field which complicates deaply the problem. In any case, the elastic energy may be written as

$$W = \frac{1}{2}\left(E-E^T\right)C\left(E-E^T\right)+H^I f^I + \frac{1}{2}H^{IJ}f^I f^J \tag{2.50}$$

where the matrixes H^I and H^{IJ} depend on the shape of domains (via the Eshelby's tensors S^{EI}) and probably on the volume fraction f^I itself.
From (2.28) and (2.50), the Helmholtz free energy is defined as :

$$\Phi\left(E,T,f^I\right) = \frac{1}{2}\left(E-E^T\right)C\left(E-E^T\right)+H^I f^I + \frac{1}{2}H^{IJ}f^I f^J + B(T-T_o)f \tag{2.51}$$

where E and T are the control variables and f^I are the internal variables describing the evolution of the microstructure for the RVE if the shape of the domains is assumed to be given. $\left(f=\sum f^I \text{ and } E^T=\sum f^I\varepsilon^{TI}\right)$. Of course, internal variables f^I are submitted to some physical constraints :

$$f^I > 0 \text{ and } \sum f^I = f \leq 1 \tag{2.52}$$

By a Legendre Transformation $\Phi+\psi=\Sigma E$, the complementary free energy $\Psi\left(\Sigma,T,\varepsilon^T\right)$ is given by

$$\Psi\left(\Sigma,T,\varepsilon^{Tr}\right) = \frac{1}{2}\Sigma M\Sigma + \Sigma E^T + \frac{1}{2V}\int_V \tau(r)\varepsilon^T(r)dV - B(T-T_0)f \tag{2.53}$$

where $M=C^{-1}$. If (2.50) is used for W, (2.53) gives :

$$\Psi\left(\Sigma,T,f^{I}\right)=\frac{1}{2}\Sigma M\Sigma+\Sigma E^{T}-H^{I}f^{I}-\frac{1}{2}H^{IJ}f^{I}f^{J}-B(T-T_{0})f \tag{2.54}$$

Equations (2.51) or (2.54) may be used to defined the behavior for grains of a polycrystal together with a scale transition model solving the interactions between the grains (see chapter 4).

2. 2. 2. Dissipation and driving forces for Martensitic transformation and reorientation

The intrinsic dissipation D of a system of volume V results from the comparison between its free energy change $\dot{\Phi}=\frac{d\Phi}{dt}$ and the external power P_{ext}. For quasi-static evolutions and isothermal approximation, D is given by

$$D=Pext-\dot{\Phi} \tag{2.55}$$

Since the fields w and φ are discontinuous across the moving internal boundaries Σ, the time derivative of Φ is given by

$$\dot{\Phi}=\frac{1}{V}\int_{V}\frac{\partial(w+\varphi)}{\partial t}dV-\frac{1}{V}\int_{\Sigma}[w+\varphi]w_{\alpha}n_{\alpha}d\Sigma \tag{2.56}$$

where the same notations as in 2.1.3 are used. Since $\frac{\partial\varphi}{\partial t}=0$ and $\frac{\partial\varepsilon^{T}}{\partial t}=0$, the first integral reduces to

$$\frac{1}{V}\int_{V}\frac{\partial(w+\varphi)}{\partial t}dV=\frac{1}{V}\int_{V}\sigma_{ij}\dot{\varepsilon}_{ij}dV \tag{2.57}$$

From the results concerning φ, the jumps $[\varphi]$ of φ across the moving boundaries are zero for martensite-martensite interfaces and equal to $\varphi_{A}-\varphi_{M}$ across martensite-austenite interfaces. This gives :

$$\frac{1}{V}\int_{\Sigma}[\varphi]\, w_\alpha\, n_\alpha\, d\Sigma \;=\; \frac{1}{V}\int_{\Sigma}(\varphi_A-\varphi_M)\, w_\alpha\, n_\alpha\, d\Sigma \tag{2.58}$$

and since $(\varphi_A-\varphi_M)$ is uniform along Σ,

$$\frac{1}{V}\int_{\Sigma}[\varphi]\, w_\alpha\, n_\alpha\, d\Sigma=(\varphi_A-\varphi_M)\frac{1}{V}\int_{\Sigma} w_\alpha\, n_\alpha\, d\Sigma=(\varphi_A-\varphi_M)\dot{f} \tag{2.59}$$

The jump [w] defined by

$$[w]=\frac{1}{2}\left(\sigma^+\left(\varepsilon^+-\varepsilon^{T_+}\right)-\sigma^-\left(\varepsilon^--\varepsilon^{T_-}\right)\right) \tag{2.60}$$

can be simplified assuming uniform elasticity and using the properties $C_{ijkl}=C_{klij}$. One gets

$$\left[w^e\right]=\frac{1}{2}\left(\sigma_{ij}^++\sigma_{ij}^-\right)\left[\varepsilon_{ij}-\varepsilon_{ij}^T\right] \tag{2.61}$$

The change of free energy is now

$$\begin{aligned}\frac{d\phi}{dt}=\frac{1}{V}\int_V \sigma_{ij}\,\dot{\varepsilon}_{ij}dV-\frac{1}{V}\int_{\Sigma}\frac{1}{2}\left(\sigma_{ij}^++\sigma_{ij}^-\right)\left[\varepsilon_{ij}-\varepsilon_{ij}^T\right]w_\alpha\, n_\alpha d\Sigma\\ -\frac{1}{V}\int_{\Sigma}(\varphi_A-\varphi_M)w_\alpha\, n_\alpha d\Sigma\end{aligned} \tag{2.62}$$

The external power P_{ext} results from the boundary conditions $T_i=\sigma_{ij}n_j$ on S and the hypothesis of zero volume forces

$$VP_{ext}=\int_S \sigma_{ij}\, n_j\, v_i\, dS \tag{2.63}$$

where v_i is the velocity of particles at S. Using the divergence theorem applied on a volume V containing moving discontinuity surfaces Σ, we get

$$V P_{ext} = \int_V (\sigma_{ij} v_i)_{,j} dV + \int_\Sigma [\sigma_{ij} n_j v_i] d\Sigma \tag{2.64}$$

At the moving surfaces Σ, the following relations result from the continuity conditions of σ and the displacement u_i

$$[\sigma_{ij}] n_j = 0 \quad \text{and} \quad [u_i] = 0 \tag{2.65}$$

and the Hadamard [23] conditions on [v]

$$[v_i] = -[u_{i,j}] n_j w_\alpha n_\alpha \tag{2.66}$$

From (2.64) (2.65) and (2.66), the external power is given by

$$V P_{ext} = \int_V \sigma_{ij} \dot{\varepsilon}_{ij} dV - \frac{1}{2} \int_\Sigma (\sigma_{ij}^+ + \sigma_{ij}^-)[\varepsilon_{ij}] w_\alpha n_\alpha d\Sigma \tag{2.67}$$

Finally, the dissipation follows from (2.62) and (2.67)

$$D = \frac{1}{V} \int_\Sigma \left\{ \frac{1}{2} (\sigma_{ij}^+ + \sigma_{ij}^-)[\varepsilon_{ij}^T] + [\varphi] \right\} w_\alpha n_\alpha d\Sigma \tag{2.68}$$

The term $F = \frac{1}{2}(\sigma_{ij}^+ + \sigma_{ij}^-)[\varepsilon_{ij}^T] + [\varphi]$ is defined as the driving force for interface motion [14]. In the case of a thermodynamically reversible system, one has to find the transformation field (active variant, position of interfaces) for which $F = 0$ on each part of Σ.

If internal stresses are neglected and a single variant with transformation strain ε^T is assumed, the driving force reduces to

$$-\Sigma_{ij} \varepsilon_{ij}^T + (\varphi_A - \varphi_M)$$

At equilibrium, the condition

$$\Sigma_{ij} \varepsilon_{ij}^T = B(T - T_0) \tag{2.69}$$

gives the relation between applied stresses Σ and temperature T for which austenite and martensite are in equilibrium.

Equation (2.69) generalizes the famous uniaxial criterion of Patel and Cohen [24]. Since the transformation condition (2.69) is independent of the amount of formed martensite and the associated internal stresses, the transformation occurs at constant stress or temperature. This is approximatively observed for single crystals but happens to be wrong for polycrystals [25].

The continuum theory developed before has to be completed by a micromechanical study on nucleation conditions. In the two next chapters, an effective and powerful thermodynamical and micromechanical framework is developed using discrete internal variables ($f, \bar{\varepsilon}^T$ or f^I) describing the transformation state for polycrystalline SMA.

3. Macroscopic two-phases approach

The previous section defines the kinematical and the thermodynamical frameworks to model the thermomechanical behavior of material undergoing a solid-solid phase transition. Depending the microstructural description used, several modelling can be defined. This section deals with a macroscopic approach where the martensitic phase is considered as an elastically homogeneous medium and the study is reduced to a two-phases material with eigenstrain where polycrystalline inhomogeneities are neglected. Kinematical analysis shows that in this case, the macroscopic transformation strain is described using only two parameters, a scalar variable (the volume fraction f of martensite) and a tensorial one (the mean value of the transformation strain $\bar{\varepsilon}^T$ defined in the martensitic domain). This class of model is commonly used in metallurgical studies [26] [27] and constitutes the basis of different phenomenological modelling developed these last years [1, 2, 28, 29, 30, 31, 32]. In most cases the volume fraction of martensite is the only internal variable considered.

3.1. Thermodynamical potential

According to the preceding section, the thermodynamical potential associated to a thermoelastic martensitic transformation is a function of the control variables Σ and T and internal variables linked to the transformation strain field $\varepsilon^T(r)$. One notes $\{\varepsilon^T(r)\}$ the set of internal variables used. This potential is composed with several contributions with different physical origine and, due to incompatibilities in the transformation strain field, an interaction energy W^{int}. from relation (2.53), complementary free energy per unit volume is expressed by:

$$\Psi(\Sigma, T, \{\varepsilon^T(r)\}) = \frac{1}{2} \Sigma_{ij} M_{ijkl} \Sigma_{kl} + \Sigma_{ij} E^T_{ij} + W^{int} - B(T - T_o) f \quad (3.1)$$

In this potential all the contributions are totally determined at the exception of W^{int} which depends strongly on the microstructural state of the material. To determine this last contribution it is necessary to describe the field $\varepsilon^T(r)$ and to estimate the related internal stress $\tau(r)$. Determination of this stress field constitutes the most complex part of this problem. From a macroscopic point of view, the internal stress field can be defined considering a transformed inclusion embedded in an infinite medium:

$$\tau_{ij} = C_{ijkl} (I_{klmn} - S^E_{klmn}) (E^T_{mn} - \varepsilon^T_{mn}) \quad (3.2)$$

In this expression, I denotes the fourth order unity tensor, S^E is the Eshelby tensor linked to the shape and the orientation of the considered domain. In the two-phase approximation, the macroscopic transformation strain depends directly on the volume fraction of martensite, f and the macroscopic average transformation strain $\bar{\varepsilon}^{TM}$ (equation (2.20)).

$$E^T_{ij} = f \, \bar{\varepsilon}^{TM}_{ij} \quad (3.3)$$

From this consideration, one chooses f and $\bar{\varepsilon}^{TM}$ as internal variables to define the thermodynamical potential. From relationship (3.3), internal stress defined by equation (3.2) turns into:

$$\tau_{ij} = -(1 - f) C_{ijkl} (I_{klmn} - S^E_{klmn}) \bar{\varepsilon}^{TM}_{mn} \quad (3.4)$$

For the sake of simplicity the transformation strain is considered as a purely deviatoric strain tensor. Considering an elastic isotropic behavior characterized by a shear modulus μ and assuming that the volume transformed into martensite constitutes a spherical inclusion, one obtains:

$$\tau_{ij} = -2\mu (1 - f)(1 - \beta) \bar{\varepsilon}^{TM}_{ij} \quad (3.5)$$

where β is the Eshelby coefficient ($\beta \approx 0.5$). In this case the interaction energy related to incompatibilities in the transformation strain field is defined by:

$$W^{int} = \mu f (1 - f)(1 - \beta)\bar{\varepsilon}_{ij}^{TM} \bar{\varepsilon}_{ij}^{TM} \tag{3.6}$$

From expression (3.6), the thermodynamical potential (3.1) becomes :

$$\Psi(\Sigma, T, f, \bar{\varepsilon}^T) = \frac{1}{4\mu}\Sigma_{ij}\Sigma_{ij} - \frac{1}{4\mu}\frac{\nu}{1+\nu}\Sigma_{kk}\Sigma_{kk} + \Sigma_{ij}\bar{\varepsilon}_{ij}^{TM} f - \mu(1 - \beta) f (1 - f)\bar{\varepsilon}_{ij}^{TM}\bar{\varepsilon}_{ij}^{TM} - B(T - T_0) f \tag{3.7}$$

This function characterizes the thermodynamical state of the austenite-martensite two-phase system. Neglecting the dissipative aspects associated to the phase transformation (hysteresis), optimisation of potential (3.7) allows to define the equilibrium state of the system with respect to the loading conditions given by Σ and T. In the following, such optimisation is realized expressing potential (3.7) as a function of the equivalent stress and strain. One denotes by S the deviatoric part of the applied stress:

$$S_{ij} = \Sigma_{ij} - \frac{1}{3}\Sigma_{kk}\delta_{ij} \tag{3.8}$$

Macroscopic Von Mises equivalent stress and strain are classically defined by:

$$\sigma = \sqrt{\frac{3}{2} S_{ij} S_{ij}} \quad \text{and} \quad \varepsilon = \sqrt{\frac{2}{3}\bar{\varepsilon}_{ij}^{TM}\bar{\varepsilon}_{ij}^{TM}} \tag{3.9}$$

To express the potential (3.7) as a function of the macroscopic scalar variables σ and ε, the modelling is restricted to radial loadings. In this case one can suppose that the transformation strain is fully oriented by the deviatoric part of the stress tensor, that means:

$$\bar{\varepsilon}_{ij}^{TM} = \frac{3}{2}\varepsilon \frac{S_{ij}}{\sqrt{\frac{3}{2} S_{kl} S_{kl}}} \tag{3.10}$$

In this assumption $\bar{\varepsilon}^{TM}$ is colinear with S and the equivalent strain ε is the only unknown. With these hypotheses and neglecting the elastic work due to volume change, potential (3.7) turns as:

$$\Psi(\sigma, T, f, \varepsilon) = \frac{1}{4\mu}\sigma\sigma + \sigma\varepsilon f - \frac{3}{2}\mu(1-\beta) f (1-f)\varepsilon\varepsilon - B(T - T_0) f \qquad (3.11)$$

the potential (3.11) only accounts for deviatoric componants of Σ, that are in fact the major contributions to control the shape memory behavior. In this description, the volume fraction of martensite f and the strain intensity ε are unknown, the strain direction is imposed a priori. One chooses f and ε as internal variables. In the two-phase approximation expression (3.11) is used as thermodynamical potential to model the shape memory alloys behavior.

3.2. *Overall behavior*

Internal variables f and ε used to describe the microstructural state of the material are submitted to physical constraints. The volume fraction f must be a positive quantity smaller than unity. Limitations applied to the strain parameter ε are more complex: it has to be positive and can not exceed a maximal value denoted by ε^{max}. In approximation (3.10) used to describe the mean transformation strain, ε^{max} is independent of the loading direction but is a function of the texture of the material. Neglecting the dissipative aspects, the material behavior is defined by optimisation of potential (3.11) respecting the following physical constraints:

$$0 \leq f \leq 1 \quad \text{and} \quad 0 \leq \varepsilon \leq \varepsilon^{max} \qquad (3.12)$$

The presence of these constraints is taken in account using the Kuhn and Tucker conditions [33] [34] [35]. One defines a Lagrangien form $\mathcal{L}(\sigma, T, f, \varepsilon)$ of the potential (3.11) from conditions (3.12). Four Lagrange multipliers are introduced: λ_o associated to condition $-f \leq 0$; λ_1 for $f \leq 1$; λ_2 for $-\varepsilon \leq 0$ and λ_3 for $\varepsilon \leq \varepsilon^{max}$. These null or positive multipliers must satisfy the following conditions:

$$\lambda_o(-f - 0) = 0 \qquad (3.13)$$

$$\lambda_1(f - 1) = 0 \qquad (3.14)$$

$$\lambda_2(-\varepsilon - 0) = 0 \qquad (3.15)$$

$$\lambda_3(\varepsilon - \varepsilon^{max}) = 0 \qquad (3.16)$$

From these expressions, the Lagrangien is expressed by :

$$\mathcal{L}(\sigma, T, f, \varepsilon) = \Psi(\sigma, T, f, \varepsilon) + \lambda_0 f - \lambda_1(f - 1) + \lambda_2 \varepsilon - \lambda_3(\varepsilon - \varepsilon^{max}) \qquad (3.17)$$

In this study, optimisation of the thermodynamical potential turns to maximize the complementary free energy (3.11), that explains the negative sign used to introduce the constraints in the Lagrangien (3.17). One determines thermodynamical forces $\mathcal{F}^f$ and $\mathcal{F}^\varepsilon$ associated to internal variables f and ε by partial derivatives of the Lagrangien according to these variables [36]. In absence of dissipation, transformation occurs when these forces are equal to zero.

$$\mathcal{F}^f(\sigma, T, f, \varepsilon) = \frac{\partial \mathcal{L}(\sigma, T, f, \varepsilon)}{\partial f} = \frac{\partial \Psi(\sigma, T, f, \varepsilon)}{\partial f} + \lambda_0 - \lambda_1 = 0 \tag{3.18}$$

$$\mathcal{F}^\varepsilon(\sigma, T, f, \varepsilon) = \frac{\partial \mathcal{L}(\sigma, T, f, \varepsilon)}{\partial \varepsilon} = \frac{\partial \Psi(\sigma, T, f, \varepsilon)}{\partial \varepsilon} + \lambda_2 - \lambda_3 = 0 \tag{3.19}$$

Introduction of expression (3.14) in the partial derivatives (3.18) and (3.19) gives :

$$\mathcal{F}^f(\sigma, T, f, \varepsilon) = -B(T - T_0) + \sigma\varepsilon - \frac{3}{2}\mu(1-\beta)(1-2f)\,\varepsilon\,\varepsilon + \lambda_0 - \lambda_1 = 0 \tag{3.20}$$

$$\mathcal{F}^\varepsilon(\sigma, T, f, \varepsilon) = \sigma f - 3\mu(1-\beta) f(1-f)\,\varepsilon + \lambda_2 - \lambda_3 = 0 \tag{3.21}$$

Conditions (3.13) to (3.16) and (3.20) (3.21) allow to define the domains of existence of the different solutions in the stress-temperature state diagram.

3.2.1. Transformation without strain

These transformations are characterized by a macroscopic transformation strain ε equal to zero. In that case constraint $\varepsilon = 0$ has to be respected for any value of f. From conditions (3.15) and (3.16) the two Lagrange multipliers associated to variable ε must satisfy:

$$\lambda_2 \geq 0 \qquad \text{and} \qquad \lambda_3 = 0 \tag{3.22}$$

Expression (3.21) for the thermodynamical forces (3.21) leads to:

$$\lambda_2 = -\sigma f \geq 0 \qquad \forall \; f \in [0;1] \tag{3.23}$$

This condition is satisfied only when the stress is equal to zero. This result is in agreement with experimental observations: the martensitic transition produces no macroscopic transformation strain when the phase change occurs in the absence of applied stress. Thermodynamical force (3.20) defines the temperature conditions for such transformation to happen. In the austenitic state, variable f must be equal to zero, so that:

$$\lambda_0 \geq 0 \qquad \text{and} \qquad \lambda_1 = 0 \tag{3.24}$$

Definition (3.20) for the thermodynamical force $\mathcal{F}^f$ assigns to multiplier λ_0 the following condition:

$$\lambda_0 = B\,(T - T_0) \geq 0 \qquad \text{that is satisfied when} \qquad T \geq T_0 \tag{3.25}$$

The same framework is applied for defining the martensitic state. In that case f has to be equal to unity, that gives :

$$\lambda_0 = 0 \qquad \text{and} \qquad \lambda_1 \geq 0 \tag{3.26}$$

Applying relation (3.18) leads to:

$$\lambda_1 = -\,B\,(T - T_0) \geq 0 \qquad \text{that is satisfied when} \qquad T \leq T_0 \tag{3.27}$$

Conditions (3.25) and (3.27) well define the austenite as the high temperature phase and the martensite as the low temperature one. However this description gives T_0, the thermodynamical equilibrium temperature, as the unique transformation temperature. That does not correspond to real transformation. Characteristic temperatures M_s and M_f do not appear in this approach. This serious shortcoming finds its origin in a microstructural description that neglects the granular structure of the material and the existence of several variants of martensite inside grains [37]. These approximations also produce some consequences for stress-induced transformations.

3.2.2 Stress-induced transformation

In this type of transformation, condition (3.23) imposes the strain parameter ε to be different from zero for any strictly positive f. To reach the maximal strain value, thermodynamical conditions required are obtained when Lagrange multipliers λ_2 and λ_3 verify:

$$\lambda_2 = 0 \qquad \text{and} \qquad \lambda_3 \geq 0 \tag{3.28}$$

Using expression (3.21) for the thermodynamical force $\mathcal{F}^\varepsilon$ leads to:

$$\lambda_3 = \sigma f - 3\mu\,(1 - \beta)\,f\,(1 - f)\,\varepsilon^{max} \geq 0 \tag{3.29}$$

When f is different to zero the applied stress must satisfy

$$\sigma \geq 3\,\mu\,(1-\beta)\,(1-f)\,\varepsilon^{max} \tag{3.30}$$

In the austenitic state, conditions (3.24) are active and equation (3.20) gives :

$$\lambda_0 = B\,(T-T_0) - \sigma\,\varepsilon^{max} + \frac{3}{2}\mu\,(1-\beta)\,(\varepsilon^{max})^2 \geq 0 \tag{3.31}$$

The critical transformation stress σ_c is then defined as:

$$\sigma_c = \frac{1}{\varepsilon^{max}}\,B\,(T-T_0) + \frac{3}{2}\mu\,(1-\beta)\,\varepsilon^{max} \tag{3.32}$$

Such transformation is thermodynamically feasible if critical stress (3.32) satisfied to condition (3.30) when f tends to zero. What imposes:

$$T - T_0 \geq \frac{3}{2B}\,\mu\,(1-\beta)\,(\varepsilon^{max})^2 \tag{3.33}$$

Using reasonable numerical values to describe polycrystalline copper-based shape memory alloys ($\varepsilon^{max} = 3\%$, $\mu = 40$ GPa, $d\sigma/dT = 2$ MPa·K^{-1}, that gived $B \approx 0.03$ MPa·K^{-1}), condition (3.33) is satisfied when the temperature gap is larger than 900 Kelvin. This result is out of any physical sense. Once again the description used for the microstructure (a unique spherical inclusion of martensite embedded into an austenitic matrix) failed. Utilization of an ellipsoïdal inclusion strongly improves this result in reducing the temperature gap (from about a factor fifty).

In this framework, the martensitic state must satisfy conditions (3.26). When f is equal to unity, thermodynamical force (3.20) leads to:

$$\lambda_1 = -B\,(T-T_0) + \sigma\,\varepsilon^{max} + \frac{3}{2}\mu\,(1-\beta)\,(\varepsilon^{max})^2 \geq 0 \tag{3.34}$$

The martensite finish stress σ_f is defined by:

$$\sigma_f = \frac{1}{\varepsilon^{max}}\,B\,(T-T_0) - \frac{3}{2}\,\mu\,(1-\beta)\,\varepsilon^{max} \tag{3.35}$$

This result leads to a new paradox, stress σ_f is smaller than σ_c, that gives an unstable behavior (figure 3.1.a). From conditions (3.31) and (3.34) the transformation temperatures when cooling under a constant imposed stress are defined by

$$M_s^\sigma = T_0 + \frac{1}{B}[\sigma\,\varepsilon^{max} - \frac{3}{2}\,\mu\,(1-\beta)\,(\varepsilon^{max})^2] \tag{3.36}$$

$$M_f^\sigma = T_0 + \frac{1}{B}[\sigma\,\varepsilon^{max} + \frac{3}{2}\,\mu\,(1-\beta)\,(\varepsilon^{max})^2] \tag{3.37}$$

The same unstable behavior is obtained (figure 3.1.b). Such result is in contradiction with experimental observations.

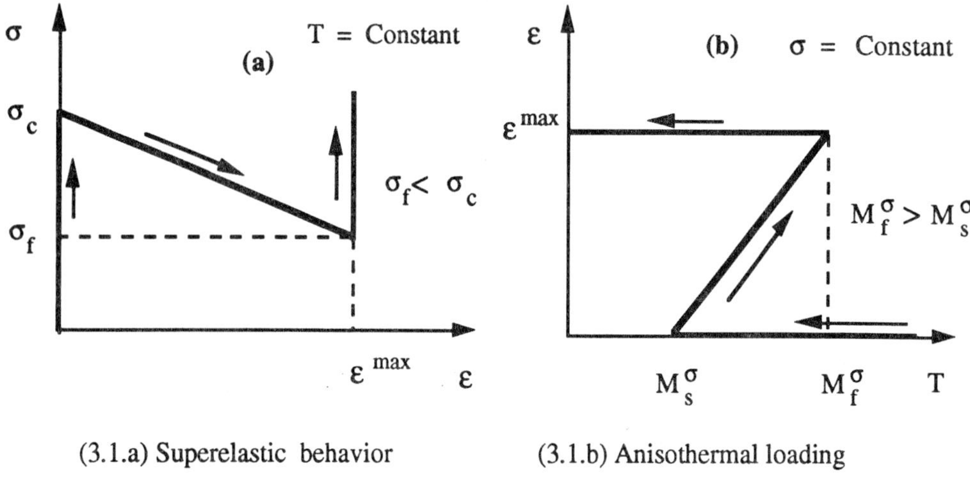

(3.1.a) Superelastic behavior (3.1.b) Anisothermal loading

Figure 3.1 : Unstable behavior determined from optimization of potential (3.17) in the macroscopic two-phases approach.

Among physical limitations on variable ε, we have to determine conditions to obtain solutions such that $0 < \varepsilon < \varepsilon^{max}$. In this case, according to (3.15) and (3.16), multipliers λ_2 and λ_3 are equal to zero. The ε value is then determined using thermodynamical force (3.21). For strictly positive f, one obtains:

$$\varepsilon = \frac{1}{(1-f)}\;\frac{\sigma}{3\,\mu\,(1-\beta)} \tag{3.38}$$

As long as the stress do not satisfy the condition (3.30) transformation strain is a linear stress function, otherwise it reaches the saturation value ε^{max} (figure 3.2). Such evolution is in agreement with the superthermoelastic behavior observed in polycrystalline materials.
Many conclusions arise from the thermodynamical study realized in this paragraph. Several results are in agreement with experimental observations, like the necessity to stress induce the transformation to produce a macroscopic transformation strain, or the dependence of this

transformation strain with the amplitude of the applied stress. But an instable behavior never observed is thus obtained. This instability arise from the f (1-f) factor in the interaction energy W^{int} of the potential (3.11). Such expression is obviously not realistic when f tends to unity. The reason is that grain to grain interactions introduce internal stresses so that even if f equals unity W^{int} cannot reach zero. A too coarse description of the microstuctural aspects linked to the transformation is responsible for this trouble. Different solutions can be used. In this section one keep the macroscopic two-phase description but a better heuristic estimation for the interaction energy W^{int} is made [38]. Results such obtained are presented in the following part.

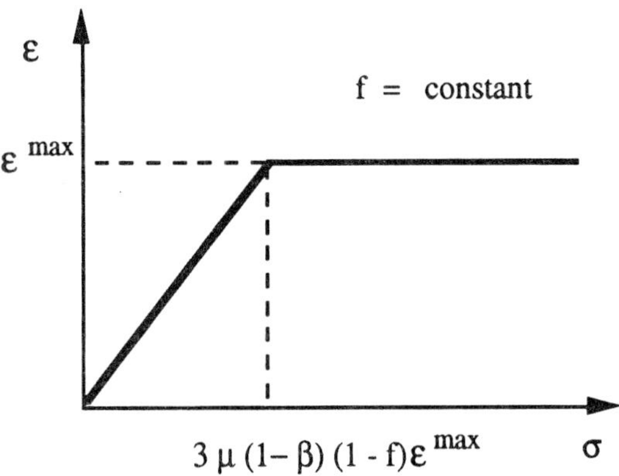

Figure 3.2: Evolution of the macroscopic transformation strain in superthermoelasticity determined from optimization of thermodynamical potential (3.17) (when the volume fraction of martensite is smaller than unity).

3.3. *Intergranular and intervariant interactions*

Interaction energy W^{int} is a complex function of the transformation strain field. This contribution is defined from:

$$W^{int}(\{\varepsilon_{ij}^{TM}(r)\}) = -\frac{1}{2V}\int_V \tau_{ij}(r)\,\varepsilon_{ij}^{TM}(r)\,dV \tag{3.39}$$

In the two-phases model, simplified microstructure leading to this expression using the inclusion problem has shown its limits. These come from a too crude description of the microstructural aspects. The dependance of the transformation strain with the amount of

martensite is neglected and the existence of two levels of strain incompatibilities is missing. The austenite/martensite and martensite/martensite interfaces are the intragranular sources of incompatibilities. Differences in the crystallographic orientations between grains give the intergranular one. From a phenomenological point of view, the choice of an expression for W^{int} without disavantages associated to relation (3.6) can improve the obtained result. This is realized in the following keeping the volume fraction f of martensite and the equivalent mean transformation strain ε as internal variables in the thermodynamical potential.
Two kinds of contributions must appear in $W^{int}(f, \varepsilon)$. In one hand, a quadratic function in f and ε, denoted by W^{gran} is rather associated to intergranular interactions. It is expressed like:

$$W^{gran} = \frac{1}{2} H \varepsilon^2 f^2 \tag{3.40}$$

In this approximation H is an intergranular interaction coefficient. On the other hand, a contribution depending on f is required, considering that a null value of variable ε is not equivalent to a null value of the interaction energy. This second term is assumed to be:

$$W^{var} = \frac{1}{2} A f^2 - K f \tag{3.41}$$

Approximation for the thermodynamical potential is given by:

$$\Psi(\sigma, T, f, \varepsilon) = \frac{1}{4\mu} \sigma\sigma + \sigma\varepsilon f - \frac{1}{2} H\varepsilon\varepsilon f^2 - \frac{1}{2} A f^2 + K f - B(T - T_0) f \tag{3.42}$$

Like in the previous scheme, dissipative aspects are neglected, the equilibrium condition for the austenite/martensite system is obtained from optimisation of potential (3.42) with respect to physical constraints (3.12). Thermodynamical forces applied on internal variables are now expressed by:

$$\mathcal{F}^f(\sigma, T, f, \varepsilon) = -B(T - T_0) + \sigma\varepsilon - H\varepsilon\varepsilon f - A f + K + \lambda_0 - \lambda_1 = 0 \tag{3.43}$$

$$\mathcal{F}^\varepsilon(\sigma, T, f, \varepsilon) = \sigma f - H\varepsilon f^2 + \lambda_2 - \lambda_3 = 0 \tag{3.44}$$

Like previously, Lagrange multipliers λ_0, λ_1, λ_2 and λ_3 are submitted to conditions (3.13) to (3.16). Resolution of the set of equations (3.43) and (3.44) defines stability domains for the different phases in a stress-temperature state diagram.

3.3.1. Transformation without strain

This kind of transformation is defined by conditions (3.22). Application to thermodynamical force (3.44) leads again to limitations (3.23). The absence of applied stress is always the criterion to obtain such transformation. The stability domain of austenite is defined from (3.24) according to:

$$\lambda_0 = B(T - T_0) - K \geq 0 \qquad \text{that gives} \qquad T \geq T_0 + \frac{K}{B} = M_s \tag{3.45}$$

Equation (3.26) defines the martensitic state from :

$$\lambda_1 = -B(T - T_0) + A - K \geq 0 \qquad \text{that gives} \quad T \leq M_s - \frac{A}{B} = M_f \tag{3.46}$$

Approximation (3.41) allows to define M_s and M_f temperatures that are different from the equilibrium temperature T_0. This result strongly improves the description obtained from potential (3.11). Nevertheless, from particular choice (3.41), there is no possible way to make a distinction between single or polycrystalline behavior. This is due to the fact that these aspects are only considered in contribution (3.40) that is vanishing when a null value is assigned to the transformation strain ε.

3.3.2. Stress-induced transformation - Superelasticity

As in paragraph 3.2.2 we determine thermodynamical conditions to obtain the maximal transformation strain. From constraints (3.28), the definition (3.44) of force $\mathcal{F}^{\varepsilon}$ imposes:

$$\lambda_3 = \sigma f - H f^2 \varepsilon^{max} \geq 0 \tag{3.47}$$

When f has a non zero value, application of this inequality gives

$$\sigma \geq H f \varepsilon^{max} \tag{3.48}$$

One notices that in the single crystal case characterized by a parameter H equal to zero, condition (3.48) is always verified. In the general case, using (3.24) the austenitic state $(f = 0)$ is defined by:

$$\lambda_0 = B(T - T_0) - \sigma \varepsilon^{max} - K \geq 0 \tag{3.49}$$

From definition (3.45) for the M_s temperature, the critical transformation stress is expressed by:

$$\sigma_c = \frac{1}{\varepsilon^{max}} B (T - M_s) \tag{3.50}$$

In that way a Clausius-Clapeyron-type equation is obtained. In the two-phases framework adopted, the martensitic state is defined using conditions (3.26). Setting f equal to unity, definition (3.43) of the thermodynamical force $\mathcal{F}^f$ leads to:

$$\lambda_1 = - B (T - T_0) + \sigma \varepsilon^{max} - H (\varepsilon^{max})^2 - A + K \geq 0 \tag{3.51}$$

From definition (3.46) the martensite finish stress σ_f is given by:

$$\sigma_f = \frac{1}{\varepsilon^{max}} B (T - M_f) + H \varepsilon^{max} \tag{3.52}$$

This stress is larger than that necessary to induce the transformation, the so determined behavior is stable. Expression (3.52) allows to distinguish the superelastic behavior in the single crystal case (H = 0). In these materials the martensite finish stress is related to the temperature by a Clausius-Clapeyron-type equation. Such a result is experimentally established, but not observed in the polycrystalline case and relationship (3.52) accounts for this difference. For H different from zero, introduction of definition (3.52) into inegality (3.48) makes this condition always satisfied in the polycrystalline case. One concludes that for superelasticity the macroscopic transformation strain is always equal to its maximal value. To characterize totally this behavior, conditions of existence of the two-phase region ($0 < f < 1$) has to be done. Application of (3.13), (3.14), and (3.43) leads to:

$$- B (T - T_0) + \sigma \varepsilon^{max} - H (\varepsilon^{max})^2 f - A f + K = 0 \tag{3.53}$$

At a given volume fraction f, the applied stress must be equal to :

$$\sigma = \frac{1}{\varepsilon^{max}} [B (T - M_s) + A f] + H \varepsilon^{max} f \tag{3.54}$$

Condition (3.48) is then always satisfied in superelasticity. In the two-phases model used, this kind of behavior is totally described from equation (3.53) and kinematical relation (3.3) taking $\varepsilon = \varepsilon^{max}$. For a given loading condition (applied stress and test-temperature) applying equation (3.53) allows to determine the volume fraction of martensite formed:

$$f = \frac{1}{H\,(\varepsilon^{max})^2 + A}\,[\,\sigma\,\varepsilon^{max} - B\,(T - M_s)] \qquad (3.55)$$

Tangent modulus $d\sigma/d\varepsilon$ associated to this transformation is given by:

$$\frac{d\sigma}{d\varepsilon} = H + \frac{A}{(\varepsilon^{max})^2} \qquad (3.56)$$

Superelastic behavior for single ($H = 0$) or polycrystalline materials are defined by relationships (3.50), (3.52) and (3.56). Figure 3.3 drawn from these results shows that the approximation used presents a good representation of this behavior. In this study hysteretic effect are ignored.

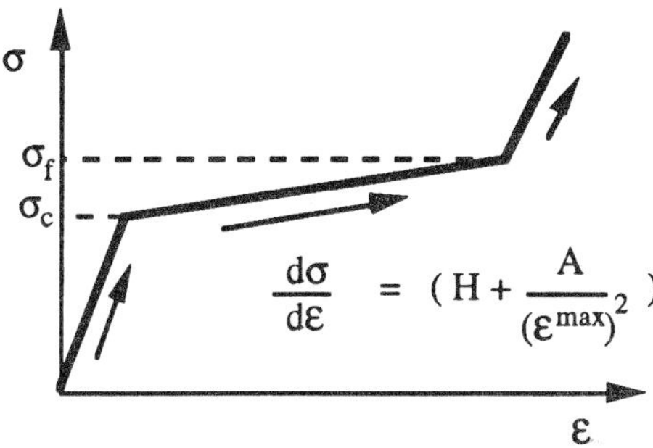

Figure 3.3: Superelastic behavior determined from optimization of potential (3.42). Parameter H is associated to intergranular interactions and parameter A to intragranular ones.

3.3.3. Superthermoelasticity

This scheme of resolution is now applied to transformations induced by cooling under constant applied stress. When the applied stress is large enough to satisfy condition (3.48), the strain parameter ε keeps its maximal value. In this case, the transformation temperatures associated to such a loading are directly defines from equation (3.53).

$$M_s^\sigma = M_s + \frac{1}{B}[\,\sigma\,\varepsilon^{max}\,] \qquad (3.57)$$

$$M_f^\sigma = M_f + \frac{1}{B}[\,\sigma\,\varepsilon^{max} + H\,(\varepsilon^{max})^2] \qquad (3.58)$$

It is important to notice that the martensitic start temperature M_s^σ, defined by equation (3.57) is directly related to the value of the applied stress σ. That is not the case for the martensitic finish temperature M_f^σ. Two cases have to be distinguished. When the applied stress satisfy condition (3.48) at any f, the transformation strain is equal to its maximal value ε^{max} and equation (3.58) can be used. In the opposite case, the transformation strain ε keeps its maximal value ε^{max} only during cooling from M_s^σ to a temperature called M_{th}^σ, which is a function of the applied stress. In further cooling below M_{th}^σ, ε becomes smaller than its maximal value and the transformation achieves at the temperature M_f. In that case, expression (3.44) of the thermodynamical force $\mathcal{F}^\varepsilon$ allows to relate ε to the applied stress.

$$\varepsilon = \frac{1}{f}\frac{1}{H}\sigma \tag{3.59}$$

Transformation strain produced during anisothermal creep depends on the applied stress. At loww stress level, a linear relationship is defined by equation (3.59). A saturation value is obtained at higher stress. Such an evolution constitutes a good approximation of experimental observations. As soon as condition (3.48) is no longer satisfied, strain ε decreases as the volume fraction of martensite increases. Critical volume fraction f_{th}^σ from which this decrease begins is defined from (3.48) as:

$$f_{th}^\sigma = \frac{\sigma}{H\,\varepsilon^{max}} \tag{3.60}$$

The M_{th}^σ temperature is then defined setting f equal to f_{th}^σ in equation (3.53).

$$M_{th}^\sigma = M_s - \frac{1}{B}\frac{A}{H\,\varepsilon^{max}}\sigma \tag{3.61}$$

Such relation is only defined in the polycrystalline case ($H \neq 0$). In these materials, the lower limit for the maximal transformation strain domain is a linear function of applied stress. The transformation is achieved when f is equal to unity. Condition (3.26) leads to the following inegality:

$$\lambda_1 = -B\,(T - M_f) + \sigma\varepsilon - H\varepsilon^2 \geq 0 \tag{3.62}$$

Substituting ε by its expression (3.59) one obtains :

$$T \leq M_f \tag{3.63}$$

Condition (3.63) is independent on the stress level imposed. Results obtained from the approximation used in this study allows to determine a schematic chart for the behavior (figure 3.4). The M_{th}^{σ} temperature line determines a limit between a transformation regime in which the martensite plates are fully oriented by the applied stress and a thermal regime in which the martensite growths in a self-accommodated way.

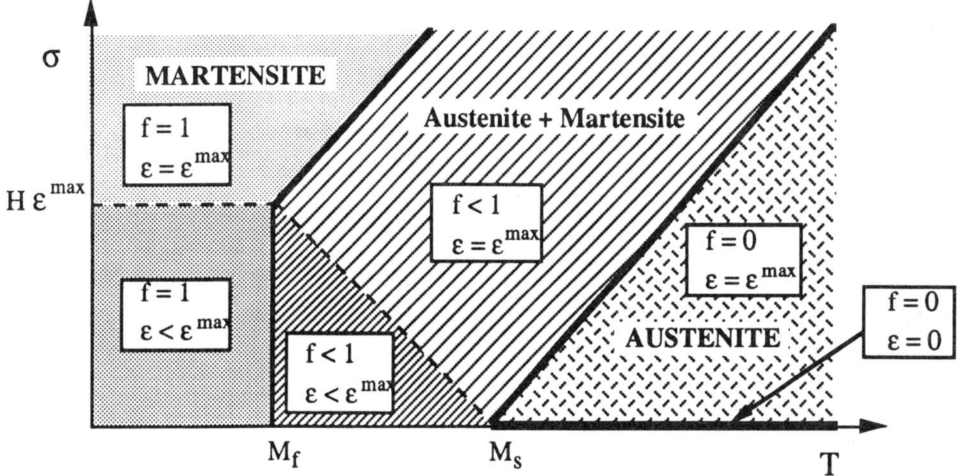

Figure 3.4: Schematic chart defined from optimization of potential (3.42). Depending on the loading state, different values are taken to internal variables f and ε.

Results obtained in this section show that an a priori estimation of the interaction energy allows to improve appreciably the two-phase model. In that way, superelastic or superthermoelastic problems can be solved for any monotonic multiaxial loading. However, such an approach presents several disadvantages. First of all, assumption (3.10) reduces the applicability of the model to radial loadings and gives no information on the evolution of strain $\bar{\varepsilon}^{TM}$ with the loading. Less brutal estimation for the interaction energy needs to introduce a priori kinetic of the transformation. This class of macroscopic approach, while being very useful in structure calculation presents in fact a weak predictive capacity. Utilization of scale transition methods allows to overcome these handicaps using a microstructural description closer to the physical strain mechanisms [39].

4. Crystallographical approach

Results obtained in the previous section using a macroscopic approach present the advantage to give analytical expression very easy to use. Nethertheless it is necessary to use a microstructural description closer to the physical strain mechanisms to overcomes the disadvantages related to this class of macroscopic approach.

In shape memory materials, transformation strain incompatibilities come from the granular structure and from the existence of several variants of martensite inside each grain. Intragranular phenomena are described considering the volume fraction of each variant of martensite as internal variables in the thermodynamical potential associated to the transformation. A local behavior is derived from this study.

Intergranular aspects are taken into account using a scale transition approach. The thermomechanical integral equation obtained considering the local behavior law and standard field equations is solved using the self-consistent approximation [40].

Numerical results are compared with experimental observations and measurement on Cu-Zn-Al superelastic alloys [41].

4.1. Single crystal behavior

The local behavior to introduced in the scale transition framework is derived considering grain inside polycrystalline material as single crystal. Particular aspects related to the difference between a grain and a real single crystal are considered in the determination of the thermodynamical potential.

4.1.1. Kinematical aspects

From kinematical equations (2.15) and (2.17) determined in section 2, the mesoscopic (at the grain level) transformation stain E^T and the global amount of martensite f can be related to the volumic fraction of each variant of martensite formed inside a grain.

$$E^T_{ij} = \sum_n \varepsilon^{Tn}_{ij} f^n \qquad \text{and} \qquad f = \sum_n f^n \tag{4.1}$$

In this description, transformation strain associated to each variant is a well-defined quantity. Variables f^n have obvious physical meaning and the following physical constraints must be respected :

$$f^n \geq 0 \quad \text{for each } n \tag{4.2.a}$$

$$\sum_n f^n \leq 1 \tag{4.2.b}$$

4.1.2. Thermodynamical potential and driving forces

Let us consider a reference austenitic crystal of volume V bounded by a surface ∂V. For temperature T larger than T_0, a transformation strain field ε^T (r) is induced appliying surface forces $T_i^d = \Sigma_{ij}\, n_j$ on the boundary ∂V. From the thermodynamical study presented in chapter 2, complementary free energy Ψ by unit volume expresses as (equation 2.53):

$$\Psi(\Sigma_{ij},T,\{\varepsilon_{ij}^T(r)\}) = \frac{1}{2}\Sigma_{ij}\, M_{ijkl}\,\Sigma_{kl} + \Sigma_{ij}\, E_{ij}^T + \frac{1}{2V}\int_V \tau_{ij}(r)\,\varepsilon_{ij}^T(r)\, dV - B(T-T_0)\, f \tag{4.3}$$

Chapter 3 has shown that free energy plays a fundamental role on the behavior. Energy W^{int} is a complex function of the transformation strain field $\varepsilon^T(r)$. This contribution is closely linked to the microstructure of the martensitic phase and can not reduce easily to a simple expression of the f^n. According to (2.45) in chapter 2, this contribution is defined as:

$$W^{int} = -\frac{1}{2V}\int_V \tau_{ij}(r)\;\varepsilon_{ij}^T(r)\; dV \tag{4.4}$$

Both topological and morphological considerations on the growing process of martensite inside grains of a polycrystal allow to evaluate this contribution. Increase of the volume fraction for a given variant is mainly produced by the growing of new plates in a limited region of the grain (see micrography 4.1). From this observation and considering internal stresses arise mainly from the martensite/martensite interactions a "cluster" description is used for the microstructure (figure 4.1) [42]. One notes V_n the domain partially filled by a variant n and v_n the volume constituted by the plates of martensite. The volume fraction f^n relative to a variant n is classically defined by:

$$f^n = \frac{v_n}{V} \qquad \text{with} \qquad \varepsilon^T_{ij}(r) = \varepsilon^{Tn}_{ij} \qquad \text{when} \qquad r \in v_n \tag{4.5}$$

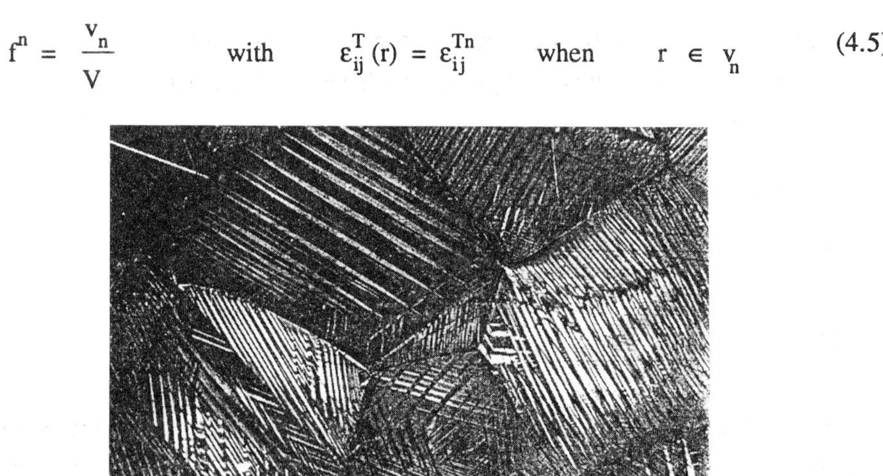

Micrography 4.1: Formation of different variants of martensite inside a grain in a Cu-Zn-Al alloy polycrystalline speciment in uniaxial tensile test.

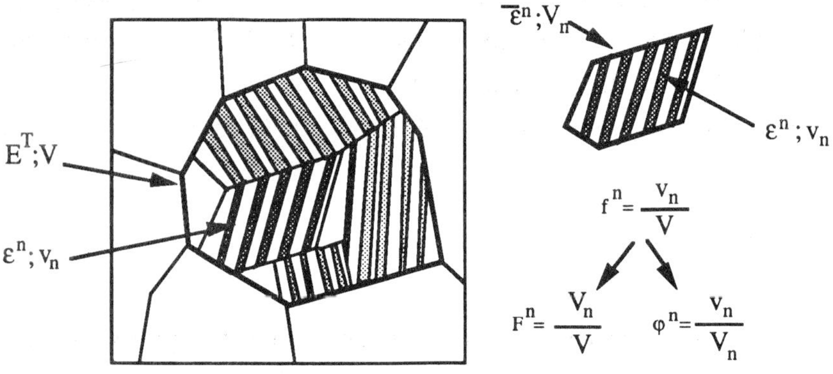

Figure 4.1 : Application of a cluster type description to represent the microstructural aspects inside a grain of the polycrystalline material. Definition of parameters V_n, v_n, F^n, f^n and φ^n.

One notes $F^n = V_n/V$ the volume fraction of the domain filled by variant n inside the grain. The amont of martensite inside V_n is characterized by the parameter $\varphi^n = v_n/V_n$. Quantities f^n, φ^n et F^n are linked by:

$$f^n = \frac{v_n}{V} = \frac{v_n}{V_n}\frac{V_n}{V} = \varphi^n F^n \tag{4.6}$$

The mean transformation strain $\bar{\varepsilon}^{Tn}$ over a domain V_n is defined as :

$$\bar{\varepsilon}_{ij}^{Tn} = \frac{1}{V_n}\int_{V_n} \varepsilon_{ij}^{T}(r)\, dV = \varepsilon_{ij}^{Tn}\frac{v_n}{V_n} = \varepsilon_{ij}^{Tn}\varphi^n \tag{4.7}$$

The transformation strain field $\varepsilon^T(r)$ is considered as piecewise uniform. Inside each volume V_n a uniform transformation strain $\bar{\varepsilon}^{Tn}$ is defined. Interaction energy (4.4) expresses then as:

$$W^{int} = -\frac{1}{2V}\int_V \tau_{ij}(r)\, \varepsilon_{ij}^{T}(r)\, dV = -\frac{1}{2V}\sum_n \bar{\tau}_{ij}^{n}\, \bar{\varepsilon}_{ij}^{Tn}\, V_n \tag{4.8}$$

where $\bar{\tau}^n$ denotes the internal stress over domain V_n. Considering these domains like ellipsoïdal inclusions characterized by an Eshelby tensor $\mathcal{S}^{En}$, equation (3.2) gives :

$$\bar{\tau}_{ij}^{n} = C_{ijkl}\,(I_{klrs} - \mathcal{S}_{klrs}^{En})\,(E_{rs}^{T} - \bar{\varepsilon}_{rs}^{Tn}) \tag{4.9}$$

Considering the mean value of $\bar{\tau}^n$ over V is equal to zero, one obtains the following property :

$$\frac{1}{V}\int_V \tau_{ij}(r)\, E_{ij}^{T}\, dV = \frac{1}{V}\sum_n \bar{\tau}_{ij}^{n}\, E_{ij}^{T}\, V_n = 0 \tag{4.10}$$

From (4.9) and (4.10) interaction energy (4.8) can express as :

$$W^{int} = \frac{1}{2}\sum_n \{\bar{\varepsilon}_{ij}^{Tn} - E_{ij}^{T}\}\, C_{ijkl}\,(I_{klrs} - \mathcal{S}_{klrs}^{n})\,\{\bar{\varepsilon}_{rs}^{Tn} - E_{rs}^{T}\}\, F^n \tag{4.11}$$

Because few variants are formed in each grain in superelasticity (see micrography 4.1), it is reasonable to limit the calculation of W^{int} to pair interactions. When only two variants p and q are taken in consideration, energy (4.11) turns then into:

$$W^{int} = \frac{1}{2}(\bar{\varepsilon}^{Tp}_{ij} - E^{T}_{ij})\, C_{ijkl}\, (I_{klrs} - S^{Ep}_{klrs})\, (\bar{\varepsilon}^{Tp}_{rs} - E^{T}_{rs})\, F^{p}$$
$$+ \frac{1}{2}(\bar{\varepsilon}^{Tq}_{ij} - E^{T}_{ij})\, C_{ijkl}\, (I_{klrs} - S^{Eq}_{klrs})\, (\bar{\varepsilon}^{Tq}_{rs} - E^{T}_{rs})\, F^{q} \quad (4.12)$$

Global transformation strain E^T is then defined by :

$$E^{T}_{ij} = \bar{\varepsilon}^{Tp}_{ij}\, F^{p} + \bar{\varepsilon}^{Tq}_{ij}\, F^{q} \qquad \text{with} \qquad F^{p} + F^{q} = 1 \quad (4.13)$$

It is worthwhile to notice that strong correlation exists between the shape of domains V_p and V_q. Considering these volumes as ellipsoïdal inclusions one obtains the following property $S^{Ep} = S^{Eq} = S^{Epq}$. Introducing (4.13) in (4.12) turns the interaction energy into:

$$W^{int} = \frac{1}{2} F^{p} F^{q}\, (\bar{\varepsilon}^{Tp}_{ij} - \bar{\varepsilon}^{Tq}_{ij})\, C_{ijkl}\, (I_{klrs} - S^{Ep}_{klrs})\, (\bar{\varepsilon}^{Tp}_{rs} - \bar{\varepsilon}^{Tq}_{rs})\, F^{p}$$
$$+ \frac{1}{2} F_{q}\, F^{p} (\bar{\varepsilon}^{Tq}_{ij} - \bar{\varepsilon}^{Tp}_{ij})\, C_{ijkl}\, (I_{klrs} - S^{Eq}_{klrs})\, (\bar{\varepsilon}^{Tq}_{rs} - \bar{\varepsilon}^{Tp}_{rs})\, F^{q}$$
$$= \frac{1}{2} F^{p} F^{q}\, (\bar{\varepsilon}^{Tp}_{ij} - \bar{\varepsilon}^{Tq}_{ij})\, C_{ijkl}\, (I_{klrs} - S^{Epq}_{klrs})\, (\bar{\varepsilon}^{Tp}_{rs} - \bar{\varepsilon}^{Tq}_{rs}) \quad (4.14)$$

Tensor S^{Epq} that appear in (4.14) is strongly linked to the choice of variants p and q. Utilization of the definition (4.7) for mean transformation strain in domains V_p and V_q allows to express (4.14) from parameters f^n et ε^{Tn}.

$$W^{int} = \frac{1}{2}\, \frac{f^{p}}{\varphi^{p}}\, \frac{f^{q}}{\varphi^{q}}\, (\varphi^{p} \varepsilon^{Tp}_{ij} - \varphi^{q} \varepsilon^{Tq}_{ij})\, C_{ijkl}\, (I_{klrs} - S^{Epq}_{klrs})\, (\varphi^{p} \varepsilon^{Tp}_{rs} - \varphi^{q} \varepsilon^{Tq}_{rs}) \quad (4.15)$$

Taking account of the physics of the transformation allows to simplify relationship (4.15). It is reasonable to assume that for every domain V_n the progress of the transformation is the same. This gives:

$$\varphi^{p} = \varphi^{q} \qquad \text{for each variant} \quad p \text{ et } q \in [0\,;n] \quad (4.16)$$

using this assumption, one obtains :

$$W^{int} = \frac{1}{2} f^{p} f^{q}\, (\varepsilon^{Tp}_{ij} - \varepsilon^{Tq}_{ij})\, C_{ijkl}\, (I_{klrs} - S^{Epq}_{klrs})\, (\varepsilon^{Tp}_{rs} - \varepsilon^{Tq}_{rs}) \quad (4.17)$$

This energy is a function of volume fractions f^p and f^q of the two considered variants. It also depends on the shape and orientation of domains V_p and V_q. Assuming penny-shape inclusions, the shape is characterized by one scalar parameter only (the c/a ratio - thickness/ diameter). To simplify furthermore this problem, in order to describe the microstructural aspects using only the volume fraction as internal variables, one determines orientation of the $\mathcal{S}^{Epq}$ tensor minimizing energy (4.17) at a given c/a ratio. Two interaction modes are distinguished corresponding to self-accommodated plates and incompatible variants. Such distinction is really observed in metallurgical studies [43]. From this result, the resistive contribution associated to interaction energy (4.4) is represented using an interaction matrix denoted by H^{pq}. Interaction energy (4.4) writes then as:

$$W_{int} = \frac{1}{2} \sum_{n,m} H^{nm} f^n f^m \tag{4.18}$$

It is necessary to note that the different assumptions used to determine this matrix limit its utilization to stress-induced transformations (superelasticity and two-way shape memory). Among these limitations, the number of variants present in a grain has to remain weak. By consequence, in a fully thermally-induced martensitic state where different self-accommodated groups coexist the use of this matrix gives a considerable overestimation of the elastic energy associated to this configuration.
Utilization of relationship (4.18) in the free energy (4.3) allows to use only the volume fraction of the different variants of martensite as internal variables :

$$\Psi(\Sigma_{ij},T, f^n) = \frac{1}{2}\Sigma_{ij} M_{ijkl} \Sigma_{kl} + \Sigma_{ij} \sum_n \varepsilon_{ij}^{Tn} f^n - \frac{1}{2} \sum_{n,m} H^{nm} f^n f^m - B(T-T_0) \sum_n f^n \tag{4.19}$$

Like in the macroscopic two-phases approach, developed in the previous chapter, the physical constraints (4.2) are taken into account using the Kuhn and Tucker conditions in order to define the thermodynamical forces $\mathcal{F}^n$ [18]. To the n + 1 conditions (4.2), n + 1 Lagrange multipliers are associated. Multiplier associated to condition (4.2.b) is denotes by λ_0. This null or positive quantity has to satisfy the following property:

$$\lambda_0 \left(\sum_n f^n - 1\right) = 0 \tag{4.20}$$

In the same way, n multipliers λ_n associated to the n inequalities (4.2.a) must verify:

$$\lambda_n \left(-f^n - 0\right) = 0 \quad \text{at each } n \tag{4.21}$$

From conditions (4.20) and (4.21) the Lagrangien of the complementary free energy is defined. This fonction is denoted by $\mathcal{L}(\Sigma_{ij},T, f^n)$:

$$\mathcal{L}(\Sigma_{ij},T, f^n) = \Psi(\Sigma_{ij},T, f^n) - \lambda_0 (\sum_n f^n - 1) - \sum_n \lambda_n (- f^m - 0) \quad (4.22)$$

Partial derivatives of Lagrangien (4.22) give the thermodynamical forces $\mathcal{F}^n$ acting on internal variables f^n [36].

$$\mathcal{F}^n = \frac{\partial \mathcal{L}}{\partial f^n} = \frac{\partial \Psi}{\partial f^n} - \lambda_0 + \lambda_n \quad (4.23)$$

Definition (4.19) of the complementary free energy leads to:

$$\mathcal{F}^n = \Sigma_{ij}\, \varepsilon_{ij}^{Tn} - \sum_m H^{nm} f^m - B(T-T_0) - \lambda_0 + \lambda_n \quad (4.24)$$

Due to the existence of hysteresis phenomena, the knowledge of thermodynamical forces (4.24) is not suffisant to derive the thermodynamical behavior associated to a stress-induced martensitic transformation. This study must be completed by the definition of a dissipative potential.

4.1.3. Dissipative potential and local transformation criteria

Hysteresis is an important phenomenon in shape memory alloys. It is linked to dissipation process occuring during the martensitic transformation which can be described by a dissipative potential W^d.

Evolution of this second potential has to respect the second principle of thermodynamic. Combined utilization of this fundamental law with the energy balance gives:

$$\dot{\Psi}\,\big|_{\Sigma,T} = \dot{W}^d \geq 0 \quad (4.25)$$

As in plasticity, one supposes that the thermodynamical driving force $\mathcal{F}^n$ has to reach a critical value denoted by F_c^n, to produce growth (or shrinkage) of a variant. This critical value is a function of the microstructural state of the material. As for the thermodynamical potential, it is assumed that these aspects can be described using internal variables f^n only. One writes then:

$$\mathcal{F}_c^n = \mathcal{F}_c^n (f^m) \quad (4.26)$$

This gives the following transformation conditions:

$$\text{when} \quad \mathcal{F}^{n} < \mathcal{F}_c^{n} \quad \text{then} \quad \dot{f}^{n} = 0 \quad \forall \quad \dot{\mathcal{F}}_c^{n} \tag{4.27.a}$$

$$\text{when} \quad \mathcal{F}^{n} = \mathcal{F}_c^{n} \quad \text{and} \quad \dot{\mathcal{F}}_c^{n} = 0 \quad \text{then} \quad \dot{f}^{n} = 0 \tag{4.27.b}$$

$$\text{when} \quad \mathcal{F}^{n} = \mathcal{F}_c^{n} \quad \text{and} \quad \dot{\mathcal{F}}_c^{n} = \dot{\mathcal{F}}^{n} \quad \text{then} \quad \dot{f}^{n} \neq 0 \tag{4.27.c}$$

This problem simplifies if one supposes that resistive forces $\mathcal{F}_c^n$ are positive material constants, identical for all the variants. In this case $\mathcal{F}_c^n$ reduces to $\mathcal{F}_c$ for each n. This hypothesis considers implicitly the dissipative potential as a function proportional to the cumulative volume fraction f_{cu} of martensite [7].

$$W^d = \mathcal{F}_c \int_0^t |df| \quad = \mathcal{F}_c f_{cu} \tag{4.28}$$

Application of inequality (4.25) produces different transformation conditions for forward and reverse transitions. Austenite to martensite transition occurs when:

$$\sum_n \mathcal{F}^n \dot{f}^n = \mathcal{F}_c \dot{f} \tag{4.29}$$

and the reverse transformation needs :

$$\sum_n \mathcal{F}^n \dot{f}^n = - \mathcal{F}_c \dot{f} \tag{4.30}$$

Assuming that the dissipation occurs without coupling effects between variants, conditions (4.29) and (4.30) turn:

$$\mathcal{F}^n \dot{f}^n = \mathcal{F}_c \dot{f}^n \quad \text{for the forward transition} \tag{4.31.a}$$

$$\mathcal{F}^n \dot{f}^n = - \mathcal{F}_c \dot{f}^n \quad \text{for the reverse transition} \tag{4.31.b}$$

This analysis allows to define local criteria. A volume of austenite transforms to a given variant of martensite when:

$$\Sigma_{ij} \varepsilon_{ij}^{Tn} - \sum_m H^{nm} f^m - B(T-T_0) + \lambda_n - \lambda_0 = \mathcal{F}_c \tag{4.32}$$

Nevertheless, this transformation will be only possible if conditions (4.20) and (4.21) are satisfied, what imposes on the Lagrange multipliers:

$$\lambda_n = -\Sigma_{ij}\,\varepsilon_{ij}^{Tn} + \sum_m H^{nm} f^m + B(T - T_0) + \lambda_0 + \mathcal{F}_c \geq 0 \qquad (4.33.a)$$

$$\lambda_0 = \Sigma_{ij}\,\varepsilon_{ij}^{Tn} - \sum_m H^{nm} f^m - B(T - T_0) + \lambda_n - \mathcal{F}_c \geq 0 \qquad (4.33.b)$$

As soon as a variant reaches a volume fraction equal to unity or if the global volume fraction reaches this threshold value, applying constraints (4.33) limits the transformation, even when condition (4.32) is verified. These restrictions constitutes a major difference from the modelling of plastic behavior by dislocations motion.

The reverse transformation happens when the following condition is satisfied:

$$\Sigma_{ij}\,\varepsilon_{ij}^{Tn} - \sum_m H^{nm} f^m - B(T - T_0) + \lambda_n - \lambda_0 = -\mathcal{F}_c \qquad (4.34)$$

in this case, Lagrange multipliers λ_0 et λ_n have to respect:

$$\lambda_n = -\Sigma_{ij}\,\varepsilon_{ij}^{Tn} + \sum_m H^{nm} f^m + B(T - T_0) + \lambda_0 - \mathcal{F}_c \geq 0 \qquad (4.35.a)$$

$$\lambda_0 = \Sigma_{ij}\,\varepsilon_{ij}^{Tn} - \sum_m H^{nm} f^m - B(T - T_0) + \lambda_n + \mathcal{F}_c \geq 0 \qquad (4.35.b)$$

Constraints (4.35) verify that it remains enough martensite for the reverse transformation to occur. This set of conditions (4.32) (4.33) (4.34) and (4.35), constitute the local criteria for thermoelastic martensitic transformation.

4.1.4. Constitutive equations

If criterion (4.32) (resp. 4.34) and conditions (4.33) (resp. 4.35) are satisfied, the utilization of the coherency relationship (4.27.c) allows to determine evolution of the transformation rates $\dot{f}^n$ on each variants:

$$\dot{\mathcal{F}}^n = \frac{\partial \dot{\mathcal{F}}^n}{\partial \Sigma_{ij}}\,\dot{\Sigma}_{ij} - \frac{\partial \dot{\mathcal{F}}^n}{\partial T}\,\dot{T} + \sum_m \frac{\partial \dot{\mathcal{F}}^n}{\partial f^m}\,\dot{f}^m = \dot{\mathcal{F}}_c = 0 \qquad (4.36)$$

Substitution of driving force (4.24) in relation (4.36) gives :

$$\varepsilon_{ij}^{Tn}\,\dot{\Sigma}_{ij} - B\,\dot{T} - \sum_m H^{nm}\,\dot{f}^m = 0 \qquad (4.37)$$

Resolution of the system equations (4.37) gives the evolution of the $\dot{f}^n$ with the loading parameters $\dot{\Sigma}$, $\dot{T}$. Using this result into the kinematical equation (4.1) defines the constitutive equation that describes the single crystal behavior.

$$\dot{E}^T_{ij} = \sum_n \varepsilon^{Tn}_{ij} \sum_m (H^{nm})^{-1} [\, \varepsilon^{Tm}_{kl} \dot{\Sigma}_{kl} - B \dot{T} \,] \tag{4.38}$$

This equation characterizes the behavior of a unit volume crystal of parent phase undergoing a stress-induced martensitic transformation. Production of several variants of martensite is taken into account. This polyvariants transformation difficult to produce in single crystal really happens in the grains of polycrystalline material. From this consideration, relationship (4.38) appears well-adapted to describe the local behavior at the grain level in a scale transition approach.

The different material parameters included in relationship (4.38) can be determined. The B-coefficient related to the chemical energy can be evaluated on single crystal from tensile tests performed at different temperatures [8]. Denoted by Σ_{33} the applied stress, condition (4.34) gives for a purely austenitic state:

$$\varepsilon^{Tn}_{33} \, \Sigma_{33} = B\,(T - T_0) + \mathcal{F}_c \tag{4.39}$$

This result is similar to that obtain using the well-known Patel and Cohen criterion [24]. For characteristic temperatures M_s and A_f, the material is in a fully parent phase state (characterized by a value f^m equal to zero for each variant m). Consequently the Lagrange multipliers λ_n and λ_0 take both a null value, that allows to define:

$$M_s = T_0 - \frac{\mathcal{F}_c}{B} \quad \text{and} \quad A_f = T_0 + \frac{\mathcal{F}_c}{B} \tag{4.40}$$

By consequence, these two temperatures are intrinsic parameters. They can be used to obtain an evaluation of the T_0 temperature and to determine critical force $\mathcal{F}_c$.

4.2. Thermomechanical integral equation

In the previous section, a constitutive relation was determined for the single crystal where the crystallographical orientation is totally defined and where mechanical loading conditions can be considered as uniform on the whole volume. Such an accurate analysis is impossible

in polycrystalline materials. Existence of the granular structure leads to strain incompatibilities. This phenomenon produces an intergranular stress field. An a priori knowledge of the local loading conditions on each grains is then impossible. Utilization of concentration and homogenization methods allows to overcome this difficulty.
Three fundamental equations govern the scale transition problem. The first is the stress equilibrium condition. According to the inelastic character of the behavior, one expresses this condition in rate form:

$$\dot{\sigma}_{ij,j}(r) = 0 \tag{4.41}$$

The second fundamental equation is the compatibility condition of the total strain field. Local quantities must satisfied the imposed boundary conditions. Let us consider a velocity field v(x) applied on the surface ∂V of the studied domain:

$$v_i(x) = \dot{u}_i(x) = \dot{E}_{ij}\, x_j \quad \text{for } x \in \partial V \tag{4.42}$$

The strain rate tensor is, by definition, equal to the symmetrical part of the gradient of the velocity field $\dot{u}$:

$$\dot{\varepsilon}_{ij}(r) = \frac{1}{2}\left(\dot{u}_{i,j}(r) + \dot{u}_{j,i}(r)\right) \tag{4.43}$$

This equation insures the strain field compatibility in the solid. The local behavior law, defined in the previous section, constitutes the third relationship. For shake of simplicity, one will assume that the temperature field is uniform and equal to the test-temperature imposed.

$$\dot{\varepsilon}_{ij}(r) = g_{ijkl}(r)\, \dot{\sigma}_{kl}(r) + n_{ij}(r)\, \dot{T} \tag{4.44}$$

The strain rate $\dot{\varepsilon}$ is composed by a thermoelastic part and an inelastic one that is directly linked to the phase change. The behavior (4.44) expressed on a dual form turns:

$$\dot{\sigma}_{ij}(r) = l_{ijkl}(r)\, \dot{\varepsilon}_{kl}(r) - m_{ij}(r)\, \dot{T} \tag{4.45}$$

Tangent moduli involve in these two expressions are linked together:

$$l_{ijkl}(r) = g^{-1}_{ijkl}(r) \qquad \text{and} \qquad m_{ij}(r) = l_{ijkl}(r)\, n_{kl}(r) \tag{4.46}$$

These moduli are determined from the constitutive equation previously defined. Introduction of the local behavior (4.45) in the equilibrium condition (4.41) gives:

$$(l_{ijkl}(r)\ \dot{\varepsilon}_{kl}(r) - m_{ij}(r)\ \dot{T})_{,j} = 0 \tag{4.47}$$

Let us consider a homogeneous reference medium characterized by two uniform tangent moduli L^o and M^o. A uniform strain rate E^o is applied to this medium. Tangent modulus in behavior law (4.45) are defined from their deviations $\delta l(r)$ and $\delta m(r)$ to these homogeneous quantities.

$$l_{ijkl}(r) = L^o_{ijkl} + \delta l_{ijkl}(r) \quad \text{and} \quad m_{ij}(r) = M^o_{ij} + \delta m_{ij}(r) \tag{4.48}$$

Equilibrium condition (4.47) turns :

$$((L^o_{ijkl} + \delta l_{ijkl}(r))\ \dot{\varepsilon}_{kl}(r) - (M^o_{ij} + \delta m_{ij}(r))\ \dot{T})_{,j} = 0 \tag{4.49}$$

Uniformity of properties in the homogeneous reference material and the hypothesis of uniformity of the temperature leads to a Navier-type equation:

$$L^o_{ijkl}\ \dot{u}_{k,lj}(r) + (\delta l_{ijkl}(r)\ \dot{\varepsilon}_{kl}(r) - \delta m_{ij}(r)\ \dot{T})_{,j} = 0 \tag{4.50}$$

Second member of this equation can be considered as the time-derivative of a volume force f(r), such as for each point r:

$$L^o_{ijkl}\ \dot{u}_{k,lj}(r) + \dot{f}_i(r) = 0 \tag{4.51}$$

setting :

$$\dot{f}_i(r) = (\delta l_{ijkl}(r)\ \dot{\varepsilon}_{kl}(r) - \delta m_{ij}(r)\ \dot{T})_{,j} \tag{4.52}$$

The Green tensor G^o of an infinite homogeneous reference medium is defined by:

$$L^o_{ijkl}\ G^o_{jn,ik}(r - r') + \delta_{ln}\ \delta(r - r') = 0 \tag{4.53}$$

In this expression δ (r-r') denotes the Dirac fonction. Setting

$$\dot{u}_n(r) = \int_V \delta_{ln}\ \delta(r - r')\ \dot{u}_l(r')dV' \tag{4.54}$$

the following integral equation is obtained from (4.53) :

$$\dot{u}_n(r) = \dot{u}_n^o + \int_V G_{in}^o(r-r')\,\dot{f}_i(r')\,dV' \tag{4.55}$$

where $\dot{u}^o$ is the solution for a homogeneous reference medium, submitted to the same boundary conditions. Using definition (4.43) leads to an integral equation in term of strain rate.

$$\dot{\varepsilon}_{mn}(r) = \dot{E}_{mn}^o + \int_V \Gamma_{mnij}^o(r-r')\,(\delta l_{ijkl}(r')\,\dot{\varepsilon}_{kl}(r') - \delta m_{ij}(r')\,\dot{T})\,dV' \tag{4.56}$$

where Γ^o denotes the symmetrical part of the modified Green tensor. Resolution of this integral equation theoretically determines macroscopic behavior. However this resolution is not trivial. Considering particular microstructures could leads to approximation of equation (4.56). Sachs-Zaoui model considers grains in serie arrangement in the polycrystal (like a bamboo structure) [44, 45]. This hypothesis leads to the homogeneity of the stress field. Such kind of simplified approach has been developed elsewhere to describe the superelastic behavior [46]. Utilization of a distribution of grains of parallel type has for corollary the uniformity of the strain field and defines the Taylor model. These two types of resolution possess an extremal character and constitute upper and lower bounds for the real behavior. However the static model underestimates the influence of the internal stress field and the Taylor model overestimates them strongly. The hypothesis of a fully disordered material also simplifies the resolution of equation (4.56) while preserving a correct estimation for the internal stress field. This last approach defining the self-consistent models, was developed for the elastoplastic behavior of microheterogeneous materials [40]. In the following we develop this kind of approach to the case of superelastic copper-based alloys. The obtained results are discussed.

4.3. Self-consistent approximation

Integral equation (4.56) is solved using the self consistent approximation. In this scheme, particular properties of the modified Green tensor are used. This tensor can be decomposed in a local part Γ^{ol} and a non local one Γ^{nl} [47] that is rapidly decreasing.

$$\Gamma^{o}_{ijkl}\,(r\text{-}r') \;=\; \Gamma^{ol}_{ijkl}\,(r)\;\delta(r\text{-}r') \;+\; \Gamma^{nl}_{ijkl}\,(r\text{-}r') \qquad (4.57)$$

The decomposition (4.57) transforms integral equation (4.56) into :

$$\begin{aligned}\dot{\varepsilon}_{mn}\,(r) = \;& \dot{E}^{o}_{mn} + \Gamma^{ol}_{mnij}\,(\delta l_{ijkl}\,(r)\;\dot{\varepsilon}_{kl}\,(r) \;-\; \delta m_{ij}\,(r)\,\dot{T}) \\ & + \int_V \Gamma^{nl}_{mnij}\,(r\text{-}r')\,(\delta l_{ijkl}\,(r')\;\dot{\varepsilon}_{kl}\,(r') - \delta m_{ij}\,(r')\,\dot{T})\;dV' \qquad (4.58)\end{aligned}$$

All the difficulties linked to the resolution of this equation come from the last contribution. This problem can be simplified by the choice of a homogeneous reference medium such that the non local contribution $(\delta l\,(r')\,\dot{\varepsilon}\,(r') - \delta m\,(r')\,\dot{T})$ fluctuates around a null average value:

$$\int_V ((l_{ijkl}\,(r) \;-\; L^{o}_{ijkl})\;\dot{\varepsilon}_{kl}\,(r) - (m_{ij}\,(r) \;-\; M^{o}_{ij})\,\dot{T})\;dV \;=\; 0 \qquad (4.59)$$

In this case, the non local contribution in relationship (4.58) can be neglected. This hypothesis imposes to choose L^{o} and M^{o} such that:

$$\int_V (l_{ijkl}\,(r)\;\dot{\varepsilon}_{kl}\,(r) - m_{ij}\,(r)\,\dot{T})\,dV \;-\; L^{o}_{ijkl}\int_V \dot{\varepsilon}_{kl}\,(r)\,dV \;+\; M^{o}_{ij}\,\dot{T} \;=\; 0 \qquad (4.60)$$

This expression is equivalent to:

$$\dot{\Sigma}_{ij} \;=\; L^{o}_{ijkl}\;\dot{E}_{kl} - M^{o}_{ij}\,\dot{T} \qquad (4.61)$$

So, the particular choice (4.59) is equivalent to choose as homogeneous reference medium the effective medium we have to determine.

$$L^{o}_{ijkl} \;=\; L^{eff}_{ijkl} \qquad \text{and} \qquad M^{o}_{ij} \;=\; M^{eff}_{ij} \qquad (4.62)$$

This assumption strongly simplifies the resolution of the integral equation (4.58) that turns now into:

$$\dot{\varepsilon}_{mn}\,(r) = \dot{E}^{o}_{mn} + \Gamma^{eff}_{mnij}\,((l_{ijkl}\,(r) - L^{eff}_{ijkl})\;\dot{\varepsilon}_{kl}\,(r) - (m_{ij}\,(r) - M^{eff}_{ij})\,\dot{T}) \qquad (4.63)$$

In this way, two localisation tensors A(r) and a(r) are defined. They connect the local quantity $\dot{\varepsilon}(r)$ to the overall ones $\dot{E}^{o}$ and $\dot{T}$ in such a way that equation (4.63) is now expressed by:

$$\dot{\varepsilon}_{ij}(r) = A_{ijkl}(r)\,\dot{E}_{kl} - a_{ij}(r)\,\dot{T} \tag{4.64}$$

These localisation tensors A(r) and a(r) are defined by:

$$A_{mnkl}(r) = \left(I_{mnkl} - \Gamma^{eff}_{mnij}\left(l_{ijkl}(r) - L^{eff}_{ijkl}\right)\right)^{-1} \tag{4.65.a}$$

$$a_{kl}(r) = A_{klmn}(r)\,\Gamma^{eff}_{mnij}\left(m_{ij}(r) - M^{eff}_{ij}\right) \tag{4.65.b}$$

Definition of these quantities allows to determine the uniform tangent moduli used in relationship (4.61) from the local ones of relation (4.45).

$$L^{eff}_{ijkl} = \frac{1}{V}\int_V l_{ijmn}(r)\,A_{mnkl}(r)\,dV \quad ; \quad M^{eff}_{ij} = \frac{1}{V}\int_V \left(l_{ijkl}(r)\,a_{kl}(r) + m_{ij}(r)\right) dV \tag{4.66}$$

Assuming polycrystal as a set of homogeneous grains allows to consider the localisation tensors as piecewise uniform functions. Denoting by l^N the uniform tangent moduli of grain N and by A^N the constant value of the localisation tensor in this grain, relationship (4.66) becomes:

$$L^{eff}_{ijkl} = \sum_N l^N_{ijmn}\,A^N_{mnkl}\,F^N \tag{4.67}$$

where F^N represents the volume fraction of grain N. In the polycrystal each grain is characterized by its crystallographic orientation (that can be determined from X-ray diffraction), its shape (from the Γ^{ol} tensor) and by F^N. The overall behavior is then totally determined using relationship (4.67) from the knowledge of the internal structure evolution in the polycrystalline material. Intergranular interactions coming from the crystallographical misorientation are obtained by the calculation of the stress in each grain.

4.4. Numerical results

A self-consistent code [48] elaborated to describe the plastic behavior in polycrystalline metallic alloy has been modified to compute the superelastic behavior of shape memory alloys. As the macroscopic reversible strain remains smaller than 10%, this code is used in small deformation and grain rotation is ignored. Variants of martensite replace slip systems and the volume fractions of these variants are now the microstructural parameters describing the internal state evolution. Physical constraints (4.2) acting on these variables have been introduced. Recovery of the inelastic strain and hysteretic effect on unloading are taken into account. Numerical results obtained are analyzed and compared to experimental observations performed on Cu-Zn-Al alloys.

4.4.1. Numerical data used

The elastic behavior is assumed to be isotropic and homogeneous in the two phases ($\mu \approx 40$ GPa and $\nu \approx 0.3$). The polycrystalline structure is schematized by one hundred spherical grains randomly oriented to produce an isotropic texture. Main crystallographic characteristics of the martensitic transformation for Cu-Zn-Al shape memory alloys are taken from De Vos et al. [49]. These characteristics are considered to be invariant during the transformation. Optical microscopy combined with image analysis give access to the shape of the different grains. X-rays measurement provides a representation of the crystallographic texture of the polycrystal. The measurement of the lattice parameters in the two phases and the mutual orientation relationships allows to determine the habit plane normals **n**, the transformation directions **m** and the strain intensity g, using the WLR phenomenological theory for the martensitic transformation [16] (table 4.1). Uniaxial tensile tests realized on single crystal at different temperatures define the B-coefficient needed in equation 4.39 [8] ($B \approx 0.23$ MPa·K^{-1} in the alloys employed).

a	n1	n2	n3		m1	m2	m3
b	n1	n3	n2		m1	m3	m2
c	- n1	n2	n3		- m1	m2	m3
d	- n1	n3	n2		- m1	m3	m2
e	- n2	n1	n3		- m2	m1	m3
f	- n3	n1	n2		- m3	m1	m2
g	- n2	- n1	n3		- m2	- m1	m3
h	- n3	- n1	n2		- m3	- m1	m2
i	n1	- n2	n3		m1	- m2	m3
j	n1	- n3	n2		m1	- m3	m2
k	- n1	- n2	n3		- m1	- m2	m3
l	- n1	- n3	n2		- m1	- m3	m2

m	n3	n1	n2		m3	m1	m2
n	n2	n1	n3		m2	m1	m3
o	n3	- n1	n2		m3	- m1	m2
p	n2	- n1	n3		m2	- m1	m3
q	n2	- n3	- n1		m2	- m3	- m1
r	n3	- n2	- n1		m3	-m2	- m1
s	- n2	n3	- n1		- m2	m3	- m1
t	- n3	n2	- n1		- m3	m2	- m1
u	- n3	- n2	- n1		- m3	- m2	- m1
v	- n2	- n3	- n1		-m2	-m3	- m1
w	n3	n2	- n1		m3	m2	- m1
x	n2	n3	- n1		m2	m3	- m1

n1 = - 0.182 n2 = 0.669 n3 = 0.721 m1= - 0.165 m2 = - 0.737 m3 = 0.655

Table 4.1: Definition of the normal to the habit plane and the direction of transformation for each martensitic variant in a Cu-Zn-Al shape memory alloys [49].

Existence of compatibility (or incompatibility) relationships between variants is taken into account by the interaction matrix H (equation 4.18). Micromechanical determination using minimisation of equation (4.17) according to the orientation of the inclusion considered [42] established that this matrix is composed using two types of terms (Table 4.2), weak interaction terms H^1 for self-accommodated variants (around $\mu/1000$) and strong interaction terms H^2 for incompatible variants (about $\mu/150$). Such distinction is in agreement with metallographical observations [43].

$$H^1 = \frac{\mu}{1000} \approx 40 \text{ MPa} \qquad H^2 = \frac{\mu}{150} \approx 270 \text{ MPa} \qquad \text{(for a Cu-Zn-Al alloy)}$$

	a	b	c	d	e	f	g	h	i	j	k	l	m	n	o	p	q	r	s	t	u	v	w	x
a	C	C	C	C			C		C		C			C						C	C			
b	C	C	C	C				C		C		C	C						C			C		
c	C	C	C	C	C				C		C					C		C					C	
d	C	C	C	C		C				C		C			C		C							C
e			C		C	C	C	C	C					C		C	C					C		
f				C	C	C	C	C		C			C		C			C			C			
g	C				C	C	C	C			C			C		C			C					C
h		C			C	C	C	C				C	C		C					C			C	
i	C		C		C				C	C	C	C				C				C	C			
j		C		C		C			C	C	C	C			C				C			C		
k	C		C				C		C	C	C	C		C				C					C	
l		C		C				C	C	C	C	C	C				C							C
m		C				C		C				C	C	C	C	C		C			C			
n	C				C		C				C		C	C	C	C	C					C		
o				C		C		C		C			C	C	C	C				C			C	
p			C		C		C		C				C	C	C	C			C					C
q				C	C							C		C			C	C	C	C		C		C
r			C			C					C		C				C	C	C	C	C		C	
s		C					C			C						C	C	C	C	C		C		C
t	C							C	C						C		C	C	C	C	C		C	
u	C					C			C				C					C		C	C	C	C	C
v		C			C					C				C			C		C		C	C	C	C
w			C					C			C				C			C		C	C	C	C	C
x				C			C					C				C	C		C		C	C	C	C

Table 4.2: Interaction Matrix H determined for a Cu-Zn-Al Shape memory alloys; C denotes the compatible interactions H^1 and blank denotes the incompatible one H^2 [41].

Hysteresis phenomenon linked to the transformation is related to the critical force $\mathcal{F}_c$. Measurement of temperature M_s and A_f allows to evaluate the amplitude of this force using relationship (4.40).

$$2\,\mathcal{F}_c = B\,(A_f - M_s)$$

Temperatures A_f and M_s strongly depend on the material composition. Difference between these two temperatures is almost constant for a given class of alloy but is very sensitive to the thermomechanical loading history. This aspect is not studied here, $\mathcal{F}_c$ is considered as being constant along the loading path. It is worthwhile to note that the different material parameters used in this work are measurable quantities, at the only exception of H^1 and H^2 that are determined from micromechanical considerations. There is no use of adjustable parameter in this approach.

4.4.2. Results

From these data the polycrystalline behavior is determined for a given thermomechanical loading path. Figure 4.2 illustrates result obtained for uniaxial tensile test at room temperature (20 °C). In this example M_s and A_f temperature are respectively taken equal to 10 °C and 15 °C. Several characteristics obtained are in agreement with experimental trends:

- the maximal transformation strain is close to three per cent, what is realistic in copper-based alloy;
- the hysteresis size determined at the polycrystalline level (45 MPa) is larger that the single crystal one used in constitutive equation (10 MPa);
- the critical transformation stress is in agreement with the well-known Patel and Cohen relationship [24].

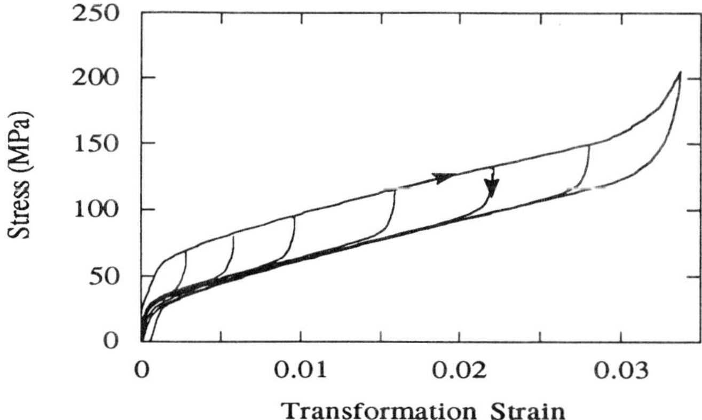

Figure 4.2: Numerical simulation obtained using the self-consistent approach for tensile test on a polycrystalline superelastic Cu-Zn-Al alloy. Computation realized considering room temperature condition (T = 20°C), M_s = 10°C and A_f = 15°C.

In this simulation, several stages may be distinguished in the course of the transformation. At the very beginning, the macroscopic transformation strain is very weak. Around 60 MPa a sharp variation is observed on the stress-strain curve which defines an apparent macroscopic critical transformation stress. The transformation proceeds then in a steady-state regime. In this second regime, the hardening rate given by the ratio $d\Sigma/dE^T$ is near 3500 MPa. This is close to usually measured value on this type of alloy. At stress larger than 150 MPa, one observes a saturation strain and the stress increases rapidly. This comes from physical condition (4.2.b) applied in each grains of the polycrystal. In this last stage, some grains totally transformed behave in a purely elastiç way.

Evolutions of the internal variables involved in the macroscopic two-phases approach developed in chapter 3 are computed. Numerical determination of the evolution of the global volume fraction f of martensite according to the loading defines the kinetic for stress induced transformation (figure 4.3). A quasi linear relationship is observed during the steady-state regime of figure 4.2. A saturation value around 80% of stress-induced martensite is observed. The end of the transformation is associated to a very large stress without physical meaning. In these conditions it is no longer possible to consider that the transformation mechanism occurs alone, other physical phenomena must appear (plasticity, crack initiation) and the transformation should remain partial in superelasticity (existence of residual austenite).

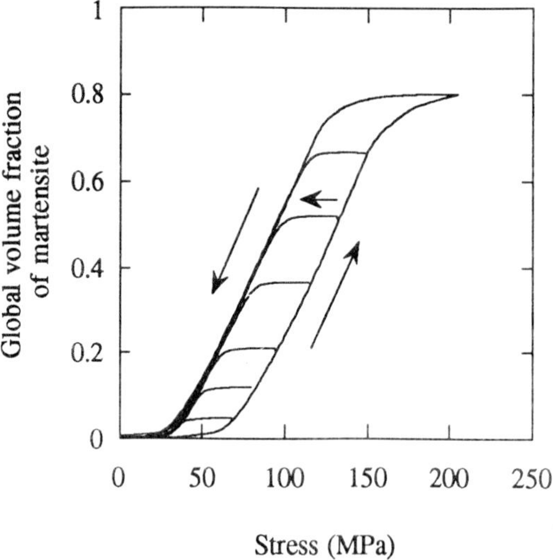

Figure 4.3: Kinetic of a stress-induced martensitic transformation numerically defined using the self-consistent simulation.

Numerical results confirm that the macroscopic mean transformation strain $\bar{\varepsilon}^{TM}$ can not be considered as a constant parameter (figure 4.4). This overall value rapidly decreases in the beginning of the loading and reached a saturation value depending of the loading state. The maximal value of $\bar{\varepsilon}^{TM}$, obtained in the very beginning of the transformation is equal to the transformation strain of the first induced well-oriented variant.

Description of the strain mechanism at a microscopic lengthscale allows to obtain evolution of microstructural parameters. In this approach, informations such as the progress of the transformation inside each individual grain and the evolution of the differents variants of martensite with respect to the loading path are obtained. This allows to put in evidence two stages in the transformation.

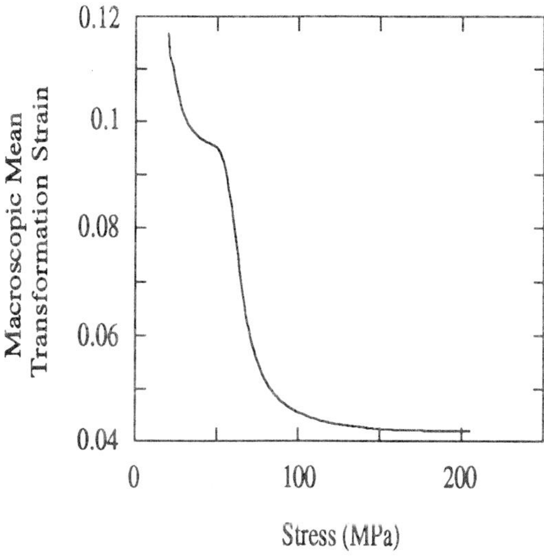

Figure 4.4: Evolution of the macroscopic mean transformation strain during a superelastic tensile test.

In the first stage, the increase of the global volume fraction of martensite is mainly due to the increase of the number of grains where the transformation takes place. In stage two, all grains transform but the number of variants per grain increases to reach an average saturation value close to three variants per grains (figure 4.5). This average trend regroups in fact very different evolution according to the grain considered. Figures 4.6 illustrates these differences giving the computed evolution in four grains [50].

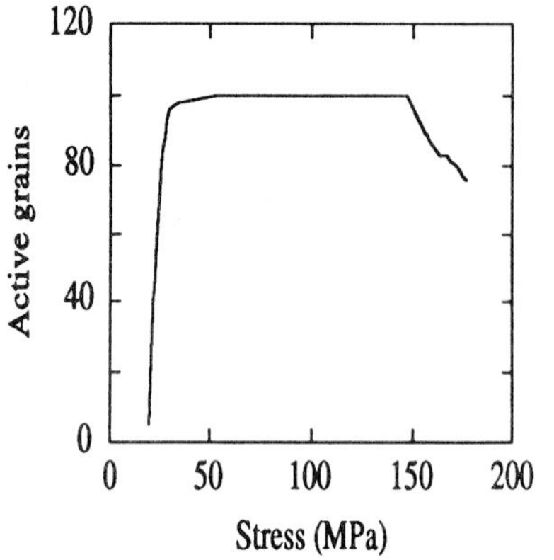

Figure 4.5.a: Evolution of the number of active grains during a tensile test.

Figure 4.5.b: Evolution of the number of active variant per grain on loading.

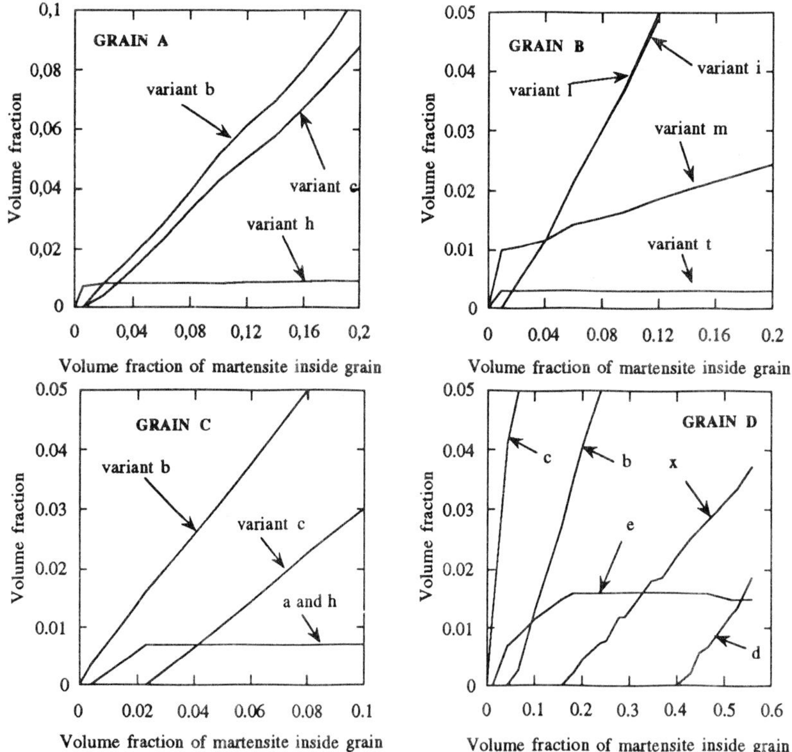

Figure 4.6: Volume fraction of the different variants of martensite versus the amount of martensite inside four grains having different crystallographic orientation. Simulation performed in tensile test condition [50]

	(a)	(b)	(c)	(d)
φ^1	354.73°	38.08°	341.26°	17.75°
ϕ	88.91°	31.90 °	83.98	69.17°
φ^2	229.61°	8.08°	34.97°	353.73°

Table 4.3: Definition of crystallographic orientation for the four grains used in figure 4.6.

One observes that the mode of selection of variants is different in the two stages. The applied stress plays a preponderant role to select the first variant induced in each grain. That is no longer the case in stage two where the local stress becomes preponderant. In some grains (figures 4.6 a and b) the first variant can be superseded by other variants. As presented in figure 4.6.d this evolution can be very complex in some grains and even show a regression of variants formed at the beginning of the transformation. Such evolution predicted by the model is in agreement with several metallurgical observations [51, 52].

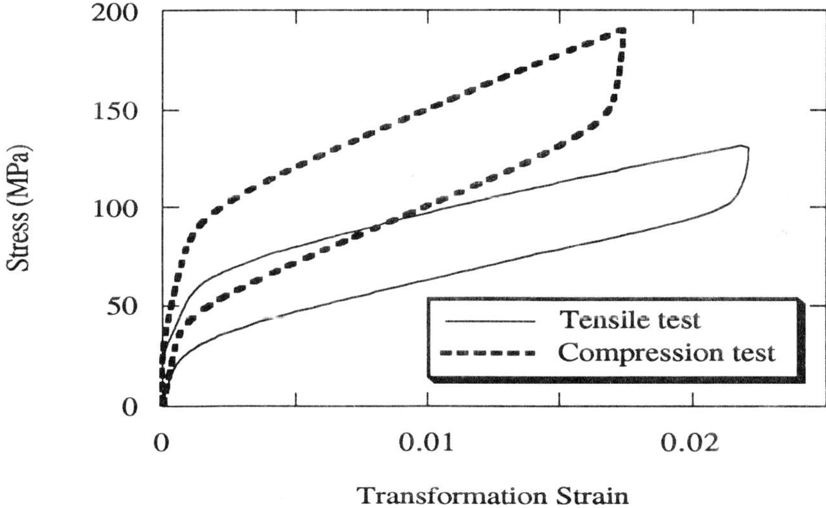

Figure 4.7: Dissymmetry observed between tensile and compression tests as obtained by numerical simulation using the self-consistent approach (T - Ms = 10°C).

The model also allows to take into account the dissymmetry observed in the behavior on these alloys between tensile and compression tests. Crystallographical origin of this phenomenon has been determined [53]. The habit plane normal and the transformation direction serving as input data, it is natural that the dissymmetry phenomenon is taken there into account. On computed curves, one observes that the critical stress, the transformation slope and the hysteresis size obtained are larger in compression that in traction (figure 4.7). This is in agreement with the experimental trends [54].

Considering different radial loading path allows to determine a transformation surface in the stress space (figure 4.8). A symmetric behavior is obtained for shearing, that is in agreement with experimental observation [55]. Analytical description of such a surface using the second and third invariant of the deviatoric stress tensor is usefull in structure calculation [56] (see chapter 8 of this volume).

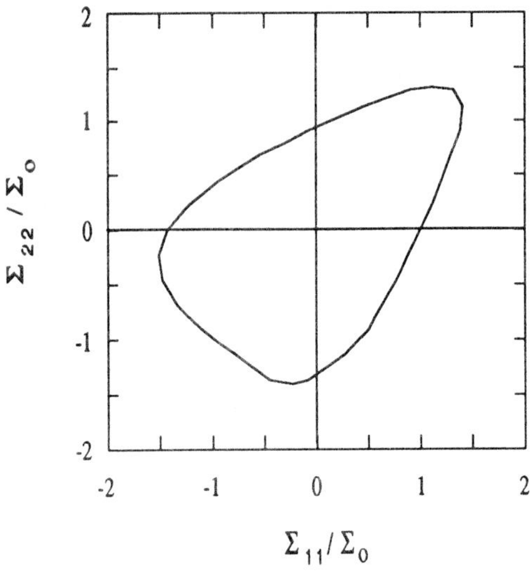

Figure 4.8: Dissymmetrical transformation surface in superelasticity numerically determined using the self-consistent approach [53]

4.4.3. Comparison with experimental observations

Numerical determination obtained using the proposed scale transition framework is now compared with experimental results on superelastic Cu-Zn-Al alloys from the literature. The alloy used in [54] was elaborated by Tréfimétaux company with a weight composition of 67.93% Cu-4% Al (including Zr, Mg, Fe as affinant). Average grain size is 0.3 mm. Thermal treatment was the following: solution heating in β-phase (10 minutes at 850 °C), water-quenched and annealed at 20 °C for 48 hours. Characteristic transformation temperatures are M_S = - 98°C and A_f = - 91°C (determined by resistivity measurement).

Uniaxial tensile test was performed on specimen of rectangular section (10 x 0.6 mm) and length 45 mm. Deformations were measured using extensometer. The imposed strain rate was 10^{-4} s^{-1}. Variations of temperature were obtained using a thermocontroled chamber in which liquid nitrogen is vaporized. Vacher and Lexcellent characterized the superelastic behavior of this alloy between - 80 °C and + 100 °C [54].

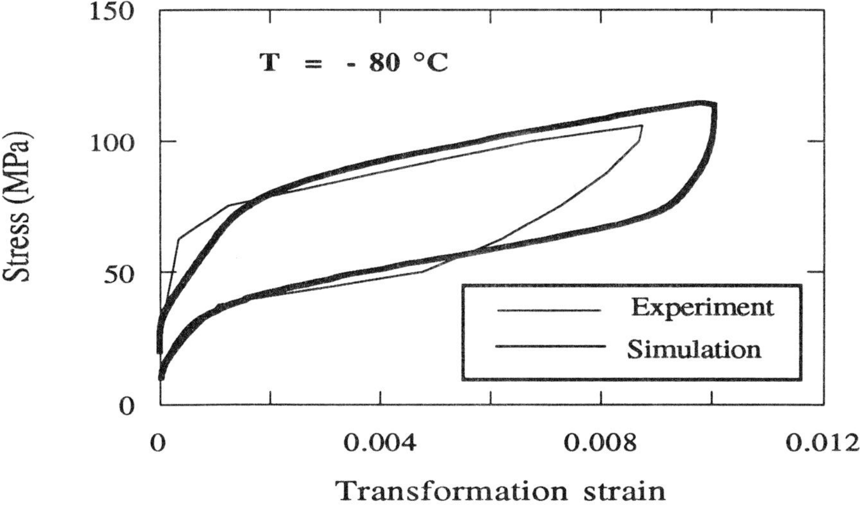

Figure 4.9: Comparaison between numerical simulation and experimental measurement [54] performed on a superelastic Cu-Zn-Al alloy ($A_f = M_s + 7$°C).

Numerical results obtained from the self-consistent model are determined using the same material parameters as those of the preceding section (**n**, **m**, g, B, μ, H^{nm}). Particularities of the studied alloy are taken into account by the T_0 and F_c parameters [41]. These quantities are determined from the experimental measurement of temperatures T, M_s and A_f given by Vacher et al. [54]. Comparison with experimental tensile stress-strain curve published by Vacher and al. for -80 °C indicates that the transformation slope and the hysteresis size are correctly evaluated by the model (figure 4.9). Computed and measured transformation stresses are in good agreement. In the same way an increase of the test temperature keep a good correspondence between computed and measured transformation stress (figure 4.10) upon a large range of test temperature.

The three dimensionnal aspects taken into account in the description of the transformation inside crystal are kept by the scale transition method, so the proposed framework can be applied to described multi-axial loading path. There are few experimental studies on the bi-axial behavior of shape memory alloys. Tensile-torsion tests were performed by Rogueda [57] on a 70.17 Cu - 25.63 Zn - 4.2 Al (wt%) alloy elaborated by Tréfimétaux company.

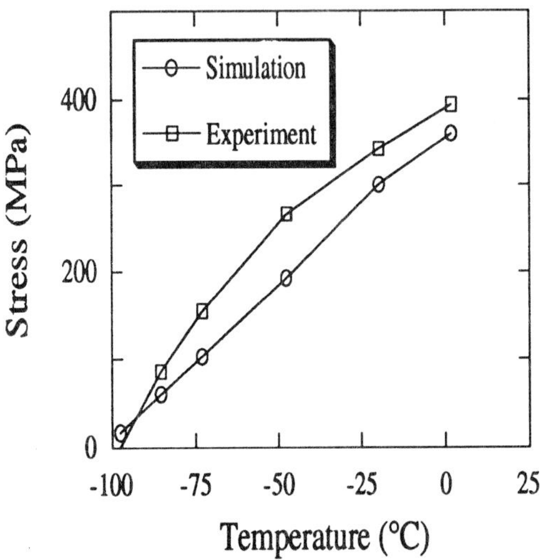

Figure 4.10: Critical transformation stress against the temperature in superelasticity for a Cu-Zn-Al alloy. Comparison between numerical results obtained using the self-consistent method and experimental data [54].

This alloy was solution treated in β-phase (15 minutes at 850 °C), quenched and annealed at 120 °C during one hour followed by a maintain for several days at room temperature. Characteristic transformation temperatures, determined by resistivity measurement, are M_S = 14°C and A_f = 20°C [57]. Superelastic tensile-torsion tests were performed on tube of 20 mm diameter, 2 mm thickness and 71 mm heigh at a temperature of 30°C. Deformations were measured using LVDT and RVDT extensometer.

Different proportionnal loadings are made from uniaxial tensile test to pure torsion keeping the maximal Von Mises equivalent stress equal to 110 MPa. These experimental conditions were simulated using the self-consistent model. A very good agreement was obtained (figure 4.11). The $r\theta$ and zz componants of the stress-strain curve are well-captured by the simulation without any fitting parameter. Only the test conditions (applied stress and test temperature) and the characteristic temperatures (M_S and A_f) are changed in the input data.

Change in the critical transformation stress due to the dissymmetrical aspects in the superelastic behavior is well captured too [58].

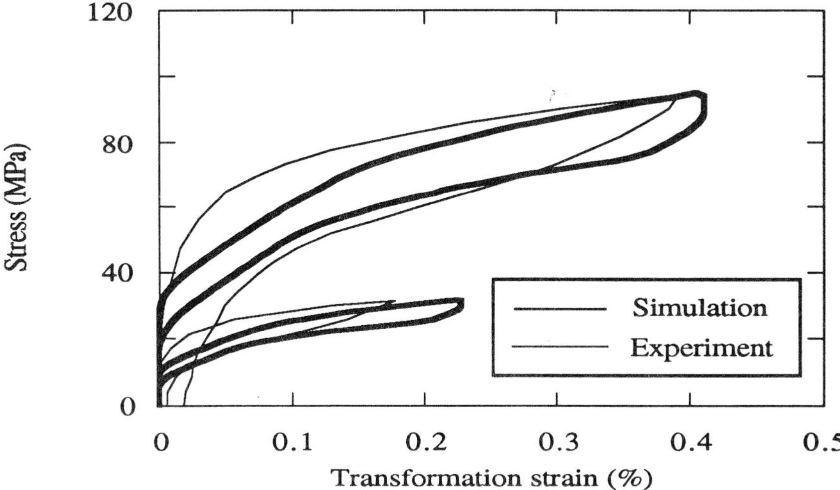

Figure 4.11 : Superelastic tensile-torsion test with $\sigma_{r\theta} / \sigma_{zz}$ ratio egal to 0.333 on a Cu-Zn-Al alloy. Comparison between experimental measurement performed by Rogueda [57] and simulation obtained using the self-consistent method.

Non isothermal behavior can be also described without adding new internal parameter. Influence of a continuous change in the imposed temperature can be accounted deriving a complete thermomechanical integral equation for the scale transition problem as exposed in [50]. In this case stress and temperature act in the same way as loading parameters. Anisothermal creep behavior of shape memory alloys is very well described using this extension of the model (see figure 4.12).

Experimental result performed by Bourbon [59] on thermal cycling of a Cu-Zn-Al shape memory alloy under a constant applied stress of 65 MPa is well captured by this scale transition approach. Numerical results put in evidence that the decrease, experimentally observed of the maximal transformation strain with the applied stress level is related to the formation of self-accommodated structure inside the martensite phase.

General feature related to the stress-induced martensitic transformation were presented in this part. For superelasticity, influence of the test temperature, hysteresis loop, tension compression dissymmetry and multiaxial loading are simulated with a good accuracy by the scale transition approach proposed. This framework is also successfully developed to describe anisothermal creep.

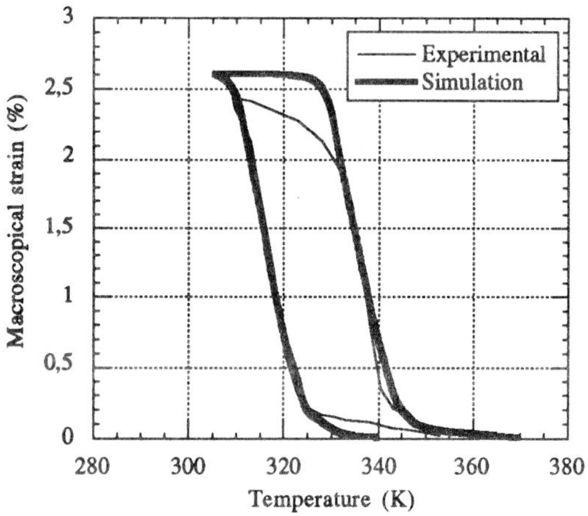

Figure 4.12 : Anisothermal creep on a Cu-Zn-Al alloy, comparison between experimental measurement and numerical simulation [50].

4.5. Conclusions

Numerical results presented in this chapter are in good agreement with experimental observations obtained on copper-based superelastic alloys. This confirms the interest of micromechanical approaches and scale transition methods for the study of behavior of materials undergoing a martensitic phase transition. The formulation used follows a kinematical description of the strain mechanism taking into account the existence of several variants of martensite and a determination of the thermodynamical potential. Polycrystalline response is derived considering single crystal as a representative volume element for the local behavior. Hysteresis associated to the phase transformation is taken into account from the definition of a pseudo-dissipative potential.

Different data needed for the computation are experimentally determined from calorimetric measurement, texture analysis and mechanical tests performed on single crystal. These determinations are independent of the loading mode envisaged.

This scheme of resolution determines the evolution of some macroscopic variables hard to obtained experimentally but very often used as internal variables in a rather macroscopic phenomenological modelling.

References

1. K. TANAKA, S. KOBAYASHI, Y. SATO, "*Thermomechanics of Transformation Pseudoelasticity and Shape Memory Effect in Alloys*", Int. J. Plasticity, Vol. 2, pp. 59-72 (1986).
2. B. RANIECKI, C. LEXCELLENT, K. TANAKA," *Thermodynamic Models of Pseudoelastic Behaviour of Shape Memory Alloys*", Arch. of Mech., Vol. 3 (1992).
3. A. L. ROYTBURD, "*Elastic Domains and Polydomain Phases in Solids*", Phase Transitions, Vol. 45, pp. 1-33 (1993).
4. K. BHATTACHARYA, "*Comparison of the geometrically nonlinear and linear theories of martensitic transformation*", Continuum Mechanics and thermodynamics, Vol.5, pp. 205-242 (1993).
5. J.M. BALL, R.D. JAMES, "*Fine phase mixtures as minimisers of elastic energy*", Archive for Rational Mechanics, Vol. 100, pp. 13-52 (1987).
6. N. ONO, A. SATO, "*Plastic Deformation Governed by the Stress Induced Martensitic Transformation in Polycristals*", Trans. Japan Inst. of Metals, Vol. 29, No. 4, pp. 267-273 (1988).
7. Q.P. SUN, K.C. HWANG, S. W. YU, "*A Micromechanics Constitutive Model of Transformation Plasticity with Shear and Dilatation Effect*", J. Mech. Phys. Solids, Vol. 9, No. 4, pp. 507-524 (1991).
8. E. PATOOR, A. EBERHARDT, M. BERVEILLER, "*Thermomechanical Behaviour of Shape Memory Alloys*", Arch. of Mech., Vol. 40, No. 5-6, pp. 775-794 (1988).
9. Y.H. WEN, S. DENIS, E. GAUTIER, "*Criterion for the progress of martensitic transformation in a finite element simulation*, Journal de Physique IV, Vol. 5 pp. C2-531-536 (1995).
10. K. SIMONSSON, "*Micromechanical finite element simulations of the plastic behavior of steels undergoing martensitic transformation*", Ph D Thesis, Linköping, Studies in Sciences and Technology Dissertations N° 362, Linköping (1994).
11. L. DELAEY, "*Diffusionless Transformations*", chapter 6 in <u>Materials Science and Technologies</u>, Vol. 5 : <u>Phase Transformations in Materials</u>, Ed. R.W. Cahn, P. Haasen, E.J. Kramen, Ed. VCH Publishers, ISBN 3-527-26818-9, pp. 339-404 (1991).
12. M. COHEN, "*Martensitic transformations in material science and engineering*", Transactions of the Japan Institute of Metals, Vol. 39, No. 8, pp. 609-624 (1988).
13. G. GUÉNIN, "*Alliages à mémoire de forme*", in <u>Techniques de l'ingénieur</u>, M 530 (1986).
14. J.D. ESHELBY, "*Elastic Inclusions and Inhomogeneities*", in <u>Progress in Solids Mechanics</u>, (Ed. I.N. Sneddon, R. Hill), Vol. 2, North-Holland, Amsterdam, pp. 87-140 (1961).
15. E. KRÖNER, "*Zur Plastischen Verformung des Vielkristalls*", Acta metall., Vol. 9, pp. 155-161 (1961).
16. M. S. WECHSLER, D. S. LIEBERMAN, T. A. READ," *On the Theory of the Formation of Martensite*", Trans. AIME, Vol. 197 , pp. 1503-1515 (1953).
17. C. M. WAYMAN, <u>Introduction to the crystallography of martensitic transformations</u>, MacMillian Series in Materials Science, New York (1966).
18. E. KRÖNER, "*Modified Green functions in the theory of heterogeneous and/or anisotropic elastic media*", in <u>Micromechanics and inhomogeneity</u>, (Ed. Weng, Taya), Vol. 2, Springer Verlag (1989).
19. M. BERVEILLER, A. ZAOUI "*Modelling of the Plastic Behavior of Inhomogeneous Media*", J. Engng. Mat. and Technology, No. 106, pp. 295-299 (1984).

20. T. MORI, K. TANAKA," *Average stress in matrix and average energy of materials with misfitting inclusions*", Acta metall., Vol. 21, pp. 571-574 (1973).
21. O. FASSI FEHRI, A. HIHI, M. BERVEILLER," *Elastic Interactions Between Variants in Pseudoelastic Single Crystals*", Scripta Met., Vol. 21, pp. 771 (1987).
22. P. H. DEDERICHS, R. ZELLER, "*Elastische konstanten vor vielkristallen*", Report of KFA, 877-FF (1972).
23. J. HADAMARD, Leçons sur la propagation des ondes et les équations de l'hydrodynamique, Cours du Collège de France, Librairie Scientifique A. HERMANN, Paris (1903).
24. J. R. PATEL,M. COHEN, "*Criterion for the action of applied stress in the martensitic. transformation*", Acta metall., Vol.1, pp. 531-538 (1953).
25. E. PATOOR, A. EBERHARDT, M. BERVEILLER, "*Potentiel pseudoélastique et plasticité de transformation martensitique dans les mono et polycristaux métalliques*", Acta metall., Vol. 35, pp. 2779-2789 (1987).
26. G. EDWARDS, J. PERKINS, "*Suggestions for Applying a Phenomenological Approach to Investigations of Mechanical Behavior in SME Alloys*", in Shape Memory Effects in Alloys, (Ed. J. PERKINS) Plenum Press, ISBN 0-306-30891-6, pp. 445-449 (1975).
27. J. BOHONG, T.Y. HSU (XU ZUYAO), "*Influence of Order, Grain Size and Pre-Strain on Shape Memory Effect in Cu-Zn-Al Alloys*", Procs. ICOMAT'89, Materials Science Forum, Vols. 56-58, pp. 457-462 (1990).
28. I. MÜLLER, K. WILMANSKI, "*A Model for Phase Transition in Pseudoelastic Bodies*", Il Nuevo Cimento, Vol. 57B, pp. 283,(1980).
29. F. FALK, "*Model Free Energy, Mechanics and Thermodynamics of Shape Memory Alloys*", Acta metall., Vol. 28, pp. 1773-1780 (1980).
30. R.D. JAMES, "*Displacive Phase Transformations in Solids*", J. Mech. Phys. Solids, Vol. 34, No. 4, pp. 359-394 (1986).
31. L.C. BRINSON, "*One-dimensional Constitutive Behavior of Shape Memory Alloys : Thermomechanical Dérivation with Non-constant Material Functions and Redefined Martensite Internal Variable*", J. of Intell. Mater. Syst. and Struct., Vol. 4, pp. 229-242 (1993).
32. J. G. BOYD, D. C. LAGOUDAS, "*A Constitutive Model for Simultaneous Transformation and Reorientation in Shape Memory Materials*", Procs. ASME, AMD-Vol. 189/PVP-Vol. 292, Mechanics of phase transformations and shape memory alloys, pp. 159-172 (1994).
33. E. PATOOR, M.O. BENSALAH, A. EBERHARDT, M. BERVEILLER, "*Détermination du comportement thermomécanique des alliages à mémoire de forme par minimisation d'un potentiel thermodynamique*", La Revue de Métallurgie, pp. 1587-1592 (1992).
34. Z. MOUMNI, "*Sur le modélisation de changement de phase solide : application aux matériaux à mémoire de forme et à l'endommagement fragile partiel*", Thèse de doctorat, E.N.P.C. (1995).
35. Z. MOUMNI, Q. S. NGUYEN, "*A Model of Materials with Phase Change and Applications*", Journal de Physique IV, Vol. 6 pp. C1-335-345 (1996).
36. J.R. RICE, " *Inelastic Constitutive Relation for Solids : an Internal Variable Theory and its Application to Metal Plasticity*," J. Mech. Phys. Solids, Vol. 19, pp. 433-455 (1971).
37. J.W. CHRISTIAN, "*Deformation by Moving Interfaces*", Metall. Trans. A, Vol. 13A, pp. 509-538 (1982).

38. E. PATOOR, M.O. BENSALAH, A. EBERHARDT, M. BERVEILLER, "*Micromechanical Aspects of the Shape Memory Behaviour*", Procs. ICOMAT'92, Monterey (CA), USA, pp. 401-406 (1993).
39. E. PATOOR, A. EBERHARDT, M. BERVEILLER, "*Micromechanical Modelling of Superelasticity in Shape Memory Alloys*", Pitman Research Notes in Mathematics Series, Vol. 296, pp. 38-54 (1993).
40. M. BERVEILLER, J. MORREALE, E. REUBREZ "*Comportement élastoplastique des aciers lors de la mise en forme : théorie micromécanique, simulations numériques et résultats expérimentaux*", Revue Européennes des éléments finis, Vol. 3, No. 4, pp. 491-514 (1994).
41. E. PATOOR, A. EBERHARDT, M. BERVEILLER, "*Micromechanical Modelling of the Shape Memory Behavior*", Proc. ASME WAM '94, Chicago, IL (USA), AMD- Vol. 189 / PVD- Vol. 292, pp. 23-37 (1994).
42. E. PATOOR, A. EBERHARDT, M. BERVEILLER, "*Micromechanical modelling of superelasticity in shape memory alloys*", Journal de Physique IV, Vol. 6 pp. C1-277-292 (1996).
43. K. ADACHI, J. PERKINS, C.M. WAYMAN, "*The Crystallography and Boundary Structure of Interplate-Group Combinations of 18R Martensite Variant in Cu-Zn-Al Shape Memory Alloy*", Acta metall., Vol. 36, No. 5, pp. 1343-1364 (1988).
44. R. HILL, "*Continuum micromechanics of elastoplastic polycrystals*", J. Mech. Phys. Solids., Vol. 13, pp. 89-101 (1965).
45. A. ZAOUI, "*Macroscopic Plastic Behaviour of Microhomogeneous Materials*", in "Plasticity Today" (Ed. A. SAWCZUK, J. BIANCHI), Ed. Elsevier, pp. 451-469 (1985).
46. E. PATOOR, M.O. BENSALAH, A. EBERHARDT, M. BERVEILLER, "*Détermination du comportement thermomécanique des alliages à mémoire de forme par minimisation d'un potentiel thermodynamique*", La Revue de Métallurgie, pp. 1587-1592 (1992).
47. P. H. DEDERICHS, R. ZELLER, "*Variational Treatment of the Elastic Constants of Desordered Materials*, Z. Phys., Vol. 259, pp. 103 (1973).
48. P. LIPINSKI, J. KRIER, M. BERVEILLER,"*Elastoplasticité des métaux en grandes déformations : comportement global et évolution de la structure interne*", Rev. Phys. Appl., Vol. 25, pp. 361-388 (1990).
49. J. DE VOS, E. AERNOUDT, L. DELAEY," *The Crystallography of the Martensitic Transformation of B.C.C into 9R a Generalized Mathematical Model*", Z. Metallkde, Bd. 69, H7, pp. 438-444 (1978).
50. D. ENTEMEYER, E. PATOOR, A. EBERHARDT, M. BERVEILLER, "*Micromechanical modelling of the superelastic behavior of materials undergoing thermoelastic phase transition*", Journal de Physique IV, Vol. 5 pp. C8-233-238 (1995).
51. K. SHIMIZU, K. OTSUKA, "*Optical and electron microscope observations of transformation and deformation characteristics in Cu-Al-Ni marmem Alloys*", in Shape Memory Effects in Alloys, (Ed. J. PERKINS) Plenum Press, ISBN 0-306-30891-6, pp.59-87 (1975).
52. L. DELAEY, F. VAN DE VOORDE, R.V. KRISHNAN, "*Martensitic formation as a deformation process in polycrystalline Copper-Zinc based Alloys*", in Shape Memory Effects in Alloys, (Ed. J. PERKINS) Plenum Press, ISBN 0-306-30891-6, pp. 351-364 (1975).
53. E. PATOOR, M. EL AMRANI, A. EBERHARDT, M. BERVEILLER, "*Determination of the Origin for the Dissymmetry Observed Between Tensile and Compression Tests on Shape Memory Alloys*", Journal de Physique IV, Vol. 5, pp. C2-495-500 (1995).

54. P. VACHER, C. LEXCELLENT, "*Study of Pseudoelastic Behaviour of Polycrystallin Shape Memory Alloys by Resistivity Measurement and Acoustic Emission*", Procs. ICM 6, Kyoto, Japan, pp. 231-236 (1991).
55. P.Y. MANACH, "*Étude du comportement thermomécanique d'alliages à mémoire de forme NiTi,*", Thèse de doctorat, Institut National Polytechnique de Grenoble, France (1992).
56. Y. GILLET, E. PATOOR, M. BERVEILLER, "*Structure calculation applied to Shape Memory Alloys*", Journal de Physique IV, Vol. 5, pp. C2-343-348 (1995).
57. C. ROGUEDA, "*Modélisation thermodynamique du comportement pseudoélastique des alliages à mémoire de forme*", Thèse de doctorat, Université de Franche-Comté, Besançon (1993).
58. M. EL AMRANI ZIRIFI, "*Contributions à l'étude micromécanique des transformations martensitiques thermoélastiques*", Thèse de doctorat, Université de Metz, France, 1994.
59. G. BOURBON,C. LEXCELLENT, S. LECLERCQ, "*Modelling of the non isothermal cyclic behaviour of a polycrystalline Cu-Zn-Al shape memory alloy*", Journal de Physique IV, Vol. 5, pp. C8-221-226 (1995).

MODELLING AND SIMULATION OF TRANSFORMATION INDUCED PLASTICITY IN ELASTO-PLASTIC MATERIALS

F.D. Fischer
University of Mining and Metallurgy, Leoben, Austria

ABSTRACT

First a phase change process in a certain microregion of a material is described in phenomenological terms by the appearance of a transformation tensor which can be interpreted as an eigenstrain tensor or a strain incompatibility. By applying the irreversible thermodynamics of solids the rates of state functions for a material specimen are derived. The integration of these rates allows to establish a condition for the transformation of a microregion by an interface movement. Further a thermodynamic condition for the sudden transformation of a certain microregion is derived. Both considerations lead to an equivalent transformation condition relating a chemical and a mechanical driving force to a transformation and mechanical barrier. This transformation condition is applied to the selection of variants in the case of a displacive transformation demonstrating the orientation effect on a global deformation. The accommodation effect resulting from the transformation volume (and shape) change is investigated for a specimen under a constant external stress state. Extended relations compared to the "classical" solution by Greenwood and Johnson are presented based on a semianalytical concept. Then a more sophisticated incremental procedure is introduced allowing to predict both the orientation effect and the accommodation effect of an ongoing transformation on the global deformation behavior. Proposals for a modified constitutive law for an elasto-plastic material considering a solid phase transformation are neglected.

Finally the concept of both a chemical and mechanical driving force is applied to derive a transformation kinetics relation for a displacive transformation. Here the "classical" phenomenological relations (e.g. by Koistinen and Marburger) are extended by a stress term based on a thermodynamical and micromechanical consideration.

INTRODUCTION

This contribution is mainly devoted to the continuum mechanics understanding of a solid phase transformation in an elasto-plastic material. Both the "restricting" effects on the transformation process due to a generation of an additional strain energy as well as the plastic dissipation process and the "favouring" effect on the global deformation behavior of loaded specimens are studied. Especially the "weakening" of the material during a displacive or diffusive transformation process (denominated as TRansformation Induced Plasticity) can then be derived. Of course, due to the nonlinear character of the material behavior such nice concepts based on Eshelby's elementary solutions for an inclusion cannot be applied directly. Therefore, semianalytical and often only numerical concepts like the finite element method must be followed to make predictions about the deformation behavior of elasto-plastic phase changing materials. It should be mentioned here that only plasticity in the sense of von Mises flow plasticity is considered. One may argue that at the level of microregions we have to deal with crystal plasticity. This may be principally correct; however, the computational efforts would be increased by such an enormous amount that even using modern high-speed computers solutions could not be achieved within a reasonable turn-around time.

This contribution needs no introduction into the physical problem since the reader is referred to the extensive presentation of both the phenomenon of solid state phase transformations itself as well as its continuum mechanics and thermo-dynamics formulation in the preceding chapters. With respect to an under-standing of the plasticity effect, the contribution is written in a self-consistent way anyway.

1. CONTINUUM MECHANICS FORMULATION OF A SOLID STATE PHASE TRANSFORMATION PROCESS

1.1 Kinematics

As mentioned in the above contributions by Gautier, Patoor and Berveiller to this book, the kinematic representation of a transformation process is given by the development of an eigenstrain tensor $\tilde{\varepsilon}^T_{ij}$. In the case of a displacive transformation it is referred to the Wechsler-Lieberman-Read tensor, see [1],

$$\tilde{\varepsilon}^T_{ij} = \begin{bmatrix} 0 & 0 & \gamma/2 \\ 0 & 0 & 0 \\ \gamma/2 & 0 & \delta \end{bmatrix} \tilde{f}(\xi) = \frac{\delta}{3}\,\delta_{ij}\,\tilde{f}(\xi) + \frac{\delta}{3} \begin{bmatrix} -1 & 0 & \frac{3\gamma}{2\delta} \\ 0 & -1 & 0 \\ \frac{3\gamma}{2\delta} & 0 & 2 \end{bmatrix} \tilde{f}(\xi)\,, \quad (1.1)$$

described with respect to a local co-ordinate system $\tilde{1}$, $\tilde{2}$, $\tilde{3}$. The $\tilde{1}$-$\tilde{2}$-base plane corresponds to the habit plane of a martensite variant with its normal $\tilde{3}$. The transformation volume change "grows" in the $\tilde{3}$-direction, too. A specific variant is related to a global co-ordinate system 1, 2, 3 by a set of Eulerian angles ϑ, φ, ψ with $0 \le \vartheta \le \pi$, $0 \le \varphi$, $\psi \le 2\pi$. ϑ is the angle between the global axis 3 and the local normal $\tilde{3}$. $\tilde{f}(\xi)$ may be interpreted as a process function $0 \le \tilde{f}(\xi) \le 1$, $\tilde{f}(0)=0$, $\tilde{f}(1)=1$, of a dimensionless process parameter $0 \le \xi \le 1$. γ is the transformation shear. δ represents the transformation volume change. Some details with respect to the derivation of $\tilde{\varepsilon}_{ij}^T$ can be found in the following chpt. 1.2.1. The transformation tensor $\tilde{\varepsilon}_{ij}^T$

$$\tilde{\varepsilon}^T \Rightarrow \tilde{\varepsilon}^T_{ij} = \frac{\delta}{3}\,\tilde{f}(\xi)\,\delta_{ij} + \frac{\delta}{3}\,\tilde{f}(\xi)\tilde{e}^T_{ij} \quad (1.2.1)$$

can now be written with respect to the global co-ordinate system via the transformation matrix Q as

$$\varepsilon^T = Q\,.\,\tilde{\varepsilon}^T\,.\,Q^t\,. \quad (1.3)$$

Q is an orthogonal matrix with its elements being functions $Q_{ij}(\vartheta, \varphi, \psi)$, for details see [2]. Thus, the component ε^T_{33} reads

$$\varepsilon^T_{33} = \frac{1}{2}\,\gamma(\sin 2\vartheta\,\sin\varphi) + \frac{1}{2}\,\delta(1 + \cos 2\vartheta)\,. \quad (1.4)$$

Relation (1.2.1) is written with respect to the global co-ordinate system as

$$\varepsilon_{ij}^{T} = \frac{\delta}{3}\, \tilde{f}(\xi)\, \delta_{ij} + \frac{\delta}{3}\, \tilde{f}(\xi)\, e_{ij}^{T} \,. \tag{1.2.2}$$

For the case of a diffusive transformation Fischer [3] presented a concept based on the assumption that the product phase grows as a sphere with radius R similar to a carbide in spheroidal graphite iron. The transformation is accompanied by a certain volume change δ. Since a spherically symmetric configuration is assumed, the transformed sphere experiences a hydrostatic stress state only and, therefore, does not plastify. The only "response" of the material to the transformation volume change δ is a plastification of the shell surrounding the sphere with radius R. Here the elastic strain components are ignored, which are usually much smaller than the plastic strain components. Therefore, the transformation process produces an eigenstrain tensor, namely the plastic strain tensor, in any material element of the surrounding shell. As a consequence of representing the transformation process by an eigenstrain tensor, this plastic strain tensor takes over the role of a transformation tensor in a microregion in the surrounding shell. The plane $\tilde{1}$, $\tilde{2}$ of the local co-ordinate system is now tangential to the sphere at a point r with $\tilde{3}$ in the direction of r. $2\tilde{R}$ is the average distance between the growing product spheres. A factor α_P, $0.524 \le \alpha_P \le 0.740$, describes the arrangement of the growing spheres in an average sense. The volume fraction f, $\dot{f} = \frac{df}{dt}$, of the transformed material is now $f = \alpha_P (R/\tilde{R})^3$.

By neglecting the elastic strains Fischer [3] has shown that $\dot{\tilde{\varepsilon}}_{ij}^{T} = \frac{d\tilde{\varepsilon}_{ij}^{T}}{dt}$ can be written as

$$\dot{\tilde{\varepsilon}}_{ij}^{T} = \frac{\delta}{3\alpha_P}\, \dot{f}\, \frac{\tilde{R}^3}{r^3} \begin{bmatrix} 1 & 0 & 0 \\ 0 & 1 & 0 \\ 0 & 0 & -2 \end{bmatrix}, \; R \le r \le \tilde{R} \,,\; f \le \alpha_P \,. \tag{1.5}$$

The model now considers that the growing spheres can impinge if $f = \alpha_P$. From that moment on, the parent phase becomes more and more an inclusion. The model meets this condition by representing the product phase, having transformed during $0 \le f \le \alpha_P$, by a hollow sphere with the inner radius R_g and the outer radius R_e. The remaining parent phase is now enclosed by this product phase hollow sphere. The transformation now progresses inside the parent phase as a sphere with the radius R,

$$R_e = R_g(1 - \alpha_P)^{-1/3}; \quad \alpha_P \le f = \alpha_P + (\frac{R}{R_e})^3 \le 1, \quad 0 \le R \le R_g \,. \quad (1.6)$$

According to this assumption the ongoing transformation process produces only a hydrostatic stress state in the enclosed sphere, $0 \le r \le R_g$, plastification happens only in the product phase shell, $R_g \le r \le R_e$. Therefore, a modified plastic strain tensor is formulated for $\alpha_P \le f \le 1$ as

$$\dot{\varepsilon}^T_{ij} = \frac{\delta}{3}\dot{f}\frac{R_e^{\,3}}{r^3}\begin{bmatrix}1 & 0 & 0\\ 0 & 1 & 0\\ 0 & 0 & -2\end{bmatrix}, \; R_g \le r \le R_e, \; \alpha_P \le f \le 1. \qquad (1.7)$$

A unique formulation after co-ordinate transformation into the global co-ordinate system delivers

$$\dot{\varepsilon}^T_{ij} = \frac{\delta}{3}\dot{f}\left(\frac{R_T}{r}\right)^3(\delta_{ij} - 3a_{ij}), \quad a_{ij} = Q_{i3}\cdot Q_{j3}\,, \qquad (1.8)$$

$$0 \le f \le \alpha_P : R_T = \tilde{R}/\sqrt[3]{\alpha_P}, \qquad \alpha_P \le f \le 1 : R_T = R_e\,.$$

Finally, it is emphasized that in the case of a diffusive transformation, the transformation tensor ε_{ij}^T is not placed in the transforming phase but in the adjacent phase as a plastic incompatibility strain tensor.

1.2 Thermomechanics

1.2.1 Introduction

Let us consider a body of elasto-plastic material whose total volume V consists of a transformed new (product) phase (label "n") with the volume V_n, external surface ∂V_n, and the old (parent) phase (label "o") with the volume V_o, external surface ∂V_0, $V_n + V_o = V$, $\partial V_n + \partial V_o = \partial V$, see Fig.1. The transformation front s_{on} (later s_{am}), r^S representing a point on s_{on}, moves with a velocity ω along the material interface normal n_i^S pointing from the product phase to the parent phase. Moreover, we define the jump of an entity h along s_{on} as $[h] = h_o - h_n$, and the average as $\langle h \rangle = 1/2\,(h_o + h_n)$.

This chapter is now mainly devoted to combine continuum mechanics and elementary thermodynamics of phase transformations to find a transformation condition.

1.2.2 Jump condition

In the following context we refer to a reference configuration with the spatial coordinates x_j and the time t. We consider a coherent phase transformation so that in an interface point r_j^S at any time t

$$[u_i\,(r_j^S, t)] = 0\,. \tag{1.9}$$

u_i is the material displacement vector adjacent to the interface point r_j^S either in the product or the parent phase. If we follow the interface point it will be located at time t +dt at the spatial position $r_j^S + \omega n_j^S\,dt$. The displacement vector there will be $u_i(r_j^S + \omega n_j^S\,dt,\ t + dt)$. Applying a Taylor series the coherency condition now reads

$$[u_i(r_j^S+\omega n_j^S\,dt,\ t+dt)] = [u_i(r_j^S, t)] + \left[\frac{\partial u_i}{\partial t}\Big|_t + \frac{\partial u_i}{\partial x_j}\Big|_{r^S}\,\omega\, n_j^S\right] dt = 0\,.$$

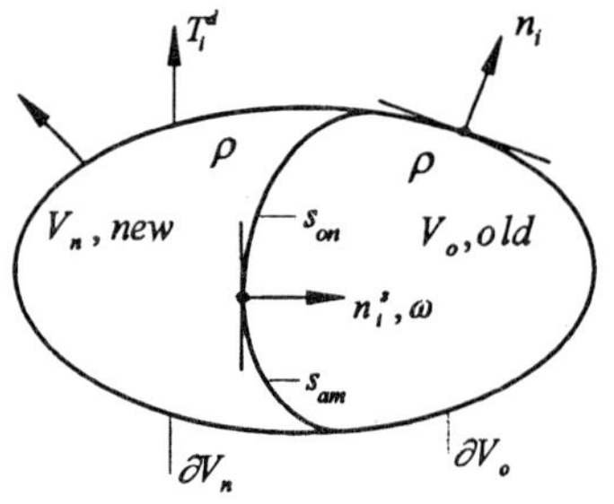

$$\partial V_n + \partial V_o = \partial V$$
$$V_n + V_o = V$$

Fig. 1: Reference configuration of a two phase material consisting of the parent phase (old) and the product phase (new).

With (1.9) and $v_i = \dfrac{\partial u_i}{\partial t}$ being the material velocity it follows that

$$[v_i]|_{r^S} = -\,\omega \left[\frac{\partial u_i}{\partial x_j}\right]\Big|_{r^S} n_j^S\,. \tag{1.10.1}$$

This is the well-known Hadamard condition which indicates a jump in the material velocity on both sides of the interface point r_j^S.

A second condition exists as

$$\left[\frac{\partial u_i}{\partial x_j}\right]\Big|_{r^S} = \left[\frac{\partial u_i}{\partial x_k}\right]\Big|_{r^S} \cdot n_k^S\, n_j^S , \tag{1.10.2}$$

which can be proved by multiplying (1.10.2) with n_j^S and considering $n_j^S \cdot n_j^S = 1$. As an immediate consequence of (1.10.2) the jump in the deformation gradient $[F_{ij}]|_{r^S}$ which is equal to $\left[\frac{\partial u_i}{\partial x_j}\right]\Big|_{r^S}$ can be written as

$$[F_{ij}]|_{r^S} = \left[\frac{\partial u_i}{\partial x_j}\right]\Big|_{r^S} = m_i^S n_j^S , \qquad m_i^S = \left[\frac{\partial u_i}{\partial x_k}\right]\Big|_{r^S} n_k^S . \tag{1.10.3}$$

Applying (1.10.3) furnishes

$$m_i^S = -\frac{1}{\omega}[v_i]|_{r^S} . \tag{1.10.4}$$

The mass flux across the interface can easily be understood considering an observer sitting on the interface, which moves with the velocity $\omega\, n_j^S$, leading to

$$\left[\rho(\omega - v_i\, n^S_i)\right]\Big|_{r^S} = 0 . \tag{1.10.5}$$

Equation (1.10.5) allows to formulate a further condition for m_i^S, namely

$$[\rho]|_{r^S} \left(\omega - \langle v_i \rangle n_i^S\right) = -\langle \rho \rangle \omega\, m_i^S\, n_i^S . \tag{1.10.6}$$

In the case of no jump in density, m_i^S is a vector in the tangent plane to the interface and, therefore, the vector v_i differs in r^S only in its tangential component. The goal of the crystallographic theory of martensite (CTM) lies now in the evaluation of the vectors m_i^S and n_i^S, for details see the pioneering papers by Wechsler, Lieberman and Read [1] and a more general and mathematically rigorous treatment by Ball and James [4]. It is important to note that the CTM ultimately follows a geometrically linear concept since only this allows to calculate from m_i^S, n_i^S the final transformation tensor

$$\tilde{\varepsilon}_{ij}^{\,T} = \frac{1}{2}\left(\left[\frac{\partial u_i}{\partial x_j}\right]\Big|_{r^S} + \left[\frac{\partial u_j}{\partial x_i}\right]\Big|_{r^S}\right) = \frac{1}{2}\left(m_i^S\, n_j^S + m_j^S n_i^S\right) . \tag{1.10.7}$$

In the case of a geometrically nonlinear concept one would need also the tensor $\left\langle\frac{\partial u_i}{\partial x_j}\right\rangle\Big|_{r^S}$ which is not available.

The geometrically linear concept further allows to write $\left[\frac{\partial u_i}{\partial x_j} a_{ij}\right] = [\varepsilon_{ij}\, a_{ij}]$ for any symmetric tensor a_{ij} and the strain tensor ε_{ij}.

From this point, we will go on only by studying the static case with the consequence that the traction vector $t_i|_{r^S} = \sigma_{ij}|_{r^S} n_j$ shows no jump,

$$[t_i]|_{r^S} = 0 \,. \tag{1.11}$$

Finally we would like to mention that due to the heat conduction the temperature T has no jump,

$$[T]|_{r^S} = 0 \,, \tag{1.12}$$

for details see eg. Nishiyama, chpt.4.6.2, [5].
As thermodynamic potentials we introduce the specific Helmholtz free energy φ and the specific Gibbs free energy ψ,

$$\varphi = (\varepsilon_{ij}, T; \varepsilon^P_{ij}, \varepsilon^T_{ij}, \alpha_i), \qquad \psi = \psi(\sigma_{ij}, T; \varepsilon^P_{ij}, \varepsilon^T_{ij}, \alpha_i) \,,$$

$$\psi = \varphi - \frac{1}{\rho} \sigma_{ij} \varepsilon_{ij} \,, \tag{1.13}$$

ε^P_{ij} is the plastic strain tensor, α_i is a set of internal variables (hardening etc.).
φ separates into a mechanical part φ_σ and a chemical part φ_{ch} which is understood as the Helmholtz free energy of the stress free body (φ_{ch} is either $\varphi_{ch,o}$ or $\varphi_{ch,n}$) and depends mainly on temperature T. φ_σ consists of the elastic energy $1/\rho\, w^e$ and some further $\varphi_D(\alpha_i)$ stored in the dislocations (hardening),

$$\varphi = \varphi_\sigma + \varphi_{ch}(T), \quad \varphi_\sigma = \frac{1}{\rho} w^e(\varepsilon_{ij} - \varepsilon^P_{ij} - \varepsilon^T_{ij}, T) + \varphi_D(\alpha_i, T) \,. \tag{1.14.1}$$

In a similar way, ψ can be written also as

$$\psi = \psi_\sigma + \varphi_{ch}(T), \quad \psi_\sigma = -\frac{1}{\rho} w^{e*}(\sigma_{ij}, T; \varepsilon^P_{ij}, \varepsilon^T_{ij}) + \varphi_D(\alpha_i) \,, \tag{1.14.2}$$

w^{e*} being the complementary elastic energy.
The surface energy along s_{on} can be neglected in the cases considered here.

The thermodynamic force F^S acting on the interface can be given as

$$F^S = \left(\rho[\varphi] - \langle\sigma_{ij}\rangle [\varepsilon_{ij}]\right) = \left(\rho[\varphi] - \langle\sigma_{ij}\rangle \left[\frac{\partial u_i}{\partial x_j}\right]\right) \,. \tag{1.15}$$

The derivation of F_s is not explained here. The reader is referred to the pioneering paper by Eshelby [6]. F^S can be considered as the normal projection of the (modified) energy momentum (or chemical) tensor. The transformation will propagate if the force F^S is at least equal to a critical value F_c^T,

$$F^S \geq F_c^T, \qquad F_c^T > 0 . \tag{1.16}$$

This critical value F_c^T has its origin both in an energy well (which must be overcome in any case) and in a dissipative process. Of course, in a material point the dissipative process depends both on the history and the actual configuration. In addition to the dissipative process a further contribution to F_c^T may occur in the case of a loss in energy due to acoustic emission.

We now suggest to estimate F_c^T from the amount of energy due to undercooling which is necessary to start the transformation at a temperature M_s in a stress-free body. M_s may be specifically lower than T_o, $\varphi_{ch,o}(T_o) = \varphi_{ch,n}(T_o)$. If this amount of energy is $\rho[\varphi_{ch}]|_{M_s}$ we fix F_c^T as

$$F_c^T = \rho \frac{T}{M_s} [\varphi_{ch}]|_{M_s} . \tag{1.17}$$

The ratio $\frac{T}{M_s}$ is employed to increase F_c^T for a temperature $T > M_s$.

1.2.3 The global transformation condition - Growth Condition

We now look at the time derivative of the global thermodynamic potentials. For this reason we apply the principle of virtual work in the form of the principle of virtual velocity for both parts of the body. With no body forces and the traction vector T_i^d on the external surface $\partial V_o + \partial V_n$ it follows that

$$V_n: \int_{\partial V_n} T_i^d \, v_i \, dA + \int_{S_{am}} t_i v_i|_n \, dA = \int_{V_n} \sigma_{ij} \, \dot{\varepsilon}_{ij} \, dV,$$

$$V_o: \int_{\partial V_o} T_i^d \, v_i \, dA + \int_{S_{am}} t_i v_i|_o \, dA = \int_{V_o} \sigma_{ij} \, \dot{\varepsilon}_{ij} \, dV .$$

The labels "n" and "o" at the integrals over the interface refer to the fact that t_i, v_i must be taken from the new and the old material side, respectively. Observing that $t_i|_o = - \sigma_{ij}|_o \, n_j^S$, and assuming $\sigma_{ij} = \sigma_{ji}$, we have

$$\int_{V_o + V_n} \sigma_{ij} \, \dot{\varepsilon}_{ij} \, dV = \int_{\partial V} T_i^d \, v_i \, dA - \int_{S_{am}} [\sigma_{ij} v_j] n_i^S \, dA . \tag{1.18.1}$$

$[\sigma_{ij} v_j] n_i^S$ can be reformulated with the Hadamard condition (1.10.1,2) and $\sigma_{ij} = \sigma_{ji}$ as

$$[\sigma_{ij} v_j n_i^S] = [\sigma_{ji} n_i^S v_j] = [t_j v_j] = \langle t_j \rangle [v_j] =$$
$$= -\omega \langle \sigma_{ji} n_i^S \rangle \left[\frac{\partial u_j}{\partial x_l} \right] n_l^S = -\omega \langle \sigma_{ij} \rangle \left[\frac{\partial u_j}{\partial x_i} \right].$$

Remark: $[ab] = [a]\langle b \rangle + \langle a \rangle [b]$, $n_i^S n_i^S = 1$.

Implementing this relation into (1.18.1) leads to

$$\int_{V_o+V_n} \sigma_{ij} \dot{\varepsilon}_{ij} dV = \int_{\partial V} T_i^d v_i \, dA - \omega \int_{s_{am}} \langle \sigma_{ij} \rangle \left[\frac{\partial u_i}{\partial x_j} \right] dA . \tag{1.18.2}$$

Next, we calculate the rate of the total Helmholtz free energy ϕ of the body as

$$\phi = \int_V \rho \varphi \, dV, \quad \dot{\phi} = \int_{V_o+V_n} \rho \overset{\circ}{\varphi} \, dV - \omega \int_{s_{am}} \rho [\varphi] \, dA . \tag{1.19}$$

For the derivation of the time derivative of an integral with a jump in the integrand along an interface s_{am}, see e.g. Chadwick, chpt.6, [7].
Irreversible thermodynamics allows us to express the local dissipation D as

$$D = \sigma_{ij} \overset{\circ}{\varepsilon}_{ij} - \rho \eta \dot{T} - \rho \overset{\circ}{\varphi} \geq 0 . \tag{1.20.1}$$

Here we assume (at least locally) a spatially constant temperature so that any contribution to the entropy rate by heat flow can be avoided.
Now $\rho \overset{\circ}{\varphi} = \sigma_{ij} \overset{\circ}{\varepsilon}_{ij} - \rho \eta \dot{T} - D$ is introduced in (1.19) together with the jump condition (1.15) and (1.16) in the form of $\rho[\varphi] = F_c^T + \langle \sigma_{ij} \rangle \left[\frac{\partial u_i}{\partial x_j} \right]$. This results in a final relation for $\dot{\phi}$ as

$$\dot{\phi} = \int_{\partial V} T_i^d v_i \, dA - \int_{V_o+V_n} \rho \eta \dot{T} \, dV - \int_{V_o+V_n} D \, dV - \omega \int_{s_{am}} F_c^T \, dA . \tag{1.21.1}$$

It can be shown that an equivalent relation can be formulated for the rate of the total Gibbs free energy as

$$\dot{\psi} = -\int_{\partial V} u_i \dot{T}_i^d \, dA - \int_{V_o+V_n} \rho \eta \dot{T} \, dV - \int_{V_o+V_n} D \, dV - \omega \int_{s_{am}} F_c^T \, dA . \tag{1.22.1}$$

It may be useful to eliminate the entropy η in the second integral on the left of (1.21.1), (1.22.1). As very often a weak thermomechanical coupling can be assumed, $\frac{\partial \varphi_\sigma}{\partial T} \sim 0$, $\frac{\partial \psi_\sigma}{\partial T} \sim 0$, thermodynamics allows to calculate η as

$$\eta = -\frac{\partial \varphi}{\partial T} = -\frac{\partial \psi}{\partial T} = -\frac{\partial \varphi_{ch}}{\partial T} .$$

Now $-\int\limits_{V_o+V_n} \rho \dot{T}\eta \, dV$ can be written as $\int\limits_{V_o+V_n} \rho \frac{\partial \varphi_{ch}}{\partial T} \dot{T} dV = \int\limits_{V_o+V_n} \rho \dot{\varphi}_{ch} \, dV$.

Using again the time derivative of an integral, see (1.19), results in

$$\int\limits_{V_o+V_n} \rho \dot{\varphi}_{ch} \, dV = \dot{\phi}_{ch} + \omega \int\limits_{S_{am}} \rho[\varphi_{ch}] \, dA .$$

Substituting this relation in (1.21.1), (1.22.1) and observing that ϕ and ψ, resp., are split into $\phi_\sigma + \phi_{ch}$ and $\psi_\sigma + \psi_{ch}$, resp., yields the modified global relations

$$\dot{\phi}_\sigma = \int\limits_{\partial V} v_i T_i^d \, dA + \omega \int\limits_{S_{am}} \rho[\varphi_{ch}] \, dA - \int\limits_{V_o+V_n} D \, dV - \omega \int\limits_{S_{am}} F_c^T \, dA , \tag{1.21.2}$$

$$\dot{\psi}_\sigma = -\int\limits_{\partial V} u_i \dot{T}_i^d \, dA + \omega \int\limits_{S_{am}} \rho[\varphi_{ch}] \, dA - \int\limits_{V_o+V_n} D \, dV - \omega \int\limits_{S_{am}} F_c^T \, dA . \tag{1.22.2}$$

A lot of conclusions can be drawn from (1.21) and (1.22): First of all, it cannot be said a priori that either ϕ or ψ must decrease during a transformation process. Due to $D > 0$, $F_c^T > 0$, only the last two integrals give rise to a decrease of $\dot{\phi}$, $\dot{\psi}$. However, if we consider an isothermal process, $\dot{T} = 0$, ϕ decreases if we fix the displacements ($v_i = 0$), and ψ decreases if we keep the tractions T_i^d constant $\left(\dot{T}_i^d = 0\right)$.

We use the relation (1.22.1-2) to explore if at a certain load level (that means with T_i^d fixed ($\dot{T}_i^d = 0$)), a microregion may transform under a varying temperature. Such a transformation process has been named "creep test" by Gautier et al [8]. We imagine that the transformation happens by interface propagation. A dimensionless process parameter ξ, $0 \le \xi \le 1$, indicates the actual state, $\xi = 0$ marks the birth and $\xi = 1$ the end of the growth of a microregion. The applied loading is assumed to be a homogeneous stress state Σ_{ij} with a local stress state $\sigma_{ij}(r,\xi) = \Sigma_{ij} + \tau_{ij}(r,\xi)$, $\tau_{ij}(r,0) = 0$, $T_i^d = \Sigma_{ij}\, n_j$. τ_{ij} is the stress fluctuation in a point r due to the growth of the microregion under consideration.

After integration with respect to the process parameter ξ, relation (1.22.2) delivers with $\int_{\partial V} \dot{T}_i^d u_i \, dA = 0$

$$\psi_\sigma(1) - \psi_\sigma(0) = \int_0^1 \omega\left(\frac{dt}{d\xi}\right) \int_{S_{am}} \rho[\varphi_{ch}] dA\, d\xi - \int_0^1 \left(\int_{V_o+V_n} D dV + \omega \int_{S_{am}} F_c^T dA \right) \left(\frac{dt}{d\xi}\right) d\xi. \quad (1.23.1)$$

If the same elastic properties exist in both phases, an exact and explicit relation for $\psi_\sigma(1) - \psi_\sigma(0)$ can be given, see e.g. [9],

$$\psi_\sigma(1) - \psi_\sigma(0) = - \int_{V_o+V_n} \left(\Sigma_{ij} + 1/2\, \tau_{ij}(r,1)\right) (\varepsilon_{ij}^P + \varepsilon_{ij}^T)\, dV\, . \quad (1.23.2)$$

Expression (1.23.2) can be reformulated by adding and subtracting the elastic strain ε_{ij}^e to the last term, leading to

$$\frac{1}{2} \int_{V_o+V_n} \tau_{ij}\, (r,1)(\varepsilon_{ij}^P + \varepsilon_{ij}^T + \varepsilon_{ij}^e - \varepsilon_{ij}^e)\, dV = \frac{1}{2} \int_{V_o+V_n} \tau_{ij}\, \varepsilon_{ij}\, dV - \frac{1}{2} \int_{V_o+V_n} \tau_{ij}\, \varepsilon_{ij}^e\, dV\, .$$

Due to the principle of virtual work the first integral on the right side is 0. The elastic strain ε_{ij}^e is now decomposed into $\varepsilon_{ij}^{e,\Sigma} + \varepsilon_{ij}^{e,\tau}$ with $\varepsilon_{ij}^{e,\Sigma}$ being the average elastic strain corresponding to the overall stress state Σ_{ij}. Since the integral $\frac{1}{2}\int_{V_o+V_n} \tau_{ij}\, \varepsilon_{ij}^{e,\Sigma}\, dV = \frac{1}{2}\, \varepsilon_{ij}^{e,\Sigma} \int_{V_o+V_n} \tau_{ij}\, dV = 0$, the final outcome with $\bar{w}_\tau^e = \int_{V_o+V_n} \tau_{ij}\, \varepsilon_{ij}^{e,\tau} dV$ is

$$\frac{1}{V} \int_{V_o+V_n} (\Sigma_{ij} + \frac{1}{2}\, \tau_{ij})(\varepsilon_{ij}^P + \varepsilon_{ij}^T)\, dV = \Sigma_{ij}\, (\varepsilon_{ij}^P + \varepsilon_{ij}^T) - \bar{w}_\tau^e\, . \quad (1.23.3)$$

$\bar{w}^e_\tau$ is the specific strain energy produced by the stress fluctuation.

For further simplification we assume that $[\varphi_{ch}(\xi)]$ is constant along the interface s_{am} with the current total area A_{am}. Since a shell with the thickness ωdt and the area A_{am} can be considered as the volume element dV of the transforming microregion, the first integral in (1.23.1) follows as

$$\int_0^1 \rho[\varphi_{ch}(\xi)]\ \omega\ A_{am}(\xi)\left(\frac{dt}{d\xi}\right)d\xi = fV\int_0^1 \rho[\varphi_{ch}(\xi)]\ d\xi\ ; \tag{1.23.4}$$

fV is the actual volume of the growing microregion.

Let us now consider the dissipative terms. Irreversible thermodynamics gives us

$$D = -\rho\frac{\partial\varphi}{\partial\varepsilon^P_{ij}}\dot{\varepsilon}^P_{ij} + \rho\frac{\partial\varphi}{\partial\varepsilon^T_{ij}}\dot{\varepsilon}^T_{ij} - \rho\frac{\partial\varphi}{\partial\alpha_i}\dot{\alpha}_i\ . \tag{1.20.2}$$

Due to the structure of φ, see (1.14), it follows that $\rho\frac{\partial\varphi}{\partial\varepsilon^P_{ij}} = \rho\frac{\partial\varphi}{\partial\varepsilon^T_{ij}} = -\sigma_{ij}$, and,therefore,

$$D = \sigma_{ij}(\dot{\varepsilon}^P_{ij} + \dot{\varepsilon}^T_{ij}) - \rho\frac{\partial\varphi_D}{\partial\alpha_i}\dot{\alpha}_i\ . \tag{1.20.3}$$

$\dot{\varepsilon}^T_{ij}$ can only exist in the product phase. We assume, however, that ε^T_{ij}, once developed, does not change (e.g. in the case of steels). Therefore, the corresponding dissipation term $\sigma_{ij}\ \dot{\varepsilon}^T_{ij}$ becomes 0!

For many materials the last term $-\rho\frac{\partial\varphi_D}{\partial\alpha_i}\dot{\alpha}_i$ can be neglected in relation to $\sigma_{ij}\dot{\varepsilon}^P_{ij}$. Then $\sigma_{ij}\dot{\varepsilon}^P_{ij}$ remains the dominant dissipation mechanism outside the interface,

$$D \rightarrow D^P = \sigma_{ij}\ \dot{\varepsilon}^P_{ij} = \sigma_{ij}\frac{d\varepsilon^P_{ij}}{d\xi}\ \dot{\xi}\ .$$

The dissipation D^P is now split into two terms and integrated over ξ, leading to

$$\bar{D}^P = \Sigma_{ij}\ \varepsilon^P_{ij}\ |_{\xi=1} + \int_0^1 \tau_{ij}\frac{d\varepsilon^P_{ij}}{dt}\frac{dt}{d\xi}d\xi = \Sigma_{ij}\ \varepsilon^P_{ij} + w^P_\tau\ . \tag{1.23.5}$$

The total specific plastic work $\bar{w}^P$ follows as

$$\bar{w}^P = \frac{1}{V}\int_{V_o+V_n}^{1} \bar{D}^P dV = \Sigma_{ij}\ \bar{\varepsilon}^P_{ij} + \bar{w}^P_\tau\ . \tag{1.23.6}$$

The second term in (1.23.5) and (1.23.6) can be considered as the specific plastic work $\bar{w}^P_\tau$ generated by the internal stress state τ_{ij} only.
The last term in (1.23.1) can be treated with (1.17.2) along the same concept as (1.23.4) yielding

$$fV\,\rho[\varphi_{ch}]|_{M_s}\int_0^1 \frac{T}{M_s}(\xi)d\xi\,. \tag{1.23.7}$$

Since we assume a constant transformation tensor ε^T_{ij} with respect to space and time in the microregion considered, insertion of (1.23.4), (1.23.6), (1.23.7) into (1.23.1) delivers with the following process averages

$$\int_0^1 \rho[\varphi_{ch}(\xi)]d\xi = \rho[\varphi ch]|_{\bar{\xi}}\,,\quad \int_0^1 T(\xi)d\xi = \bar{T}\,, \tag{1.24}$$

a relation

$$\rho\,[\varphi_{ch}]|_{\bar{\xi}} + \Sigma_{ij}\,\varepsilon^T_{ij} = \rho\frac{\bar{T}}{M_s}[\varphi_{ch}]|_{M_s} + \frac{1}{f}(\bar{w}^e_\tau + \bar{w}^P_\tau)\,. \tag{1.25}$$

Equation (1.25) can be considered the global transformation condition of a microregion V_n which has been generated after its nucleation by a growth process. On the right side, the strain energy $\bar{w}^e_\tau$ due to the stress fluctuation as well as the plastic work $\bar{w}^P_\tau$ due to the stress fluctuation appear as additional restricting entities (not forces !), since they are positive. It is interesting to note that Levitas showed in his papers, e.g. [10], that starting from the dissipation D (1.20.1), an interface propagation criterion can be derived which is exactly equivalent to the above transformation condition.

Remark 1:

The temperature $T(\xi)$ must be higher than M_s, $T(\xi) > M_s$, $0 \leq \xi \leq 1$, otherwise a sudden transformation could happen. For $T(\xi) > M_s$ it can immediately be seen, e.g. from the contribution by Ortin in this book, that the difference of the transformation barrier and the chemical driving force is always greater than 0. Therefore, one needs a (positive) mechanical driving force to promote the transformation.

Remark 2:

As explained in detail e.g. by Patoor and Berveiller in this book, a distinct number of martensitic variants (≤ 24!) may develop. It seems natural that the variant with the largest mechanical driving force will be the most probable one. To check this statement we insert (1.23.3) into (1.23.2) and use (1.23.6) yielding

$$\frac{1}{V}\left(\psi_\sigma(1) - \psi_\sigma(0)\right) = - \Sigma_{ij}\, \varepsilon^T_{ij} - \frac{1}{V}\int_{V_o+V_n} \bar{D}^P \, dV + \bar{w}^e_T + \bar{w}^P_T \, .$$

The difference of the mechanical part of the specific Gibbs free energy is $-\Sigma_{ij}\, \varepsilon^T_{ij} + \bar{w}^e_T$ in a nondissipative process. Since $\bar{w}^e_T$ is independent of both Σ_{ij} and the number of variants, it is obvious that a maximum mechanical driving force $\Sigma_{ij}\, \varepsilon^T_{ij}$ leads to a minimum possible Gibbs free energy.
The same conclusion can be drawn also in the case of a dissipative process, at least qualitatively, taking into account that $\bar{D}^P$ has reached a maximum value due to the principle of maximum dissipation (rate), and that $\bar{w}^e_T$, $\bar{w}^P_T$ are only weakly dependent on both Σ_{ij} and the number of the variants.

Remark 3:
We now derive the transformation condition for a microregion V_μ which grows in a load stress field Σ_{ij} as well as in an internal stress field T_{ij} which may have developed by the stress fluctuation due to some earlier transformed microregions. The local stress state is, therefore, described by $\sigma_{ij} = \Sigma_{ij} + T_{ij}$. If T^d_i is kept constant now and if we look for the formation of a further transformed microregion V_μ with a transformation tensor $_\mu\varepsilon^T_{ij}$, an additional stress fluctuation τ_{ij} with corresponding strains $_\mu\varepsilon^e_{ij}$, $_\mu\varepsilon^P_{ij}$ develops. The difference in the mechanical part of the Gibbs free energy now yields

$$\psi_\sigma(1) - \psi_\sigma(0) = -\int_V (\Sigma_{ij} + \frac{1}{2}\, T_{ij} + \frac{1}{2}\, \tau_{ij})(\varepsilon^T_{ij} + \theta_\mu\, {}_\mu\varepsilon^T_{ij} + \varepsilon^P_{ij} + {}_\mu\varepsilon^P_{ij})\, dV +$$

$$+ \int_V (\Sigma_{ij} + \frac{1}{2}\, T_{ij})(\varepsilon^T_{ij} + \varepsilon^P_{ij})\, dV \, .$$

θ_μ is the indicator function with respect to V_μ. Decomposition of the right side of the above equation delivers

$$\psi_\sigma(1) - \psi_\sigma(0) = - \int_V \Sigma_{ij}\, (\theta_\mu\, {}_\mu\varepsilon^T_{ij} + {}_\mu\varepsilon^P_{ij})\, dV - \int_V \frac{1}{2}\, T_{ij}\, (\theta_\mu\, {}_\mu\varepsilon^T_{ij} + {}_\mu\varepsilon^P_{ij})\, dV -$$

$$- \int_V \frac{1}{2} \tau_{ij} (\varepsilon_{ij}^T + \varepsilon_{ij}^P) \, dV - \int_V \frac{1}{2} \tau_{ij} (\theta_\mu \, {}_\mu\varepsilon_{ij}^T + {}_\mu\varepsilon_{ij}^P) \, dV .$$

For this integral, adding and subtracting of ${}_\mu\varepsilon_{ij}^e$ to the fourth integral leads to
$\int_V \frac{1}{2} \tau_{ij} \, {}_\mu\varepsilon_{ij}^e \, dV$.

Moreover it can be shown that both the second and third integral are equivalent due to the principle of virtual work and considering uniform elastic behavior represented by C_{ijkl},

$$\int_V \frac{1}{2}\tau_{ij}(\varepsilon_{ij}^T + \varepsilon_{ij}^P + \varepsilon_{ij}^e - \varepsilon_{ij}^e) \, dV = -\int_V \frac{1}{2}\tau_{ij} \, \varepsilon_{ij}^e \, dV = -\int_V \frac{1}{2} C_{ijkl} \, {}_\mu\varepsilon_{kl}^e \, \varepsilon_{ij}^e \, dV,$$

$$\int_V \frac{1}{2} T_{ij}(\theta_\mu \, {}_\mu\varepsilon_{ij}^T + {}_\mu\varepsilon_{ij}^P + {}_\mu\varepsilon_{ij}^e - {}_\mu\varepsilon_{ij}^e) \, dV = -\int_V \frac{1}{2} T_{ij} \, {}_\mu\varepsilon_{ij}^e dV = -\int_V \frac{1}{2} C_{ijkl} \varepsilon_{kl}^e \, {}_\mu\varepsilon_{ij}^e dV.$$

Finally it follows for

$$\psi_\sigma(1) - \psi_\sigma(0) = -\int_V (\Sigma_{ij} + T_{ij})(\theta_\mu \, {}_\mu\varepsilon_{ij}^T + {}_\mu\varepsilon_{ij}^P) \, dV + \frac{1}{2} \int_V \tau_{ij} \, {}_\mu\varepsilon_{ij}^e \, dV .$$

The last integral on the right side of $\psi_\sigma(1) - \psi_\sigma(0)$ represents again the internal strain energy $V\bar{w}_\tau^e$.

According to the derivation of (1.25) the global transformation condition now reads

$$\rho[\varphi_{ch}]|_\xi + (\Sigma_{ij} + T_{ij}) {}_\mu\varepsilon_{ij}^T = \rho \frac{\overline{T}}{M_s} [\varphi_{ch}]|_{M_s} + \frac{1}{f} (\bar{w}_\tau^e + \bar{w}_\tau^P) \, , \; V_\mu = fV .$$

Finally it follows that, instead of Σ_{ij}, the local stress state before transformation, $\Sigma_{ij} + T_{ij}$, produces the mechanical driving force. This corresponds also with experimental observations, see eg. Gautier et al. [11]. These authors initially observed a favoured orientation of the martensite islands with respect to Σ_{ij} and, with the ongoing transformation, to $\Sigma_{ij} + T_{ij}$.

1.2.3 The local transformation condition - Nucleation Condition

In addition to an interface propagation criterion, Levitas proposed in [10] a nucleation criterion for a sudden generation of a new phase in a material point without observation of an interface movement. Following this line a nucleation condition is derived by first introducing a process parameter ξ, indicating in a material point by $\xi = 0$ the parent phase and by $\xi = 1$ the product phase. The specific Helmholtz free energy φ now contains ξ as an internal variable,

$$\varphi = \frac{1}{\rho} w^e (\varepsilon_{ij} - \varepsilon^P_{ij} - \varepsilon^T_{ij}, T; \xi) + \varphi_D(\alpha_i) + \varphi_{ch}(T; \xi) . \tag{1.14.3}$$

It is very important to consider in this case that the chemical part φ_{ch} depends also on the process variable ξ. This variable allows to describe the change of φ_{ch} from the parent to the product phase, also in the isothermal case.
Based on (1.20.3), the dissipation D is calculated as

$$D = \sigma_{ij} \dot{\varepsilon}^P_{ij} + \sigma_{ij} \dot{\varepsilon}^T_{ij} - \frac{\partial w^e}{\partial \xi} \dot{\xi} - \rho \frac{\partial \varphi_{ch}}{\partial \xi} \dot{\xi} - \rho \frac{\partial \varphi_D}{\partial \alpha_i} \dot{\alpha}_i . \tag{1.20.4}$$

$\left(\sigma_{ij} \dot{\varepsilon}^T_{ij} - \frac{\partial w^e}{\partial \xi} \dot{\xi} - \rho \frac{\partial \varphi_{ch}}{\partial \xi} \dot{\xi}\right)$ is interpreted as that energy rate $\tilde{F}^T_c$ per unit volume which is "extracted" from the system to produce (nucleate) the new phase (remembering the Bain strain and shearing). It follows that

$$\sigma_{ij} \dot{\varepsilon}^T_{ij} - \frac{\partial w^e}{\partial \xi} \dot{\xi} - \rho \frac{\partial \varphi_{ch}}{\partial \xi} \dot{\xi} = \tilde{F}^T_c , \qquad D = \sigma_{ij} \dot{\varepsilon}^P_{ij} - \rho \frac{\partial \varphi_D}{\partial \alpha_i} \dot{\alpha}_i + \tilde{F}^T_c . \tag{1.20.5}$$

As in the case of interface propagation, see (1.16), (1.17), we suggest also for this process to estimate $\tilde{F}^T_c$ by

$$\tilde{F}^T_c = \rho \frac{T}{M_s} [\varphi_{ch}] \mid M_s . \tag{1.20.6}$$

The time derivative of φ follows as

$$\dot{\varphi} = \frac{1}{\rho} \dot{w}^e + \frac{\partial \varphi_D}{\partial \alpha_i} \dot{\alpha}_i + \dot{\varphi}_{ch} ,$$

and with $\frac{1}{\rho} \dot{w}^e = \frac{1}{\rho} \sigma_{ij} \dot{\varepsilon}^e_{ij}$

$$\dot{\varphi} = \frac{1}{\rho} \sigma_{ij} \dot{\varepsilon}^e_{ij} + \frac{\partial \varphi_D}{\partial \alpha_i} \dot{\alpha}_i + \dot{\varphi}_{ch} . \tag{1.26.1}$$

Inserting D now in the form of (1.20.5) and $\dot{\varphi}$ in the form of (1.26.1) into (1.20.1) yields

$$\sigma_{ij} \dot{\varepsilon}^T_{ij} - \rho \dot{\varphi}_{ch} - \rho \eta \dot{T} = \tilde{F}^T_c . \tag{1.26.2}$$

Introducing for $\eta = -\frac{\partial \varphi_{ch}}{\partial T}$ and keeping in mind the total differential $\dot{\varphi}_{ch} = \frac{\partial \varphi_{ch}}{\partial T}\dot{T} + \frac{\partial \varphi_{ch}}{\partial \xi}\dot{\xi}$, it follows for (1.26.2) that

$$\sigma_{ij}\,\dot{\varepsilon}^{T}_{ij} - \rho\,\frac{\partial \varphi_{ch}}{\partial \xi}\,\dot{\xi} = \tilde{F}^{T}_{c}\,. \tag{1.26.3}$$

Integration with respect to the process variable ξ yields with $\sigma_{ij} = \Sigma_{ij} + \tau_{ij}(r,\xi)$

$$\rho[\varphi_{ch}]|_{r} + \Sigma_{ij}\,\varepsilon^{T}_{ij} + \int_{0}^{1} \tau_{ij}(r,\xi)\,\frac{d\varepsilon^{T}_{ij}(\xi)}{d\xi}\,d\xi = \int_{0}^{1} \tilde{F}^{T}_{c}\left(\frac{dt}{d\xi}\right)d\xi = \rho\,\frac{\overline{T}}{M_{s}}[\varphi_{ch}]|_{M_{s}}\,. \tag{1.26.4}$$

$\overline{T}$ is the process average of T, $\overline{T} = \int_{0}^{1} T(\xi)\left(\frac{dt}{d\xi}\right)d\xi$.

At this stage it seems to be appropriate to consider the birth of a certain microregion V_{μ}. Therefore, relation (1.26.4) will be integrated over V_{μ}, leading to the stress-dependent part

$$\int_{V_{\mu}}\int_{0}^{1} \tau_{ij}\,(r,\xi)\,\frac{d\varepsilon^{T}_{ij}}{d\xi}\,d\xi dV = \int_{V}\int_{0}^{1} \tau_{ij}\,(r,\xi)\;\theta_{\mu}\,\frac{d\varepsilon^{T}_{ij}}{d\xi}\,d\xi dV\,, \tag{1.26.5}$$

θ_{μ} being the indicator function with respect to V_{μ}. If we add again the plastic strain ε^{P}_{ij} and the elastic strain ε^{e}_{ij} to (1.26.5) and subtract them, the same consideration that led to (1.23.3-6) yields

$$\int_{V_{\mu}}\int_{0}^{1} \tau_{ij}\,(r,\,\xi)\,\frac{d\varepsilon^{T}_{ij}}{d\xi}\,d\xi dV = -V\,(\bar{w}^{e}_{T} + \bar{w}^{P}_{T})\,, \tag{1.26.6}$$

for the definition of $\bar{w}^{e}_{T}$, $\bar{w}^{P}_{T}$ see above.

If (1.26.6) is inserted into (1.26.4), again a transformation condition can be formulated for the sudden birth of a microregion V_{μ} situated around the location vector r_{i} as

$$\rho[\varphi_{ch}]|_{r_{i}} + \Sigma_{ij}\,\varepsilon^{T}_{ij} = \frac{\overline{T}}{M_{s}}\rho\,[\varphi_{ch}]|_{M_{s}} + \frac{1}{f}\left(\bar{w}^{e}_{T} + \bar{w}^{P}_{T}\right),\; f = V_{\mu}/V\,. \tag{1.27}$$

The transformation (1.27) can be seen as a local condition, since only the microregion V_μ and not the specimen has been investigated. Of course, the plastic strain ε^P_{ij} and, therefore, w^P_T depend on the history and the actual configuration of the specimen.
Again on the right side two restricting entities appear as in the case of (1.25).

1.2.4 Conclusion

A transformation condition for a (transformed) microregion is derived which applies both in the case of

* an initial nucleation followed by a growth process by interface movement to a "full" microregion (the corresponding condition is denoted as "growth condition") and
* a sudden nucleation (burst) of a "full" microregion (the corresponding condition is denoted as "nucleation condition").

The condition reads

$$\rho[\varphi_{ch}(\xi)] + \left(\Sigma_{ij}+T_{ij}\right)\varepsilon^T_{ij} = \frac{\overline{T}}{M_S}\rho\,[\varphi_{ch}]|_{M_S} + \frac{1}{f}\left(\bar{w}^e_T + \bar{w}^P_T\right). \qquad (1.28)$$

Here, $\left(\Sigma_{ij}+T_{ij}\right)$ represents the stress state before transformation and can be the sum of the global load stress Σ_{ij} and a local stress fluctuation T_{ij} before transformation of the considered microregion, see Remark 3 of chpt.1.2.2. $\xi = 1$ in the case of the nucleation condition.

The left side of (1.28) can be interpreted as the "driving" term and the right side as the "restricting" term. Two further conclusions can be drawn immediately:

* Since 24 variants are physically possible, obviously that variant will develop which leads to a maximum mechanical driving force $(\Sigma_{ij}+T_{ij})\varepsilon^{n*}_{ij}$, $(\Sigma_{ij}+T_{ij})\varepsilon^{n*}_{ij} = \max\left((\Sigma_{ij}+T_{ij})\varepsilon^{n}_{ij}\right)$, $n=1, \ldots 24$.
* Obviously a variant will appear with such a shape that $\bar{w}^e_T$ and $\bar{w}^e_p$ become as small as possible. This means, of course, that the martensite phase forms plates or, even, blades instead of spheroids.

For further literature on transformation conditions the reader is referred to [9].

2. CONTINUUM MECHANICS FORMULATION OF THE DEFORMATION PROCESS DURING AND AFTER A SOLID STATE PHASE TRANSFORMATION

2.1 The orientation process

If we consider the transformation condition (1.28) for a microregion we distinguish between a chemical and a mechanical driving force. Obviously, only the mechanical driving force $(\Sigma_{ij}+T_{ij})\varepsilon^T_{ij}$ includes the transformation tensor ε^T_{ij}. T_{ij} is the stress fluctuation which has developed before the transformation of the microregion considered. Since up to 24 martensitic variants with corresponding transformation tensors ε^n_{ij}, n=1, ... 24, may exist (see the Wechsler-Lieberman-Read theory), the according driving forces F^n_σ, n=1 ... 24, are

$$F^n_\sigma = (\Sigma_{ij} + T_{ij})\, \varepsilon^n_{ij} \,. \tag{2.1.1}$$

The label "σ" refers to the mechanical part of the total driving force F^n.
As already mentioned, that variant will be the most likely one whose F^n_σ reaches a maximum $F^{n*}_{\sigma,max}$, n* being the corresponding variant number. Patel and Cohen [12] were, to the knowledge of the author, the first to recognize the influence of F^n_σ on increasing the martensite start temperature from M_s to M_σ. They assumed an ideal (monocrystal) configuration, that means, a sudden change of the whole specimen from the parent to the product phase without developing any stress fluctuation (T_{ij}=0) or dissipation. In this case the nucleation condition (1.27) reads as

$$\rho[\varphi_{ch}]|_{M_\sigma} + F^{n*}_{\sigma,\,max} = \frac{M_\sigma}{M_s}\,\rho[\varphi_{ch}]\,|_{M_s}, \quad F^{n*}_{\sigma,max} = \Sigma_{ij}\varepsilon^{n*}_{ij} \,. \tag{2.1.2}$$

By making an approximation of a linear relation $\rho[\varphi_{ch}]_T = \rho\big([\varphi_{ch}]|_{M_s}\big)(T-T_0)/(M_s - T_0)$, for T_0 see e.g. the contributions by Ortin and Stüwe in this book and the explanations above, the shift in temperature will be

$$M_\sigma - M_s = \big(F^{n*}_{\sigma,max}/\rho\,[\varphi_{ch}]|_{M_s}\big)(T_0-M_s)\,(M_s/T_0) \,. \tag{2.1.3}$$

Two comments should be made here:
Since Patel and Cohen [12] assumed $\rho[\varphi_{ch}]|_{M_s}$ instead of $\frac{T}{M_s}\rho[\varphi_{ch}]|_{M_s}$, the last expression (M_s/T_0) does not appear in their equation for M_σ - M_s.
The critical force $\rho[\varphi_{ch}]|_{M_s}$ may be lower in the ideal (monocrystal) case

compared to its value for a polycrystal which obviously contains a certain amount of dissipation due to the inhomogeneity of the material.

If we consider a uniaxial loading Σ_{33} in the global 3-direction, F^n_σ, given by (1.4), will be

$$F^n_\sigma = \frac{\Sigma_{33}}{2} [\gamma \sin 2\vartheta \sin 2\varphi + \delta(1 + \cos 2\vartheta)] . \tag{2.1.4}$$

The maximum possible value $\max(F^n_\sigma)$ can be found as

$$\max(F^n_\sigma) = \Sigma_{33}\varepsilon^T_+ \text{ for } \Sigma_{33} > 0,\ \varepsilon^T_+ = \frac{1}{2}\left(\delta + \sqrt{\delta^2 + \gamma^2}\right),$$

$$\min(F^n_\sigma) = \Sigma_{33}\varepsilon^T_- \text{ for } \Sigma_{33} > 0,\ \varepsilon^T_- = \frac{1}{2}\left(\delta - \sqrt{\delta^2 + \gamma^2}\right), \tag{2.1.5}$$

and the corresponding angles $\vartheta = \vartheta_{PC}$, $\varphi = \varphi_{PC}$ as

$$\vartheta_{PC} = \frac{1}{2} \arctan(\gamma/\delta),\quad \varphi_{PC} = \pi/2 . \tag{2.1.6}$$

Of course, $\max(F^n_\sigma) = F^{n^*}_{\sigma,max}$ exists only if the crystal lattice is oriented in such a way to the global load that a martensite variant n* is possible. In this case any tension (compression) load would generate a longitudinal strain ε^T_+ (ε^T_-) in addition to the elastic deformation. Since $\gamma \sim 0{,}2$ for steels, a longitudinal strain of ca. 10% would appear!

Marketz and Fischer [13] now study extrema for a three-dimensional principal stress state $\Sigma_{11} \neq 0$, $\Sigma_{22} \neq 0$, $\Sigma_{33} \neq 0$. The necessary conditions $\frac{\partial F^n_\sigma}{\partial \vartheta} = \frac{\partial F^n_\sigma}{\partial \varphi} = = \frac{\partial F^n_\sigma}{\partial \psi} = 0$ lead to a highly nonlinear system of algebraic equations in the Eulerian angles.The result is surprisingly simple:

$$\max(F^n_\sigma) = \Sigma_{33}\,\varepsilon^T_+ + \Sigma_{11}\,\varepsilon^T_- \text{ for } \Sigma_{33} > \Sigma_{22} > \Sigma_{11}, \tag{2.2.1}$$

$$\max(F^n_\sigma) = \Sigma_{33}\,\varepsilon^T_+ + \Sigma_{22}\,\varepsilon^T_- \text{ for } \Sigma_{33} > \Sigma_{11} > \Sigma_{22}, \tag{2.2.2}$$

$$\max(F^n_\sigma) = \Sigma_{22}\,\varepsilon^T_+ + \Sigma_{11}\,\varepsilon^T_- \text{ for } \Sigma_{22} > \Sigma_{33} > \Sigma_{11}. \tag{2.2.3}$$

For the remaining three cases, one only has to exchange the corresponding indices of the maximum and minimum principle stresses.

Subsequently, they studied the effect of an arbitrary triaxial stress state on the mechanical driving force and the global deformation in two simulations which are briefly repeated below. The components of Σ_{ij} are generated in the first simulation as random numbers in the range -500[MPa] < Σ_{ij} < 500 [MPa]. For the sake of comparison we perform a normalization by introducing

$$\|\Sigma_{ij}\| = (\Sigma_{ij}\Sigma_{ij})^{1/2}, \quad \|\Sigma_{ij}\| = \left(3p^2 + \frac{2}{3}\Sigma_v{}^2\right)^{1/2}, \tag{2.2.4}$$

$p = \frac{1}{3}\,tr(\Sigma_{ij})$, Σ_v is the von Mises equivalent stress. It should be mentioned that $\|\Sigma_{ij}\|$ is used instead of Σ_v for normalization to avoid the discarding of the influence of the hydrostatic part of the stress tensor.

We consider 24 variants ε_{ij}^n with the according habit plane family $(3\ 14\ 10)_{fcc}$ as typical for a Fe-Ni alloy. With respect to the kinematics explained in chpt.1.1.1, δ = 0,0515, γ = 0,22 is assumed. The habit plane normals are vectors of the type $[0,1848\ 0,7823\ -0,5948]^t$, the shear directions $[-0,2090\ 0,7090\ 0,6736]^t$, for details see Nishiyama [5].

The normalized mechanical driving force $\tilde{F}_\sigma^n$,

$$\tilde{F}_\sigma^n = \Sigma_{ij}\,\varepsilon_{ij}^n \,/\, \|\Sigma_{ij}\|, \tag{2.2.5}$$

can be written with (1.2.1) for $\xi = 1$ as

$$\tilde{F}_\sigma^n = \left(S_{ij}\,e_{ij}^n + p\delta\right) / \|\Sigma_{ij}\|,$$

S_{ij} is the deviator of Σ_{ij} and is now calculated for more than 10^8 stress tensors.

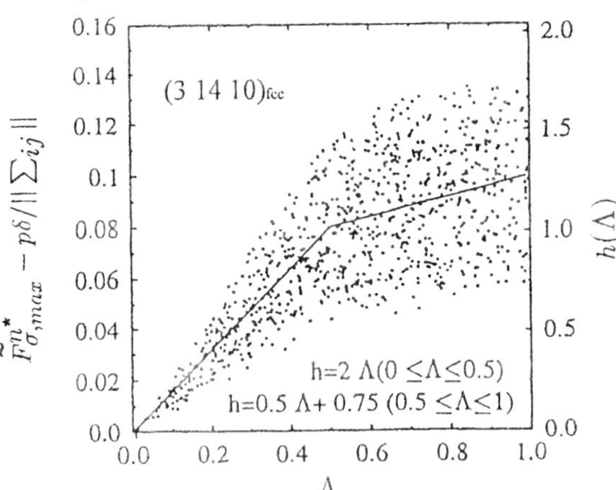

Fig. 2: Normalized maximum mechanical driving force $\tilde{F}^{n*}_{\sigma,max}$, reduced by $p\delta / \|\Sigma^n_{ij}\|$, in dependence on the multiaxiality factor Λ.

Its maximum value $\tilde{F}^{n*}_{\sigma,max}$, reduced by $p\delta/\|\Sigma_{ij}\|$, is now depicted, see Fig.2, in relation to a multiaxiality factor Λ,

$$\Lambda = \Sigma_v/(\Sigma_v + 3|p|) \ . \tag{2.2.6}$$

Note that $\Lambda = 0$ for $\Sigma_v = 0$, $p \neq 0$, $\Lambda = 1/2$ for uniaxial loading and $\Lambda = 1$ for pure shear ($\Sigma_v \neq 0$, $p = 0$).

The following conclusions can be drawn from Fig.2:

* The higher the multiaxiality factor Λ, the higher the scatter of the normalized maximum mechanical driving force.
* There is a clear tendency for the normalized maximum mechanical driving force to increase with the multiaxiality factor Λ.
* There are both an upper and a lower bound for the normalized maximum mechanical driving force.
* The average of $\tilde{F}^{n*}_{\sigma,max} - p\delta$ can be approximated by a function $h(\Lambda)$ given in Fig.2.

This simulation allows at least the following summarizations:

* In addition to the mechanical driving force term $p\delta$, there will always exist a positive maximum mechanical driving force $S_{ij}\ e^{n*}_{ij}$ which gives rise to a favorite orientation of a variant.
* The higher the degree of multiaxiality of the stress state occurs, expressed by Λ, the higher is the value of the mechanical driving force for a distinct optimal variant.

A second simulation can shed some light on the global deformation of a specimen caused by the selection of locally optimal variants. Marketz and Fischer [13] took for this purpose a mesodomain consisting of 960 crystallites with orientations g_k, $k = 1, \dots 960$. Up to 24 transformation tensors ${}^k\varepsilon^n_{ij}$ may appear in each crystallite whose orientation is selected by a random distribution. We assume a homogeneous stress state $\Sigma_{33} = \Sigma$, otherwise $\Sigma_{ij} = 0$, which is kept constant during the following simulated cooling process starting with a homogeneous temperature distribution T_s.

It should be emphasized here that this study is purely devoted to demonstrating the effect of the developing variants on the total deformation of a specimen. It does not include any actual interaction of the grains during the transformation process. This is treated within a further analysis, see chpt.2.2.2.

$\rho[\varphi_{ch}(T)]$ can be taken from experimental data published by Kaufman and Cohen [14] as

$$\rho\,[\varphi_{ch}(T)] = a(T - T_o) + b(T - T_o)^2\,,$$

$$a = -0.904\ [\mathrm{MPa\ K^{-1}}],\ b = -3.356\ .\ 10^{-4}\ [\mathrm{MPa\ K^{-2}}]\,, \tag{2.3.1}$$

$$T_o = 440[\mathrm{K}],\ M_s = 220\ [\mathrm{K}]\,.$$

As the transformation kinetics we chose the proposal by Magee, see [15],

$$f(T, \Sigma) = 1 - \exp\left\{-\tilde{b}\rho\left[\varphi_{ch}(T) - \varphi_{ch}(M_s)\right]\right\}, \quad \tilde{b} = 0.025\,. \tag{2.3.2}$$

The influence of the stress state on kinetics is incorporated by applying in equation (2.3.2) M_σ instead of M_s, see (2.1.3); here an average value of M_σ- - M_s=12[K] is assumed.

With (2.3.2) an increment of Δf is related to a temperature increment ΔT as

$$\Delta T = \left|\left(\frac{\Delta f}{1-f}\right) \Big/ \left(\tilde{b}\rho\,\frac{[\partial\varphi_{ch}(T)]}{\partial T}\right)\right|. \tag{2.3.3}$$

This type of kinetics would lead to an infinite increment ΔT for f approaching 1. Therefore, the procedure is "cut" at f = 0.99, for a discussion see [16].

Let us now assume that a total volume fraction f has been reached during a cooling process starting at T_s and arriving at the current temperature T. An increase of f by Δf will now be produced by a cooling step from T to T-ΔT, see (2.3.3). The product phase increment Δf is now distributed to each crystallite k with a volume ${}^kV$ due to its volume fraction ${}^kV/V$, leading to

$$\Delta^k f = \Delta f \cdot {}^kV/V\,. \tag{2.4.1}$$

We now apply the transformation condition (1.28) neglecting $\bar{w}^e_\tau$ and $\bar{w}^p_\tau$ and calculate the force term

$${}^kF^n = \Sigma_{ij}\,{}^k\varepsilon^n_{ij} + \rho[\varphi_{ch}\,(\bar{\xi})] - \frac{\bar{T}}{M_s}\,[\varphi_{ch}]|_{M_s}\ ; \tag{2.4.2}$$

$\bar{\xi}$ corresponds to an average temperature $\bar{T} = T - \alpha\Delta T$, $\alpha \sim 1/2$.

For ${}^kF^n \geq 0$ the corresponding variant is activated, for ${}^kF^n < 0$ it is discarded. The total force term ${}^kF$ is now the sum of the force terms of the "surviving" variants, let us say n^+ with ${}^kF^{n^+} \geq 0$, ($n^+ \leq 24$),

$$ {}^{k}F = \sum_{n^+} {}^{k}F^{n^+} . \qquad (2.4.3) $$

The incremental volume fraction of a variant, $\Delta {}^{k}f^{n^+}$, is now assumed to be proportional to the corresponding force term ${}^{k}F^{n^+}$, thus

$$ \Delta {}^{k}f^{n^+} = \frac{{}^{k}F^{n^+}}{\sum_{n^+} {}^{k}F^{n^+}} \cdot \Delta f \cdot {}^{k}V/V . \qquad (2.4.4) $$

This simple weighting procedure allows also to calculate the increment of the macroscopic transformation tensor $\bar{\varepsilon}^{T}_{ij}$ as

$$ \Delta \bar{\varepsilon}^{T}_{ij} = \sum_{k} \sum_{n^+} {}^{k}\varepsilon^{n^+}_{ij} \, \Delta {}^{k}f^{n^+} . \qquad (2.4.5) $$

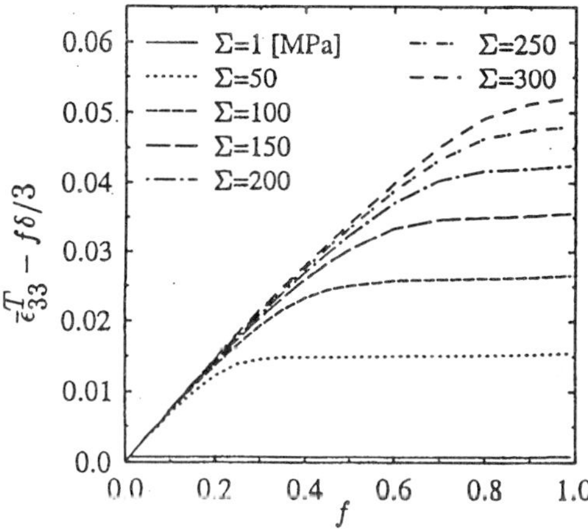

Fig. 3: Overall transformation strain $\bar{\varepsilon}^{T}_{33}$ reduced by δ/3, in direction of the applied stress Σ, dependent on the martensite volume fraction f.

Fig. 3 demonstrates the total transformation strain component $\bar{\varepsilon}^{T}_{33}$, reduced by f δ/3, due to the orientation effect. It can clearly be seen that in the range where the orientation effect stops playing a role, the tangent of the curve becomes horizontal. The smaller the applied stress Σ, the earlier this "saturation" effect occurs.

A generalization of the current study for a three-dimensional stress state leads to a proposal for the deviator $\Delta\bar{e}^T_{ij}$ of the increment of the macroscopic transformation tensor $\tilde{\varepsilon}^T_{ij}$ by use of the multiaxiality function $h(\Lambda)$,

$$\Delta\bar{e}^T_{ij} = \Delta f \left| c_1 \exp(-c_2 f^{c_3}) - \delta/3 \right| h(\Lambda) \frac{S_{ij}}{\max(S_{kl})} , \tag{2.4.6}$$

c_1, c_2 and c_3 are constants, $\max(S_{kl})$ is the maximum value of the elements of the deviator and ensures a normalization to 1 of the largest element of S_{ij} (here, Marketz and Fischer [13] applied $\frac{\Sigma_{ij}}{\|\Sigma_{ij}\|}$ instead of $\frac{S_{ij}}{\max(S_{kl})}$).

Of course, relation (2.4.6) is something like a provisional proposal.

2.2 The deformation process

2.2.1 A simplified analytical model

2.2.1.1 Assumptions and definitions

In the following chapter a semianalytical procedure is outlined which allows a basic understanding of the TRIP strain. The concept goes back to the book by Mitter [3] and is based on the following assumptions as well as simplifications:

* The transformation strain tensor ε^T_{ij} as well as its rate $\dot{\varepsilon}^T_{ij}$ are known in each microregion located around a position vector γ_i, defined by a set of Eulerian angles ϑ, φ, ψ, with their components in a global cartesian co-ordinate system, see (1.2.2) and (1.7). An orientation function $g(\xi, \vartheta, \varphi, \psi)$ shall describe the spatial distribution of the microregions in the case of a martensitic transformation.
* The transformation process is homogeneous, meaning that in every unit volume the same specific amount of transformed particles (described by ξ in case of martensite and f in the case of diffusive transformation) exists.
* The elastic strain component is neglected leading to

$$\varepsilon_{ij}|_{r_i} = \varepsilon^T_{ij}|_{r_i} + \varepsilon^P_{ij}|_{r_i} . \tag{2.5.1}$$

* The total strain component is homogeneously distributed, $\varepsilon_{ij}|_{r_i}$ = constant. This corresponds to the famous Taylor-Lin assumption in crystal plasticity. In this assumption the solution of the field equations is omitted. One is then free to concentrate only on the calculation of $\varepsilon^P_{ij}|_{r_i}$ (abbreviated as ε^P_{ij}) .

* Standard plasticity leads to the following expressions for the stress deviator $s_{ij}|_{r_i} = s_{ij}(r_i)$ (abbreviated as s_{ij}),

$$\dot{\varepsilon}^P_{ij} = \dot{\lambda}\, s_{ij}, \quad \frac{3}{2} s_{ij}\, s_{ij} = \sigma_y \, . \tag{2.5.2}$$

σ_y is the yield stress and may be a function of ξ or f.
$\dot{\lambda}$ is expressed by $\dot{\varepsilon}^P_{ij} = \dot{\varepsilon}_{ij} - \dot{\varepsilon}^T_{ij}$ as

$$\frac{3}{2\sigma_y^2}(\dot{\varepsilon}_{ij} - \dot{\varepsilon}^T_{ij})(\dot{\varepsilon}_{ij} - \dot{\varepsilon}^T_{ij}) = \dot{\lambda}^2 \, . \tag{2.5.3}$$

From this relation s_{ij} can be calculated as

$$s_{ij} = \sigma_y(\dot{\varepsilon}_{ij} - \dot{\varepsilon}^T_{ij}) \Big/ \left\{ \frac{3}{2} (\dot{\varepsilon}_{kl} - \dot{\varepsilon}^T_{kl})(\dot{\varepsilon}_{kl} - \dot{\varepsilon}^T_{kl}) \right\}^{1/2} . \tag{2.5.4}$$

* The global equilibrium condition must be fulfilled. To calculate this we have to define the average value $\bar{a}$ of an entity a by integration in the Eulerian space as

$$\bar{a} = \int_{\vartheta,\varphi,\psi} g\, a \sin\vartheta \, d\vartheta \, d\varphi \, d\psi \Big/ \int_{\vartheta,\varphi,\psi} g \sin\vartheta \, d\vartheta \, d\varphi \, d\psi \, , \tag{2.5.5}$$

$0 \le \vartheta \le \pi, \;\; 0 \le \varphi, \psi \le 2\pi$.

Therefore, the volume average $\bar{s}_{ij}$ must be equal to the global values S_{ij} being the deviator to Σ_{ij},

$$\bar{s}_{ij} = S_{ij} \, . \tag{2.5.6}$$

If one inserts (2.5.4) into (2.5.5) and keeps in mind that S_{ij} is given as the deviator to the load stress tensor Σ_{ij}, relation (2.5.5) yields a set of 5 differential equations for the five deviatoric components of $(\varepsilon_{ij} - \varepsilon^T_{ij})$.
These differential equations can be solved only by numerical methods since the unknown tensor $\dot{\varepsilon}_{kl}$ appears in the integral over the Eulerian angles in the numerator as well as in the denominator of (2.5.4).

2.2.1.2 Displacive (martensitic) transformation
According to (1.1), (1.2.1) the transformation tensor is written as

$$\varepsilon^T_{ij} = \tilde{f}(\xi)\, {}_1\varepsilon^T_{ij} \, , \quad {}_1\varepsilon^T_{ij} = \frac{\delta}{3}\delta_{ij} + {}_1e^T_{ij} \, . \tag{2.6.1}$$

The label "1" refers to the final value of ε_{ij}^{T} at $\xi = 1$, since $0 \le \tilde{f}(\xi) \le 1$, $\xi = 0 : \tilde{f}(\xi) = 0$, $\xi = 1 : \tilde{f}(\xi) = 1$.

Further we introduce

$$\varepsilon_{ij} = \frac{\delta}{3}\,\tilde{f}(\xi)\,\delta_{ij} + \varepsilon_{ij}^{TP} \quad , \varepsilon_{ij}^{TP} \text{ is denominated as the TRIP strain.}$$

The time derivatives in (2.5.4) are substituted by

$$\frac{d\varepsilon_{ij}^{TP}}{dt} = \frac{d\varepsilon_{ij}^{TP}}{d\xi}\,\dot{\xi} = \varepsilon_{ij}^{TP'}\dot{\xi} \quad \text{etc.}$$

It follows for (2.5.4)

$$s_{ij} = \sigma_y(\xi)\,(\varepsilon_{ij}^{TP'} - \tilde{f}'\cdot{}_1e_{ij}^{T}) \Big/ \left\{\frac{3}{2}\,(\varepsilon_{kl}^{TP'} - \tilde{f}'\cdot{}_1e_{kl}^{T})\,(\varepsilon_{kl}^{TP'} - \tilde{f}'\cdot{}_1e_{kl}^{T})\right\}^{1/2}$$

and with (1.1)

$$s_{ij} = \sigma_y(\xi)\,(\varepsilon_{ij}^{TP'} - \tilde{f}'\cdot{}_1e_{ij}^{T}) \Big/ \left\{\frac{3}{2}\,\varepsilon_{kl}^{TP'}\varepsilon_{kl}^{TP'} - 3\tilde{f}'.\varepsilon_{kl}^{TP'}\,{}_1e_{kl}^{T} + \tilde{f}'^2.(\delta^2 + \frac{3}{4}\gamma^2)\right\}^{1/2} . \quad (2.6.2)$$

The evaluation of $\varepsilon_{ij}^{TP'}$ from (2.5.6) now depends on the assumption of $g(\xi,\vartheta,\varphi,\psi)$. If we assume $g \equiv 1$, which means no orientation effect, S_{ij} can be derived analytically by a Taylor series

$$S_{ij} = S_{ij}\Big|_{\varepsilon_{kl}^{TP'}=0} + \frac{\partial S_{ij}}{\partial\,\varepsilon_{kl}^{TP'}}\Big|_{\varepsilon_{kl}^{TP'}=0} \cdot \varepsilon_{kl}^{TP'} + \ldots\ldots .$$

Since $\varepsilon_{kl}^{TP'}=0$ to $S_{ij}=0$ (due to ${}_1\bar{e}_{ij}^{T}=0$) it follows with truncation after 2 terms that

$$S_{ij} = \frac{\partial S_{ij}}{\partial\varepsilon_{kl}^{TP'}}\Big|_{\varepsilon_{kl}^{TP'}=0} \cdot \varepsilon_{kl}^{TP'} . \quad (2.6.3)$$

One has to calculate the derivative

$$\frac{\partial S_{ij}}{\partial\varepsilon_{kl}^{TP'}}\Big|_{\varepsilon_{kl}^{TP'}=0} = \sigma_y(\xi)\left\{\delta_{ik}\delta_{jl} - \frac{3}{2}(\varepsilon_{ij}^{TP'} - \tilde{f}'\cdot{}_1e_{ij}^{T})(\varepsilon_{kl}^{TP'} - \tilde{f}'\cdot{}_1e_{kl}^{T})/N^2\right\}\Big/N\,\Bigg|_{\varepsilon_{kl}^{TP'}=0} .$$

N is the denominator in (2.6.2) and is $\tilde{f}'\left(\delta^2 + \frac{3}{4}\gamma^2\right)^{1/2}$ for $\varepsilon_{kl}^{TP'} = 0$. (2.6.3) can be calculated as

$$S_{ij} = \frac{\sigma_y(\xi)}{\tilde{f}'\left(\delta^2 + \frac{3}{4}\gamma^2\right)^{1/2}}\left\{\delta_{ik}\delta_{jl} - \frac{3}{2(\delta^2 + \frac{3}{4}\gamma^2)}\,\overline{({}_1e_{ij}^{T}\;{}_1e_{kl}^{T})}\right\}\varepsilon_{kl}^{TP'} . \quad (2.6.4)$$

$\overline{({}_1e^T_{ij}\;{}_1e^T_{kl})}$ means the Eulerian space volume average of $({}_1e^T_{ij}\;{}_1e^T_{kl})$ and is a fourth-order tensor whose elements are functions of δ, γ. The components of ${}_1e^T_{ij}$ can be taken from [2], [17]. Although the products ${}_1e^T_{ij}\;{}_1e^T_{kl}$ are very lengthy expressions in the Eulerian angles, their volume averages can be calculated exactly for $g \equiv 1$. They yield

$$\overline{({}_1e^T_{ij}\;{}_1e^T_{kl})} = 8\pi^2 \cdot \frac{4}{45}\left(\delta^2+\frac{3}{4}\gamma^2\right) \qquad \text{for } i = j = k = l\,,$$

$$\overline{({}_1e^T_{ij}\;{}_1e^T_{kl})} = 8\pi^2 \cdot -\frac{2}{45}\left(\delta^2+\frac{3}{4}\gamma^2\right) \qquad \text{for } i=j,\ k=l,\ i\neq k\,,$$

$$\overline{({}_1e^T_{ij}\;{}_1e^T_{kl})} = 8\pi^2 \cdot \frac{3}{45}\left(\delta^2+\frac{3}{4}\gamma^2\right) \qquad \text{for } i\neq j,\ k=i,\ l=j\,,$$

all others are 0.

The insertion of these integrals into (2.6.4) allows an exact inversion of (2.6.4) leading to the surprisingly simple relation

$$\varepsilon^{TP'}_{ij} = \frac{5}{4}\,\frac{\tilde{f}'}{\sigma_y(\xi)}\left(\delta^2+\frac{3}{4}\gamma^2\right)^{1/2} S_{ij}\,. \qquad (2.6.5.1)$$

It is reasonable to assume that $\sigma_y(\xi)$ varies with ξ in an analogous way to (2.6.1), $\sigma_y(\xi) = \sigma_{y_o} + \tilde{f}(\xi)\,(\sigma_{y_n} - \sigma_{y_o})$. Then the integration of (2.6.5.1) with respect to ξ yields

$$\varepsilon^{TP}_{ij} = \frac{5}{4}\,\frac{(\delta^2+3/4\gamma^2)^{1/2}}{\sigma_y^*}\,S_{ij}\,, \qquad \sigma_y^* = \sigma_{y_n}\left(\frac{1-\sigma_{y_o}/\sigma_{y_n}}{\ln(\sigma_{y_n}/\sigma_{y_o})}\right). \qquad (2.6.5.2)$$

If we concentrate on uniaxial loading with

$$S_{11} = -\Sigma/3 = S_{22},\ S_{33} = \frac{2}{3}\Sigma,\ S_{ij} = 0\ \forall\, i \neq j\,,$$

it follows for ε^{TP}_{33}

$$\varepsilon^{TP}_{33} = \frac{5}{6}\left(\delta^2 + \frac{3}{4}\gamma^2\right)^{1/2} \Sigma/\sigma_y^*\,. \qquad (2.6.6)$$

This relation corresponds to the famous "Greenwood and Johnson relation", cf. [18]. They delivered a relation of the type $\varepsilon^{TP} = \frac{5}{6}\delta\Sigma/\sigma_y$ considering only a volume change during a time-dependent transformation. It should be repeated that relations (2.6.4-6) have been derived for a uniform distribution of microregions and, therefore, only reflect an "accommodation" effect.

We now investigate some specific distribution functions $g \neq 1$ for the uniaxial case. Here we refer to chpt.2.1 and equations (2.1.1-6) since now, in addition to the accommodation, also an orientation effect is introduced.

We define the term "φ-restriction" to mean that $\varphi = \pi/2$ for $\Sigma > 0$ and $\varphi = -\pi/2$ for $\Sigma < 0$. This condition extremizes F_σ^n with respect to φ.

Moreover, "ϑ-restriction" means that under tension ($\Sigma > 0$) only variants with a length elongation $\left(\varepsilon_{33}^T > 0\right.$, see (1.4) and $\left.(2.1.3)\right)$ and under compression ($\Sigma < 0$) only variants with a length contraction ($\varepsilon_{33}^T < 0$) contribute. This leads to the following relations for ϑ:

$$\Sigma > 0: 0 \leq \vartheta \leq \pi/2, \pi - \tilde{\vartheta} \leq \vartheta \leq \pi\,, \qquad \Sigma < 0: \tilde{\vartheta} \leq \vartheta \leq \pi/2\,,$$

$$\tilde{\vartheta} = \arctan(\delta/\gamma)\,.$$

This relation now defines g as

$$\Sigma > 0:\ g = 0 \text{ for } \frac{\pi}{2} \leq \vartheta \leq \pi - \tilde{\vartheta}, \text{ otherwise } 1.0$$

$$\Sigma < 0:\ g = 0 \text{ for } 0 \leq \vartheta \leq \tilde{\vartheta} \text{ and } \frac{\pi}{2} \leq \vartheta \leq \pi, \text{ otherwise } 1.0\,.$$

A study of ε_{33}^{TP} in dependence of $\Sigma/\sigma_y{}^*$ for the various above mentioned restrictions is depicted in Fig.4 which is taken from [2]. Observe that Fig.4 demonstrates $\varepsilon_{33}^{TP} + \delta/3$ instead of ε_{33}^{TP}!

The following conclusions can be drawn:

* A linear "Greenwood and Johnson-type" relation like (2.6.6) describes very well the relation between ε and $\Sigma/\sigma_y{}^*$.
* A slight difference occurs for the tension and compression regime.
* The experimental data by Sattler and Wassermann [19] are well matched by the linear relation (2.6.6) with $\gamma = 0.2$, $\delta = 0.04$ and $\sigma_y{}^* = 750$[MPa].
* Application of the "ϑ-restriction" as well as the "φ-restriction" will lead to a drastic overestimation of the orientation effect in comparison with experiments.

With respect to the overestimation of the orientation effect by the "ϑ-restriction" a more realistic function $g(\xi)$ is investigated, taking into account that in the initial phase of martensitic transformation an optimal orientation of the variants prevails (represented by the "ϑ-restriction") while, with ongoing transformation, a more and more uniform distribution of the variants emerges ($g \equiv 1$).

For g, this restriction reads as "ξ-restriction"

$$\Sigma > 0 \quad g = 0 \text{ for } \frac{\pi}{2} + \xi\left(\frac{\pi}{4} - \frac{\tilde{\vartheta}}{2}\right) \leq \vartheta \leq \pi - \tilde{\vartheta} - \xi\left(\frac{\pi}{4} - \frac{\tilde{\vartheta}}{2}\right),$$

$$\text{otherwise } 1.0; \; \varphi = \pi/2 .$$

$$\Sigma < 0 \quad g = 0 \text{ for } 0 \leq \vartheta \leq \tilde{\vartheta}\,(1-\xi) \text{ and } \frac{\pi}{2}(1+\xi) \leq \vartheta \leq \pi,$$

$$\text{otherwise } 1.0; \; \varphi = -\pi/2 .$$

Calculation of the volume average (2.5.5) for s_{ij}, (2.5.4), with g due to the "ξ-restriction" and integration with respect to ξ, but with constant $\sigma_y=\sigma_y{}^*$ for simplicity, leads to

$$\varepsilon^{TP} = \varepsilon_0^{TP} + k\,\Sigma/\sigma_y{}^* . \tag{2.6.7}$$

This relation is valid for $\Sigma/\sigma_y{}^*$ being not too large (e.g. $\Sigma/\sigma_y{}^* < 0{,}5$). The results of a numerical study can be taken from Fig.5 for a tension load $\Sigma>0$; slightly differing results occur in the case of $\Sigma<0$. $k\Sigma/\sigma_y{}^*$ represents the plastic accommodation of the transformation incompatibility. ε_0^{TP} reflects the favoured orientation of martensite variants, however, in an averaged sense. As investigated by Marketz et al. [13] and explained in chpt.2.1, the amount of the orientation effect depends both on the loadstress level and the progress of transformation, f.

As discussed in the review paper on TRIP by Fischer et al. [20], in the past various authors proposed, based on the "Greenwood and Johnson-relation" (2.6.6), a TRIP-strain rate term in the form of

$$\dot{\varepsilon}_{ij}^{TP} = \frac{3}{2}\,K\,\frac{d}{df}\left(\Phi(f)\right)\,\dot{f}\,S_{ij} . \tag{2.7}$$

K is often denominated the "Greenwood and Johnson"-constant and corresponds to $\frac{5}{6}\left(\delta^2+\frac{3}{4}\gamma^2\right)^{1/2}\left(\frac{1}{\sigma_y{}^*}\right)$. Experimentally found K-values for steel are typically in the range of $4 - 5.0\cdot 10^{-5}$[1/MPa]. $\Phi(f)$ describes the dependence of ε_{ij}^{TP} on the martensite volume fraction for constant S_{ij}. Several proposals for $\Phi(f)$ exist in the literature and are briefly discussed in [20], too.

Relation (2.7) can be seen as a "phenomenological" anology to (2.6.5.1). To the knowledge of the author Franitza was the first in 1972 [21] who came up with a (2.7)-type relation.

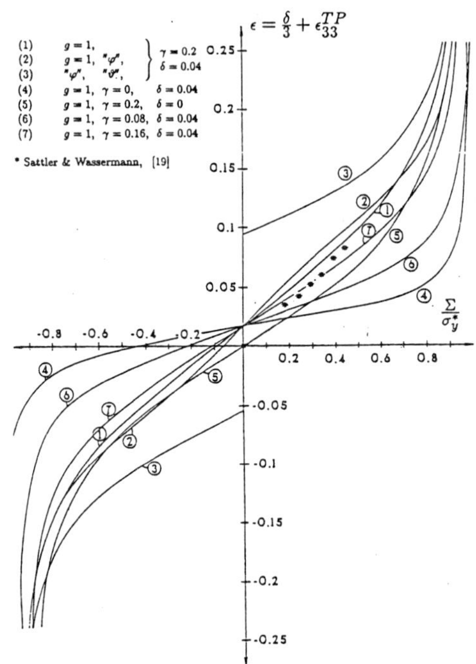

Fig. 4: Overall TRIP strain ε_{33}^{TP} plus δ/3 dependent on the normalized load stress Σ/σ_y^* for various γ, δ and restrictions.

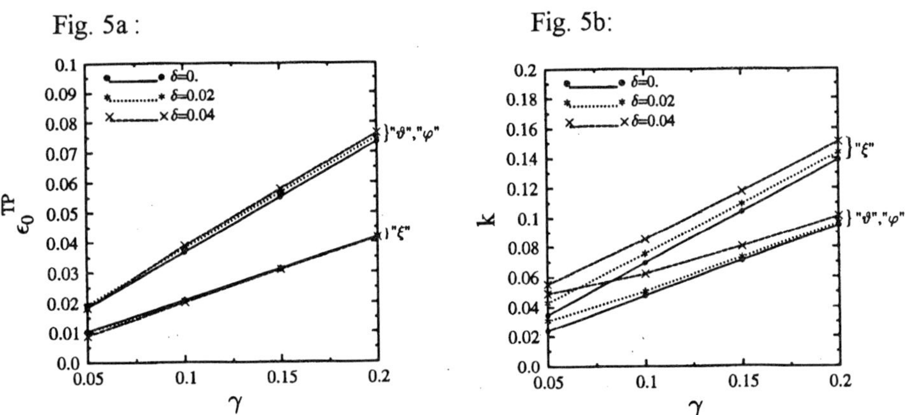

Fig. 5a: Stress independent part ε_0^{TP} for either the "φ -" and "ϑ-restriction" or the "ξ-restriction", dependent on γ (tension).

Fig. 5b: Inclination k of the stress dependent part k Σ/σ_y^* for either the "φ-" and "ϑ-restriction" or the "ξ-restriction", dependent on γ (tension).

Remark 1:

Relations of the type (2.7) only take into account the plastic accommodation of the transformation strain. They do not reflect any orientation process as discussed in detail in chpt.2.1. The author, therefore, recommends to represent the effect of transformation by an extended TRIP term consisting of an accommodation term (2.7) and an orientation term of the type (2.4.6). Of course, in addition to the TRIP term the transformation volume change $\frac{\delta}{3}\delta_{ij}$ must not be forgotten.

Remark 2:

Recently a study has been finished, [17], investigating the TRIP effect in an anisotropic material. The interest in this research stems from the experimental evidence that steel with a certain texture (e.g. forging texture) may change its shape during transformation without any external load, see e.g. [19]. The plastic anisotropy assumed is a so-called "transverse" isotropy with a yield stress R_l (R_{ln} in the product phase, R_{lo} in the parent phase) in the longitudinal (3) direction of the specimen and the yield stress R_q in the transverse 1-2 plane. The relevant yield condition is as follows:

$$\frac{3}{2}\, s_{ij}\left(s_{ij} + \frac{2}{3}\, a\, \tilde{s}_{ij}\right) = R_l^2 , \quad \tilde{s}_{ij} = \begin{bmatrix} s_{11}-s_{22} & 2s_{12} & \frac{1}{2}s_{13} \\ 2s_{12} & -(s_{11}-s_{22}) & \frac{1}{2}s_{23} \\ \frac{1}{2}s_{13} & \frac{1}{2}s_{23} & 0 \end{bmatrix} , \tag{2.8.1}$$

a is an anisotropy parameter, $a = (R_l/R_q)^2-1$, $-0{,}75 \leq a \leq 3{,}0$.
According to the normality rule the plastic strain rate works out as

$$\dot{\varepsilon}^P_{ij} = \dot{\lambda}\left(s_{ij} + \frac{2}{3}\, a\, \tilde{s}_{ij}\right) . \tag{2.8.2}$$

An analogous procedure is followed as outlined at the beginning of this chapter (from (2.5.1) to (2.6.6)). The denominator of (2.5.4) is now supplemented by a term multiplied by a. Finally a linearization procedure leads with $R_l^* = R_{ln}\left(\frac{1-R_{lo}/R_{ln}}{\ln\,(R_{ln}/R_{lo})}\right)$ for the uniaxial load case to the relation

$$\varepsilon^{TP} = \varepsilon^{TP}_{33} = \varepsilon^{TP}_{0,3} + k_3\Sigma/R_l^* , \tag{2.8.3}$$

similar to relation (2.6.7), however, for $g \equiv 1$. This means that a certain longitudinal strain $\varepsilon_{0,3}^{TP}$ does occur in the case of $\Sigma = 0$ without any activation of an orientation process. Numerical data for $\varepsilon_{0,3}^{TP}$ and k_3, including the case of the activation of the "φ-restriction", can be taken from [17].

Further a numerical study has been performed for a triaxial principle stress state Σ_{ij}, $\Sigma_{ij} = 0$ for $i \neq j$. It is interesting to note that a linearization leads to the following TRIP strain for $g \equiv 1$,

$$\varepsilon_{ij}^{TP} = \varepsilon_{0,ij}^{TP} + \frac{3}{2} k_3 S_{ij}/R_1{}^* + (k_1 - k_3) \tilde{S}_{ij}/R_1{}^* . \tag{2.8.4}$$

The tensor $\tilde{S}_{ij}$ possesses the same arrangement of elements as $\tilde{s}_{ij}$, see (2.8.1), e.g. $\tilde{S}_{11} = S_{11} - S_{22}$ etc.. $\varepsilon_{0,ij}^{TP}$ is a diagonal tensor with the components $\varepsilon_{0,1}^{TP}$, $\varepsilon_{0,2}^{TP}$, $\varepsilon_{0,3}^{TP}$. Data for $\varepsilon_{0,ij}^{TP}$, k_3 and k_1 can be taken from [17] for various parameters a. If $a = 0$, then $k_3 = k_1$ and $\varepsilon_{0,ij}^{TP} = 0$.

Again it is repeated that an irreversible strain may appear for a transformation under no external load. Fischer et al. [17] report e.g. on experiments with a low carbon (C < 0,03%), 11% Ni-steel, $M_s \sim 320°C$, $M_f \sim 20°C$ showing a 0,2% elongation strain $\varepsilon_{0,3}^{TP} = 0{,}002$ per transformation cycle.

Remark 3:

It must be emphasized again that the derivation of the TRIP strain ε_{ij}^{TP} has been performed under the assumption of a constant loadstress Σ_{ij} and a transformation driven by a cooling process. The TRIP-term (2.7), however, is used in practice in addition to the classic plasticity term in its rate form, for details see again the review report [20]. The main argument is that the TRIP strain rate can only exist with a change of f. However, recent experiments by Videau et al. [22] with steel specimens under tension and/or torsion of a very similar composition as that mentioned in Remark 2 have shown that a change of the load stress level (or even unloading) may also change the TRIP strain rate. E.g., if a specimen is loaded under tension or torsion to a loadstress level of 120 MPa, an unloading to a 60 MPa level will stop a further TRIP-strain. Even a complete unloading drives the TRIP-strain into the opposite direction! This encourages to introduce a backstress α_{ij} into a TRIP-strain rate formulation,

$$\dot{\varepsilon}_{ij}^{TP} = \frac{3}{2} K \frac{d}{df} \left(\Phi(f) \right) \dot{f} \, (S_{ij} - \alpha_{ij}) . \tag{2.9}$$

Of course, this formulation makes it necessary to implement an evolution law for α_{ij}, too. Such a proposal by the same group has been presented in [23]. An

American group implemented two backstress terms arguing that one belongs to "long range" stress and the other one to a "short range" stress effect, see Bammann et al. [24]. Further experimental research and theoretical justification seem to be absolutely necessary to establish such a formulation.

2.2.1.3 Diffusive transformation

In the case of a diffusive transformation the transformation strain tensor ε_{ij}^{T} is given in rate form as explained with respect to (1.8) and depends on $\dot{f}$, for repetition

$$\dot{\varepsilon}_{ij}^{T} = \dot{f}\,\frac{\delta}{3}\left(\frac{R_T}{r}\right)^3 (\delta_{ij} - 3a_{ij}),\quad 0 \le f \le \alpha_P : R_T = \tilde{R}/\sqrt[3]{\alpha_P},\quad \alpha_P \le f \le 1 : R_T = R_e. \quad (2.10.1)$$

α_P describes the arrangement of the growing spheres in an average sense, $\alpha_P \sim$ ~ 0.632. With respect to the other variables, see chpt.1.1., (2.10.1) is now inserted into (2.8.4). Since no orientation effect has to be considered, the volume average (2.5.6) is calculated with $g \equiv 1$. Details on the numerical procedure as well as results can be taken from [4], where the uniaxial load case was treated. In the case of linearization the following relation can be derived analytically for ε_{33}^{TP},

$$\varepsilon_{33}^{TP} = \frac{5\delta}{6}\,\psi(f)\,\Sigma/\sigma_{y_o}, \quad (2.10.2)$$

$$0 \le \xi \le \alpha_P : \quad \psi(f) = \ln\left(1 + \frac{f^2}{\alpha_P}\right),$$

$$\alpha_P \le \xi \le 1 : \quad \psi(f) = \frac{\sigma_{y_o}}{\sigma_{y_n}}\left[\frac{\sigma_{y_o}}{\sigma_{y_n}}\,\frac{2(f-\alpha_P)}{2-\alpha_P} + \ln 4\right].$$

For $\alpha_P = 0.632$ and $f = 1.0$ it follows

$$\varepsilon_{33}^{TP} = k\,\Sigma/\sigma_{y_o}, \quad k = \frac{5\delta}{6}\left(1 + 0.439\,\sigma_{y_o}/\sigma_{y_n}\right). \quad (2.10.3)$$

Relation (2.10.3) agrees surprisingly well with the "Greenwood and Johnson relation" in the case of small $\sigma_{y_o}/\sigma_{y_n}$. The exact, although somewhat lengthy solution of the problem gives slightly different results for tension and compression. The same generalization with respect to a triaxial loadstress state as (2.7) is now used in the literature. However, the function $\Psi(f)$ in (2.7) can directly be calculated instead of being heuristically assumed.

2.2.2 An enhanced numerical model

2.2.2.1 Some general aspects

Since a transformation process is described within continuum mechanics by an incompatibility, namely the transformation tensor, and a thermodynamic force driving the transformation process, the field equations must be solved to present a compatible strain state ε_{ij} in relation to a load stress Σ_{ij} and a (homogenous) temperature T. Various approaches exist to derive such a solution ε_{ij}. In the case of a purely elastic material such as a shape memory alloy, the Green's function method in combination with Eshelby's solution of an inclusion problem can be an appropriate tool, see e.g. the contribution by Patoor and Berveiller in this book. The problem becomes much more complicated if plastic accommodation must be considered, too, at least in the parent phase. In chpt.2.2.1 the most simple concept has been proposed assuming an overall homogenous strain state (Taylor-Lin assumption). In the following context we apply now the finite element method in an incremental and iterative concept to simulate both the "Creep-Test" (Σ_{ij} is kept constant and the transformation is driven by cooling/heating) and the "Tension-Test". The finite element method allows to follow the elasto-plastic behaviour in each integration point checking from increment to increment if unloading may happen or not. With respect to this aspect the finite element method differs from a secant modulus concept with a quasi-linear description of the material. G. Weng and collaborator recently applied this secant modulus concept in [25] to model the tension test for a TRIP-steel. In this case the martensitic transformation is driven by straining in the plastic range only. Further, they do not take into account any orientation effect during their incremental procedure. The author can show, however, that the orientation effect must not be neglected in the case of martensitic transformation.

2.2.2.2 Displacive (martensitic) transformation

The author would like to mention the results of two working groups which proceeded in a similar manner although they worked independently. The first group is the one guided by Prof. Sjöström in Sweden. Their main results are published in the thesis by Simonsson [26].

The second group is the Leoben group. The author refers here to a set of three papers. The first paper [27] studies in detail the growth of a martensitic plate (lense) as a first variant until the distribution of the mechanical driving force urges the nucleation and growth of a second variant. On the one hand, the interface condition applying the energy momentum tensor (1.15) together with (1.16) is used to check the growth of the variant. On the other hand, the transformation condition (in a somewhat simplified form compared with (1.28)) is applied to the nucleation of a further variant. Although, in the opinion of the author, the concept simulates the actual physical process, it cannot be followed to "fill" a volume element by martensitic variants since the calculation efforts are too high, even with today's high speed workstations. Therefore, Marketz et al. moved to an averaging concept, reported in [28] and [29]. The following assumptions and definitions are introduced:

* A "non-isothermal Creep-Test" is simulated as performed by Gautier et al. [7], [11] and in the contribution by Gautier in this book. During these tests the uniaxial loadstress $\Sigma_{33} = \Sigma$ is kept constant, and the transformation is driven by cooling.
* A Fe-31% Ni-steel has been tested. All data with respect to the transformation process are explained in chpt.2.1. The following data are assumed with respect to the deformation behaviour:
  ** Elastic behaviour for both phases γ and α':
  As elastic compliances S_{11}, S_{12}, S_{44}, those for α-Fe are taken,
  $S_{11}=7.57\cdot10^{-6}[1/\text{MPa}]$, $S_{12}=-2.82\cdot10^{-6}[1/\text{MPa}]$,
  $S_{44}=8.62\cdot10^{-6}[1/\text{MPa}]$.
  ** Plastic behavior:
  γ: $\sigma_{y,\gamma} = 250$ MPa, $E_{p,\gamma} = 2000$ MPa,
  α': $\sigma_{y,\alpha} = 800$ MPa, $E_{p,\alpha} = 8000$ MPa .
  Isotropic linear hardening with a tangent modulus E_p is, therefore, taken into account. Both σ_y and E_p are assumed to vary with the process parameter ξ linearly,
  $$\sigma_y(\xi) = \sigma_{y,\gamma} + \xi(\sigma_{y,\alpha} - \sigma_{y,\gamma}), \quad E_p(\xi) = E_{p,\gamma} + \xi(E_{p,\alpha} - E_{p,\gamma}) .$$
* As mentioned above, the process is driven by cooling. The increment Δf of the newly formed product phase is calculated to the increment ΔT as outlined in chpt.2.1 (2.3.3). The variant selection procedure corresponds fully with the one outlined in chpt.2.1 starting with (2.3.3). Only the thermodynamic driving

force is repeated here as

$$^{k}F^{n} = (\Sigma_{ij} + T_{ij})^{k}\varepsilon_{ij}^{n} + \rho\left[\varphi_{ch}(T - \xi\Delta T)\right] - \frac{T-\xi\Delta T}{M_s}[\varphi_{ch}]|_{M_s},$$

ξ is set to 1/2. It should be noted that the mechanical barrier is neglected.
In brief, the algorithm can be described as follows:
Let us assume that we know all field entities as $\sigma_{ij}^{*} = \Sigma_{ij} + T_{ij}^{*}$, ε_{ij}^{*} and the various transformed microregions represented by f^{*} at a certain temperature T^{*}. Now we lower the temperature to $T^{*}-\Delta T$, at the same time changing the volume fraction f^{*} to $f^{*}+\Delta f$. During this increase various martensite variants develop, driven by the thermodynamic driving force $^{k}F^{n}$ depending on Σ_{ij}^{*}, T_{ij}^{*}, T^{*} according to the selection procedure mentioned in relation with (2.4.3-5).
The newly developed variants serve to update the current fields, that means T_{ij}^{*} is updated to $T_{ij}^{*}-\Delta T_{ij}^{*}$ etc. After this updating has been finished, the algorithm starts with the next step as described at the state σ_{ij}^{*}, T^{*}.

* A unit cell with 192 equisized grains in the 1-3 plane is assumed along with generalized plain strain conditions. Each grain itself consists of 6 triangular elements with a quadratic displacement field. To each grain a crystal lattice of the fcc-parent phase is attached by means of a random (uniform) distribution. The selection of the variants is performed with respect to this lattice. The load Σ in the 3-direction remains constant. Appropriate antisymmetric boundary conditions are attached to the unit cell that allow a periodic repeating of the unit cell preventing gaps and overlapping as well as assuming a "minimum of symmetry" (-for details see chpt.4.3 of [27]). As a first step, the local stress state in each grain is calculated to the external stress state Σ. Since elastic anisotropy is considered, a remarkable fluctuation of the stress state σ_{ij} can be observed. Thus the stress component σ_{33} corresponding to the load stress $\Sigma_{33} = \Sigma$ varies from 60% to 150% of Σ. All finite element calculations are performed by the ABAQUS program [30]. The author acknowledges here the provision of ABAQUS within an academic license agreement.
The most important results of this simulation can be described in the following two figures:

Figure 6 depicts ε_{33}^{TP} for various levels of the loadstress $\Sigma_{33} = \Sigma$ as a function of the martensite volume fraction f. The results agree surprisingly well with the experimental data which are marked by o in the diagram.
The orientation effect is represented by ε_{33}^{T}. Its contribution to ε_{33}^{TP} in per cent is depicted in Figure 7. As also reported by Gautier et al. in [11], the orientation effect is the dominating mechanism for small fractions f<0.2. For higher f values the orientation effect is nearly constant. The accommodation effect increases with increasing level of the applied stress but remains smaller than the orientation effect, which does not even vanish if a transformation-induced microstress state becomes more and more dominant. The origin of a continuing orientation effect of the microstress state lies in its deviator. Arguments from the past that only a hydrostatic stress state would develop during transformation are not correct. If the loadstress Σ reaches the yield stress of the austenitic parent phase, plastic deformations occur even at very low f-values. The accommodation effect decreases with increasing f since the yield stress of the two-phase system increases with increasing f.

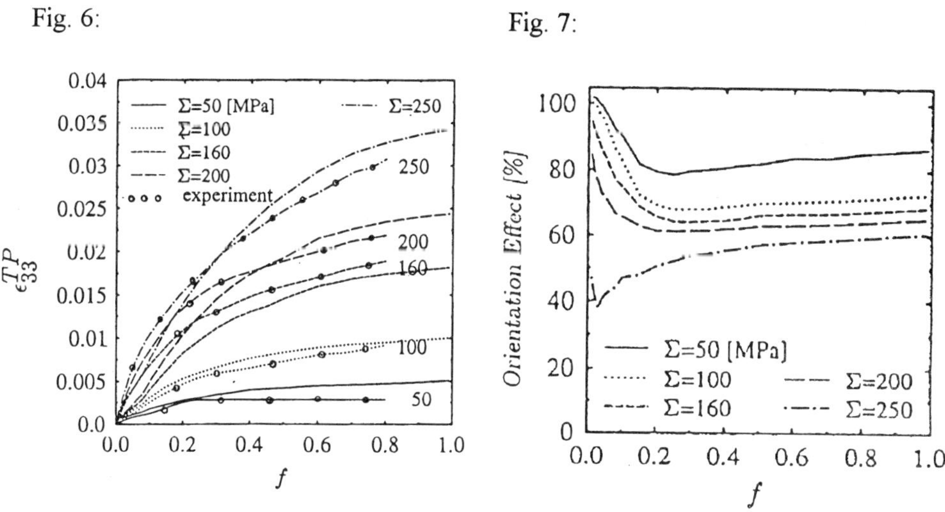

Fig. 6: TRIP strain ε_{33}^{TP} in dependence of martensite volume fraction f for different levels of uniaxial load tensile stress Σ, ∘∘∘ experimental data by Gautier et al., [11].

Fig. 7: Orientation effect contributing to the TRIP strain in dependence of martensite volume fraction f for different levels of uniaxial load tensile stress Σ.

Further interesting information can be gathered concerning the development of each of the 24 variants during the transformation history. As known from the crystallographic theory of martensitic transformation, the family of habit planes $(3\ 14\ 10)_{fcc}$ can be divided into 6 groups of habit planes with 4 variants, each with (nearly) the same habit plane normal. Of course, within the habit plane the shear direction may vary. Let us consider the following 4 variants as given in the table below:

Variant Index	Habit Plane normal indices	Components of habit plane normal in the 1,2,3-space			Schmid factor of the transformation shear
n	(. . .)	n_1^n,	n_2^n,	n_3^n	m^n
2	$(\bar{3}\ 14\ \overline{10})$	-0.185	0.782	-0.595	0.22
5	$(3\ 14\ \overline{10})$	0.185	0.782	-0.595	0.33
18	$(3\ 10\ \overline{14})$	0.185	0.595	-0.782	0.10
24	$(\bar{3}\ 10\ \overline{14})$	-0.185	0.595	-0.782	0.12

The Schmid factor m^n for the transformation shear γ is defined as $m^n = \cos\vartheta\ \cos\lambda_n$ with λ_n being the angle between the transformation shear and the 3-axis (load axis).

The angles between the normal vectors can be read from the following matrix:

	n^2	n^5	n^{18}	n^{24}
n^2	0	21,4°	26,3°	15,3°
n^5		0	15,3°	26,3°
n^{18}			0	21,4°
n^{24}				0

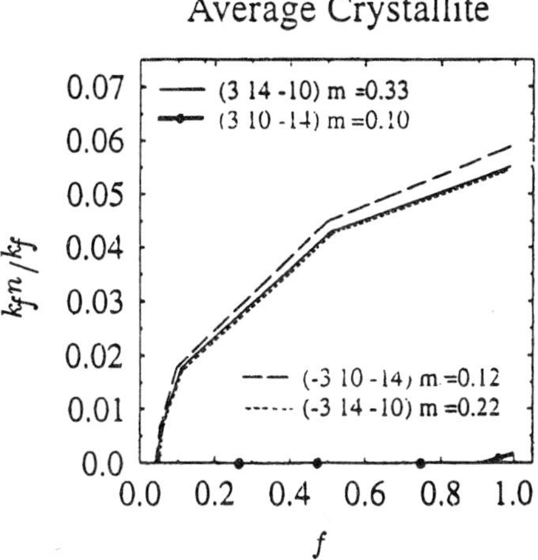

Fig. 8: Relative microfraction of the variants of a group.

Fig. 8 depicts the relative microfraction ${}^{k}f^{n}/{}^{k}f$ of the variants n = 2, 5, 18, 24 with the increasing martensite volume fraction f. It is interesting to note that the variants 5, 18 and 24 start at the same time and show nearly the same volume fraction. Variant 2, however, appears only for a high value of f ~ 0.9 with almost no volume fraction. If one calculates the average transformation strain in the load direction as

$$\sum_n {}^{k}f^{n}\, \varepsilon_{33}^{n} / \sum_n {}^{k}f^{n}\,, \qquad n=2,5,18,24\,,$$

one obtains a value of 0.048. A comparison with the maximum possible transformation strain ε_{+}^{T}, see (2.1.5), $\varepsilon_{+}^{T}=0.139$, leads to the conclusion that a significant amount of self-accommodation has occurred, reducing the influence of the transformation shear on the elongation of the specimen.

2.3 Kinetics

This chapter proposes a derivation of martensite kinetics, expressed as f being a function of Σ_{ij} and T. Since the martensitic transformation is a displacive one with the atoms only moving over very short distances, the time must not be included as an explicit parameter. However, in the case of a diffusive transformation the time plays an explicit role; here the reader is referred to standard textbooks on physical metallurgy.

The author distinguishes between two approaches to achieve martensite kinetics:

* The phenomenological approach giving an equation of the following type,

$$f = 1 - \exp\left[-\alpha(M_s - T)\right], \ \alpha > 0, \text{ const.}, \tag{2.11}$$

introduced by Koistinen and Marburger [31] in 1959. Also polynomial relations like available as that by Besserdich [32]. Relation (2.11) have been extended by Inoue and Tanaka, see e.g. [33], by an additional phenomenological stress term.

* The thermodynamically based approach.
In this case thermodynamic forces are introduced to achieve a certain type of martensite kinetics with the overall stress Σ_{ij} included. For a review of the relevant literature, see e.g. [33], [34]. We distinguish here between a statistical approach as followed by Olson and collaborators, see e.g. [35] and a deterministic approach as suggested by Magee [15] in 1959. He argued that the number of new plates nucleated per unit volume of the parent phase, dN, is proportional to the increase of the thermodynamic force in the parent phase. Considering the nucleation condition (1.27) for a microregion situated around the location r_i (the label $|_{r_i}$ is avoided in the following text) for a stress free state, we define as the thermodynamic force F the difference between the chemical driving force and the transformation barrier,

$$F = \rho[\varphi_{ch}] - F_c^T, \quad F_c^T = \frac{\overline{T}}{M_s}\rho\,[\varphi_{ch}]|_{M_s}\,. \tag{2.12.1}$$

F can also be seen as a global entity since Magee assumed a homogeneous temperature field.
We now calculate dF as

$$dF = \rho[d\,\varphi_{ch}] - dF_c^T \tag{2.13}$$

and assume furthermore that the critical force increment dF_c^T can be ignored $\left(\frac{\overline{T}}{M_s} \text{ changes only slightly}\right)$.
Magee's proposal now yields

$$dN = k\,dF = k\,\rho\,[d\,\varphi_{ch}] = k\,\rho\left|\frac{\partial\varphi_{ch}}{\partial T}\right| dT\,, \quad k \text{ .... constant}\,. \tag{2.14}$$

If V_μ is the average volume of a microregion, the total volume increment of the product phase, Vdf, follows as

$$Vdf = \big(dN\ V(1\text{-}f)\big)V_\mu\ , \qquad dN = \frac{df}{V_\mu(1\text{-}f)}\ . \tag{2.15}$$

Inserting (2.15) into (2.14) yields

$$\frac{df}{1\text{-}f} = V_\mu\ k\ \rho \left[\frac{\partial \varphi_{ch}}{\partial T}\right] dT\ , \tag{2.16}$$

and with $f = 0$ at $T = M_s$,

$$f = 1 - \exp\left\{-V_\mu\ k\rho\ \left[\varphi_{ch}(T) - \varphi_{ch}\ (M_s)\right]\right\}, \tag{2.17}$$

as stated also in (2.3.2).

If $\rho[\varphi_{ch}]$ is a linear function of T, e.g. the linear part of (2.3.1), $\rho[\varphi_{ch}] =$ $= a(T - T_0)$, it follows that

$$f = 1 - \exp\left[V_\mu\ k\rho a(M_s - T)\right],\ a < 0\ . \tag{2.18}$$

In this case the constant α in (2.11) corresponds to $V_\mu k\rho a$ in (2.18).
It seems now to be quite natural to introduce a stress term, too, in $dN = kdF$, however, with F according to (1.28) as

$$F|_{r_i} = \rho[\varphi_{ch}] + (\Sigma_{ij} + T_{ij})\ \varepsilon_{ij}^{n} - F_c^T\ . \tag{2.12.2}$$

Notice that now an upper label "n" is set to ε_{ij} instead of "T" to refer to a specific variant n in a microregion located around r_i.
For the sake of simplicity, let us skip T_{ij}. One must keep in mind that $\varepsilon_{ij}^{n}|_{r_i}$ and, therefore, $F|_{r_i}$, are local entities. To get to a global entity $\bar{F}$ we assume (- only as an imaginative concept -) that all transformation tensors are defined a priori and that they are activated if $F|_{r_i} \geq 0$. If we now apply an external loadstep, $d\Sigma_{ij}$, we increase $F|_{r_i}$ by $d\Sigma_{ij}\ \varepsilon_{ij}^{n}|_{r_i}$. Different approaches may now be followed for reaching a global measure $d\bar{F}$.
We suggest as $d\bar{F}$

$$d\bar{F} = d\Sigma_{ij}\tilde{\varepsilon}_{ij}^{T} \tag{2.19}$$

with $\tilde{\varepsilon}_{ij}^{T}$ being estimated by $\frac{\Delta\varepsilon_{ij}^{T}}{\Delta f}$, see Remark 1.
Now according to (2.12-16) and (2.19), the following differential equation can be written as an extended statement of Magee's proposal,

$$\frac{df}{1-f} = V_\mu\, k \left\{ \rho \left[\frac{\partial \varphi_{ch}}{\partial T} \right] dT + \tilde{\varepsilon}^T_{ij} d\Sigma_{ij} \right\} . \tag{2.20}$$

The integration of this differential equation must be done numerically, for detailed results see [36]. However, if $\tilde{\varepsilon}^T_{ij}$ does not depend on f, an integral of (2.20) can be given as

$$f = 1\text{-}\exp \left\{ -V_\mu k \rho \, [\varphi_{ch}(T) - \varphi_{ch}(M_s)] - V_\mu \, k \, \tilde{\varepsilon}^T_{ij} \, \Sigma_{ij} \right\} . \tag{2.21}$$

This can be considered as the simplest form of a thermodynamically based extension of the Koistinen and Marburger relation (2.11) by a stress term.
Finally it should be mentioned that a combined integration of a constitutive law, relating the load stress Σ_{ij}, the temperature T and the transformed volume fraction f, and a kinetic law of the type (2.20), could describe very well in [37] the thermomechanical behavior of a shape memory alloy under various load stress-temperature paths.

Remark 1:

Let us assume that a load increment $\Delta\Sigma_{ij}$ gives rise to the development of m microregions V^l_μ, $l=1, \ldots m$, with the transformation tensors ε^l_{ij}. The average of these newly developed tensors is

$$\sum_{l=1}^{m} \varepsilon^l_{ij} \, V^l_\mu / \sum_{l=1}^{m} V^l_\mu \, .$$

Further, Δf is equivalent to $\sum_{l=1}^{m} V^l_\mu / V$, and the increment $\Delta\varepsilon^T_{ij}$ of ε^T_{ij} is $\sum_{l=1}^{m} \varepsilon^l_{ij} \, V^l_\mu / V$.

The resulting average incremental driving force is

$$\Delta\Sigma_{ij} \left\{ \sum_{l=1}^{m} \varepsilon^l_{ij} \, V^l_\mu / \sum_{l=1}^{m} V^l_\mu \right\} = \Delta\Sigma_{ij} \frac{\Delta\varepsilon^T_{ij}}{\Delta f} \, .$$

The ratio $\frac{\Delta\varepsilon^T_{ij}}{\Delta f}$ is denominated as $\tilde{\varepsilon}^T_{ij}$.

Remark 2:

Specifically in the case of shape memory alloys, a transformation tensor ε^T_{ij} of the type

$$\varepsilon_{ij}^{T}|_{\xi} = f^{c}\,\varepsilon_{ij}^{T}|_{1}\ ,\ c < 1\ ,$$

can often be observed. In this case $\tilde{\varepsilon}_{ij}^{T}$ can be estimated as

$$\tilde{\varepsilon}_{ij}^{T} \sim \frac{c}{f^{1-c}}\,\varepsilon_{ij}^{T}|_{1} = \frac{c}{f}\,\varepsilon_{ij}^{T}|_{\xi}\ .$$

Remark 3: Tanaka et al were able to prove in [34] that the increment df must, indeed, be of the structure

$$df = \tilde{\beta}\,dT + \tilde{\tilde{\beta}}_{ij}\,\Sigma_{ij}\ , \qquad \tilde{\beta} = \tilde{\beta}(\xi,T)\ , \qquad \tilde{\tilde{\beta}}_{ij} = \tilde{\tilde{\beta}}(\xi,T)\ .$$

The proof starts by defining the transformation state of a microregion (described by the set of Eulerian angles $\omega=(\vartheta,\varphi,\psi)$) subjected to a local thermomechanical load (σ_{ij},T) with a micro-fraction $\tilde{f}$ defined as

$$\tilde{f} = \tilde{f}(\sigma_{ij},T,\omega) = 1 \text{ for } F \geq 0\ , \quad \tilde{f} = 0 \text{ for } F < 0\ ,$$

for F see (2.12.2).

The total volume fraction f is now given as

$$f = \int_{\omega} g(\omega)\,\tilde{f}\,d\omega\ , \quad 0 \leq f \leq 1.$$

$g(\omega)$ represents the normalized distribution density function of the microregions. The differential df can be calculated as

$$df = \int_{\omega} g(\omega)\,d\tilde{f}\,d\omega\ .$$

Proper differentiation of the above jump condition for $\tilde{f}$ finally leads to df as written at the beginning of this remark. Both $\tilde{\beta}$ and $\tilde{\tilde{\beta}}_{ij}$ are presented in [34] and evaluated for different types of $\varepsilon_{ij}^{n}|_{r_i}$. Kinetic equations of the discussed type can be verified by somewhat simplified but still reasonable assumptions.

ACKNOWLEDGEMENT

First and foremost, the author is indebted to his former PhD student Dr. F. Marketz, now Shell International Exploration and Product BV, Research and Technology Services, Netherlands, both for the extensive numerical studies published in context with chpts 2.1 and 2.2.2 and the careful reading of the manuscript as well as for the numerous hints concerning the formulations.

The author expresses his thanks to Prof. V.I. Levitas, formerly Ukraine, now University of Hannover, Germany, for his helpful discussions and formulations, specifically with respect to chpts 1.2.2, 1.2.3, 1.2.4.

Further the author appreciates the long-standing cooperation with Prof. K. Tanaka, Tokyo, Japan, who significantly contributed to chpt.2.3.

Thanks are also due to Doz. E.R. Oberaigner for working out the integrals following relation (2.6.4).

Finally the author is very grateful to Prof. L. Truskinovsky, University of Minnesota, USA, for his comments, specifically with respect to the physics of martensitic transformation.

REFERENCES

1. Wechsler, M.S., Lieberman, D.S. and T.A. Read: On the theory of the formation of martensite, AIME Trans. J. Metals, 197 (1953), 1503-1515.
2. Fischer, F.D.: Transformation induced plasticity in triaxially loaded steel specimens subjected to a martensite transformation, Eur. J. Mech. A/Solids, 11 (1992), 233-244.
3. Fischer, F.D.: A micromechanical model for transformation plasticity in steels, Acta metall. mater., 38 (1990), 1535-1546.
4. Ball, J.M. and R.D. James: Fine phase mixtures and minimizers of energy, Arch. Rat. Mech. Anal., 100 (1987), 13-52.
5. Nishiyama, Z.: Martensitic Transformation, Academic Press, New York et al. 1978.
6. Eshelby, J.D.: Energy relations on the energy-momentum tensor in continuum mechanics, in: Inelastic Behavior of Solids (Eds M.F. Kanninen, W.F. Adler, A.R. Rosenfield and R.I. Joffee), McGraw-Hill, New York 1970, 77-115.

7. Chadwick, P.: Continuum Mechanics, George Allen & Unwin Ltd., London 1976.
8. Gautier, E. and A. Simon: Role of internal stress state on transformation induced plasticity and transformation mechanisms during the progress of stress induced phase transformation, in: International Conference on Residual Stresses - ICRS2 (Eds G. Beck, S. Denis and A. Simon), Elsevier Applied Science, London and New York 1989, 777-783.
9. Fischer, F.D., Berveiller, M., Tanaka, K. and E.R. Oberaigner: Continuum mechanical aspects of phase transformations in solids, Arch. Appl. Mech., 64 (1994), 54-85.
10. Levitas, V.I.: Thermomechanics of martensitic phase transitions in elastoplastic materials, Mech. Res. Commun., 22 (1995), 87-94.
11. Gautier, E. and A. Simon: Transformation plasticity mechanisms for martensitic transformation of ferrous alloys, in: Phase Transformations '87 (Ed. G.W. Lorimer), Institute of Metals, London 1988, 285-287.
12. Patel, J.R. and M. Cohen: Criterion for the action of applied stress in the martensitic transformation, Acta metall., 1 (1953), 531-538.
13. Marketz, F. and F.D. Fischer: A mesoscale study on the thermodynamic effect of stress on martensitic transformation, Met. Trans., 26A (1995), 267-278.
14. Kaufman, L. and M. Cohen: Thermodynamics and kinetics of martensitic transformations, in: Progress in Metal Physics (Eds B. Chalmers and R. King), Pergamon Press, London et al. 1958, 165-246.
15. Magee, C.L.: The nucleation of martensite, in: Phase Transformations (Ed. H.I. Aaronson), ASM, Metals Park 1969, 115-156.
16. Tanaka, K.: Analysis of recovery stress and cyclic deformation in shape memory alloys, in: Advances in Continuum Mechanics (Eds O. Brüller, V. Mannl and J. Najar), Springer-Verlag, Berlin et al. 1991, 441-451.
17. Fischer, F.D. and S.M. Schlögl: The influence of material anisotropy on transformation induced plasticity in steel subject to martensitic transformation, Mech. Mat., 21 (1995), 1-23.
18. Greenwood, G.W. and R.H. Johnson: The deformation of metals under small stress during phase transformation, Proc. Royal Soc. London, A 283 (1965), 403-422.

19. Sattler, H.P. and G. Wassermann: Transformation plasticity during martensitic transformation of iron with 30% Ni, J. Less-Common Met., 28 (1972), 119-140.
20. Fischer, F.D., Sun, Q.-P. and K. Tanaka: Transformation-induced Plasticity (TRIP), Appl. Mech. Rev., 49 (1996), 317-364.
21. Franitza, S.: Zur Berechnung der Wärme- und Umwandlungsspannungen in langen Kreiszylindern, PhD Thesis, Techn. Univ. Braunschweig 1972.
22. Videau, J.-Chr., Cailletaud, G. and A. Pineau: Experimental study of the transformation-induced plasticity in a Cr-Ni-Mo-Al-Ti steel, to be published Proc. MECAMAT 95, J. de Physique IV, Colloque C1, supplément au J. de Physique III, 6 (1996), C1-465 - C1-474.
23. Videau, J.-Chr., Cailletaud, G. and A. Pineau: Modélisation des effets mécaniques des transformations de phases pour le calcul des structures, J. de Physique IV, Colloque C3, Supplément au J. de Physique III, 4 (1994), C3-227 - C3-232.
24. Bammann, D.J., Prantil, V.C. and J.F. Lathrop: A model of phase transformation plasticity, in: Modelling of Casting, Welding and Advanced Solidification Processes (Eds M. Cross and J. Campbell), TMS, Warrendale 1995, 275-285.
25. Bhattacharyya, A. and G.J. Weng: An energy criterion for the stress-induced martensitic transformation in a ductile system, J. Mech. Phys. Solids, 42 (1994), 1699-1724.
26. Simonsson, K.: Micro-mechanical FE simulations of the plastic behavior of steels undergoing martensitic transformation, PhD Thesis, Linköping Studies in Science and Technology, Dissertations, No. 362, Linköping 1994.
27. Marketz, F. and F.D. Fischer: Micromechanical modelling of stress-assisted martensitic transformation, Modelling Simul. Mater. Sci. Eng., 2 (1994), 1017-1046.
28. Marketz, F. and F.D. Fischer: A micromechanical study of the deformation behavior of Fe-Ni alloys under martensitic transformation, in: Solid→Solid Phase Transformations (Eds W.C. Johnson, J.M. Howe, D.E. Laughlin and W.A. Soffa), TMS, Warrendale 1994, 785-790.
29. Marketz, F. and F.D. Fischer: Micromechanics of transformation-induced plasticity and variant coalescence, J. de Physique IV, Colloque C1,

supplément au J. de Physique III, 6 (1996), C1-445 - C1-454.

30. Hibbitt, Karlsson & Sorensen Inc., ABAQUS User Manual, Version 5.4, Pawtucket, R.I. 1994.

31. Koistinen, D.P. and R.E. Marburger: A general equation prescribing the extent of the austenite-martensite transformation in pure iron-carbon alloys and plain carbon steels, Acta metall., 7 (1959), 59-60.

32. Besserdich, G., Scholtes, B., Müller, H. and E. Macherauch: Consequences of transformation plasticity on the development of residual stresses and distorsions during martensitic hardening of SAE 4140 steel cylinders, steel research, 65 (1994), 41-46.

33. Tanaka, K.: Analysis of transformation superplasticity and shape memory effect, in: Computational Plasticity (Eds T. Inoue, H. Kitigawa and S. Shima), The Society of Mat. Sci., Japan, Current Japanese Materials Research, Vol. 7, Elsevier Applied Science, London 1991, 43-60.

34. Tanaka, K., Oberaigner, E.R. and F.D. Fischer: Kinetics on the micro- and macro-levels in polycrystalline alloy materials during martensitic transformation, Acta Mech., 116 (1996), 171-186.

35. Olson, G.B., Tsuzaki, K. and M. Cohen: Statistical aspects of martensitic transformation, Proc. Mat. Res. Soc. Symp., 57 (1987), 129-147.

36. Oberaigner, E.R., Fischer, F.D. and K. Tanaka: A new micromechanical formulation of martensite kinetics driven by temperature and/or stress, Arch. Appl. Mech., 63 (1993), 522-533.

37. Oberaigner, E.R., Tanaka, K. and F.D. Fischer: The influence of transformation kinetics on stress-strain relations of shape memory alloys in thermomechanical processes, J. Intell. Mat. Sys. Struct., 5 (1994), 474-486.

THEORY OF PRESSURE SOLUTION CREEP IN WET COMPACTING SEDIMENTS

F.K. Lehner
University of Bonn, Bonn, Germany

Many rocks and especially sedimentary rocks contain water as a pore fluid. It has long been recognized that the mineral grains forming the rock matrix tend to dissolve in water preferentially along highly stressed, fluid-permeated grain–to–grain contacts. The dissolved material is then transported by molecular diffusion along wet grain boundaries and pores to low-energy precipitation sites. Since this 'pressure solution' process changes the rock fabric, it can lead to the accumulation of substantial macroscopic creep strain. Geologists specializing in materials science are engaged in formulating and testing appropriate macroscopic 'pressure solution creep laws', starting from a pore-scale description of stress-enhanced solution-precipitation and diffusive mass transfer processes. In this manner, they are seeking to constrain the rate and extent of important geological processes, e.g. the loss of porosity of sedimentary rocks in the course of their burial.

This chapter sketches a line of theoretical research, rooted in the thermodynamic theory of irreversible processes, that can provide a framework for the discussion of creep phenomena caused by changes of phase and of 'intergranular pressure solution' (IPS) in particular. The analysis of this phenomenon confronts the student with a fundamental question in thermodynamic theory, i.e., that of a generalized chemical potential suitable for nonhydrostatic systems. Here, the reader is directed to recent work, in which tensorial chemical potentials are introduced as the appropriate concept for discussing first-order phase transformations under nonhydrostatic stress and pressure solution phenomena in particular.

The 'creep law' obtained in this chapter describes the response of an elementary volume of a porous, fluid-permeated rock which, at the grain scale, is modeled by a simple cubic packing of spheres. On a macroscale, such a volume element behaves as an open sytem. Creep rates therefore depend, not only on the kinetics of grain boundary solution and diffusion processes, but also on coupled transport and deposition processes, thus illustrating the complexity of chemomechanical processes in geology.

1. Introduction

Anyone with an interest in nature's workings, on travelling through one of the Earth's mountain ranges, will on occasion have marvelled at the striking evidence of permanent deformation exposed by seemingly solid rocks on all scales—from tiny wrinkles on a hand specimen or measurably deformed fossiles to grandious folded structures the size of entire mountains. Geologists have been engaged in a continuing effort, rich in controversy, to explain such observations in terms of certain dynamical process that affect the crust of the Earth, but also in terms of requisite microscale deformation mechanisms that will enable rocks to accumulate large permanent strains. They have recognized that these aspects tend to represent but 'opposite sides of one medal'; this also meant that the hypothesis of continental drift, which in its modern form of "plate tectonics" forms the central unifying concept of present-day geology, could be opposed effectively until the early sixties by a geophysical orthodoxy, who declared solid state creep of mantle rocks—and hence continental drift—an impossibility.

Today, geologists distinguish at least three groups of processes by which rocks deform permanently, namely (1) cataclasis, involving fracture and frictional sliding between particles, (2) intracrystalline plasticity through dislocation motion, and (3) creep caused by diffusive mass transfer through grains and along grain boundaries and by phase transformations, including solution and reprecipitation processes. By comparing microtextures in naturally deformed rocks with those obtained under well-controlled experimental conditions, dominant deformation mechanisms can be associated with regions in pressure-temperature-strain rate space. The understanding of rock deformation processes gained in this manner from extensive research over the last decades is gradually enabling earth scientists to come to grips with quantitative questions about our planet that may be of fundamental as well as practical interest to men. Among the latter we mention the prediction and mitigation of natural hazards such as earthquakes or volcano eruptions, the prediction of land subsidence induced by underground excavations or fluid extraction, and the design of safe underground storage facilities for hazardous waste products.

In the following, we wish to discuss a particular deformation mechanism that is most often termed 'pressure solution' by geologists (Rutter, 1983), but is also known as 'solution-precipitation' or 'solution transfer creep'. It belongs into the third of the above groups of processes and provides an important example of creep deformation caused by a change of phase. We shall first describe the most familiar manifestations of pressure solution in natural rocks and thereafter sketch the theoretical basis and derivation of a typical pressure solution 'creep law' as may be used to constrain deformation rates in certain geological materials.

2. Pressure solution phenomena in rocks

That rocks can deform by a chemomechanical process of stress-enhanced dissolution in an aqueous pore fluid had been observed by geologists as early as in 1863, when Sorby first explained the phenomenon of 'pitted pebbles' (cf. Fig. 1) in terms of a selective removal of material from the more soluble of two impinging pebbles (references to early work on pressure solution and related theories may be found in a review article by Durney (1978). Sorby also hypothesized that the (limestone) material he studied dissolved preferentially "where the pressure is greatest, and crystallized, where it is least".

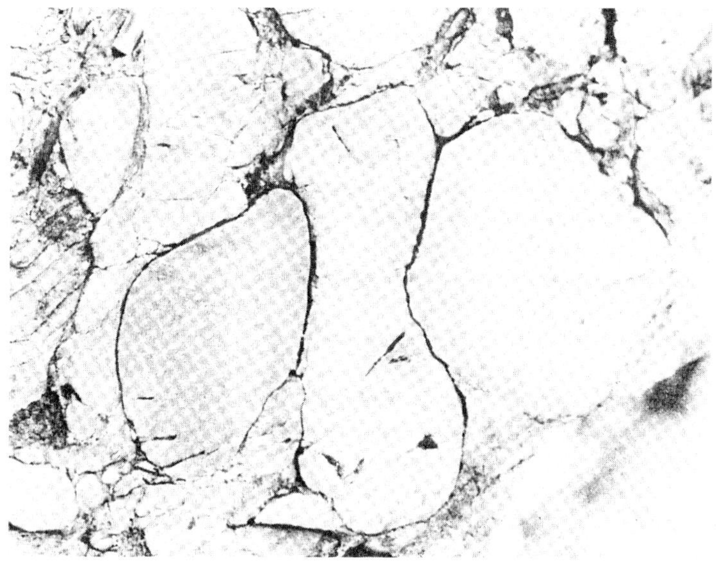

Figure 1: Pitted quartzite pebbles, as observed on an exposed joint face that cuts through a quartzite conglomerate (taken from Mosher, 1976). Pencil end in upper left corner for scale.

A characteristic feature of pressure solution are indeed the microscopic (i.e. grain-scale) as well as macroscopic solution seams or sutures which tend to assume a preferred orientation perpendicular to the largest compressive stress. These appear most conspicuos in the form of 'stylolites', the digitated solution seams most frequently formed in porous, water-saturated carbonate rocks (cf. Fig. 2). Stylolites owe their name to the stylo-like shapes that are seen in cases of extreme column-and-socket interdigitations, when the solution surface is exposed in three dimensions. These solution seams appear in cross-section as serrated veins and are easy to spot by their dark colour which they owe to less soluble, usually clay-type minerals that are deposited by the dissolved rock matrix. The stylos, columns, or 'teeth' of a stylolite can reach lengths up to a meter and the seam thickness itself can reach similar dimensions in rare cases, suggesting an enormous reduction in layer thickness by dissolution. However, the stylolites easily spotted in every-day life in marble tables or floors have typical seam thicknesses of the order of a millimeter and typical overall lengths (in cross-section) of the order of a meter.

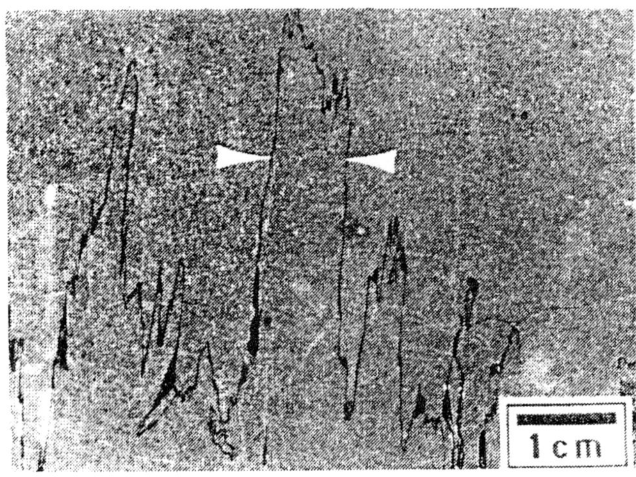

Figure 2: Typical irregular solution seam, known as 'stylolite', in a wet limestone.

The kinematics of stylolites can be explained in terms of spatially varying dissolution rates due to material heterogeneity. Thus, if two blocks of rock are imagined to converge with uniform velocities $+v$ and $-v$ towards a common solution interface, different rates r^+ and r^- of removal of material on either side of the interface will necessitate a compensatory displacement $w = \frac{1}{2}(r^+ - r^-)$ of the interface into the more rapidly dissolving material, where w may vary from point to point along the solution seam. In this manner a highly irregular stylolitic solution seam can develop, whose shape (as seen in cross-section in Fig. 2) will reflect the spatial variability of the solution rate of the material in the pore fluid.

It is more difficult to understand why the solution process tends to localize macroscopically along discrete and often quite evenly spaced, preferentially oriented solution seams. Here we refer the interested reader to a stimulating article by Merino (Merino, 1992), on "selforganization in stylolites". Merino proposes a simple and therefore attractive model for the spontaneous generation of evenly spaced solution seams in porous rocks, based on a positive feedback mechanism in the form of a stress-enhanced porosity increase which in turn magnifies the local stress in the dissolving solid matrix. Nevertheless, the validity of some of the assumptions underlying Merino's model remains difficult to judge in the absence of pertinent experimental data or theoretical developments shedding light on the relevant microscale process.

Indeed, the fundamental phenomenon that must be studied first is a pervasive mode of pressure solution which operates on a grain scale throughout a certain volume of rock and has been documented in thin sections, notably from sandstones, by several investigators (Fig. 3). On the basis of a more quantitative understanding of the microscale processes underlying this phenomenon, one can then attempt to model permanent volume and shape changes in rocks due to stress-enhanced solubility and the rates at which these can take place in a given rock type; and on can finally enquire into the conditions under which a pervasive mode of pressure solution will switch to a localized mode, giving rise to the development of macroscopic solution seams.

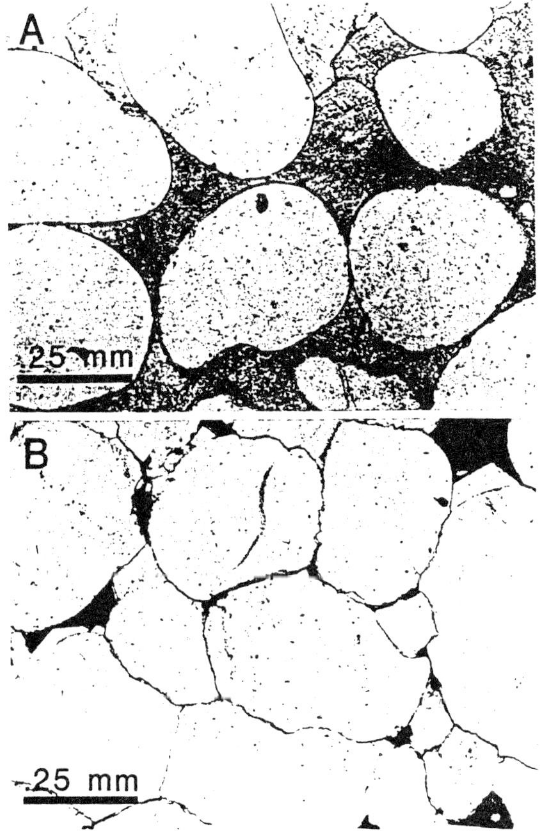

Figure 3: Photomicrographs showing intergranular pressure solution (IPS) in a sandstone. (A) Pore-filling calcite cement has preserved original rounded quartz grains by inhibiting IPS. (B) Sample from same depth as (A) without cement (dark areas are porosity) has undergone significant IPS (taken from Houseknecht & Hathon, 1987).

In this chapter we shall focus on a pervasive mode of pressure solution, also called 'intergranular pressure solution' (IPS), for reasons apparent from Figure 3. While studying IPS *sensu strictu* hereafter, we remain aware of the possibility that macroscopically pervasive pressure solution of a volume of rock may be accomplished in distinctly different ways on a pore scale, depending upon the properties of grain boundaries, pore geometry, and mineral composition of a rock. Conceivably, pressure solution could operate by 'free-face dissolution', i.e. erosion of the pore walls, without the involvement of grain boundaries, yielding a porosity increase as required by Merino's model, but in sharp contrast with IPS which tends to reduce the porosity; indeed, the possible existence and consequences of distinct pore-scale modes of pressure solution remain an important research topic.

Figure 3, apart from documenting IPS, also illustrates the following important point. If a sequence of porous and permeable rocks is open to advective fluid transport and is susceptible to pervasive pressure solution, then each of its volume elements constitutes an open thermodynamic system that will allow 'export' of soluble matter from dissolution sites or 'import' and deposition of such material as cement at precipitation sites. In such a context, pressure solution will be of interest to geologists primarily as a process that can explain large-scale mass movements, accounting for a substantial loss in layer thickness in some location and providing a source of cement at a different location (Tada & Siever, 1989). Alternatively, a sequence of porous, fluid-infiltrated rocks and every volume element therein may behave as a closed system on a macroscopic scale. Material dissolved at grain-to-grain contacts will then be transported by grain boundary diffusion and diffusion through an open pore space to nearby precipitation sites, e.g. at low-stress pore walls, giving rise to a characteristic microtexture with overgrowth of the original grains in so-called 'pressure shadows'.

This mass transfer from high energy dissolution sites to low energy precipitation sites enables an aggregate of grains to accumulate macroscopic creep strains in response to an applied load. Under closed-system conditions, the process closely resembles a type of grain boundary diffusion creep known as 'Coble creep' in the materials science literature (Coble, 1963). Its importance as a deformation mechanism in virtually all rock types is gradually being appreciated as a corollary of the presence of aqueous fluids throughout large parts of the Earth's crust.

In the following, we shall derive a pressure solution creep law assuming, for convenience, a pure solid substance forming a simple cubic packing of spherical grains. Our derivation follows the pioneering work by Weyl (1959) and Rutter (1976, 1983) up to a point, where the need is felt for a deeper understanding of thermodynamics of nonhydrostatic heterogeneous systems; from there on our approach parallels that by Lehner (1995). For simplicity, we assume linear phenomenological force-flux relations for the three distinct kinetic processes of intergranular dissolution, grain boundary diffusion, and free-face precipitation. Under closed system conditions these must occur in series; the rate-controlling step is then readily determined by comparing the characteristic times associated with each kinetic process. Free-face precipitation, in particular, may control the rate of bulk deformation under closed system conditions (Mullis, 1991). In the following we shall not examine that limiting case, however, but concentrate more on the coupling of transport and deformation under open-system conditions.

3. Model assumptions

Consider the idealized model of a flat contact between two identical spherical grains composed of a single-component, isotropic solid substance. The grains are in contact with water and initially the solid phase is assumed to be in chemical equilibrium with its own aqueous solution at a uniform pressure p and temperature T. We represent the initial state of a macroscopic rock sample by a simple cubic packing of identical spheres, whose diameter d_0 determines a cube of volume $V_0 = {d_0}^3$ as the smallest representative volume element or 'unit cell'. We now postulate that the pack of spheres is subjected to a macroscopically uniform compressive stress $\overline{\sigma}$, equal in all directions, and that as a consequence of this extra load beyond the ambient pressure p, the grains will be pressed together and will start to dissolve along their contacts. We shall denote the resulting grain shortening perpendicular to the faces of the cube by the shortening ratio λ_n; the unit cell will thus have shrunk from $V_0 = {d_0}^3$ to $V = \lambda_n^3 {d_0}^3$ at a certain moment.

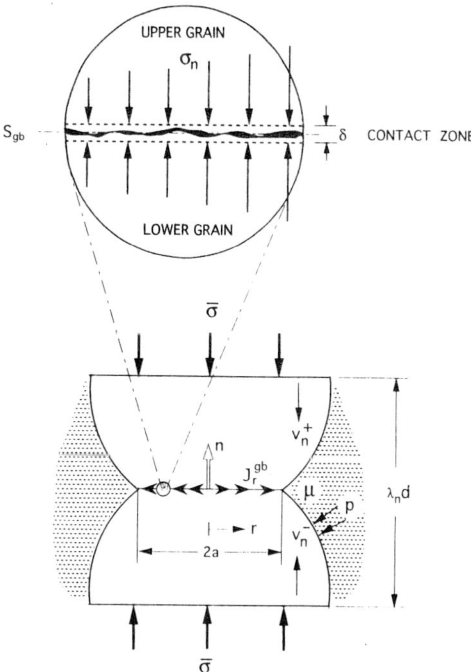

Figure 4: Intergranular pressure solution (IPS), affecting two identical spherical grains; nominally flat, fluid-permeated contact zone shown in magnification.

If elastic shape and volume changes of the grains are ignored and if the grain contacts remain flat, then the grains converge with uniform velocities v_n^- and $v_n^+ = -v_n^-$, measured relative to a stationary grain-to-grain contact (Fig. 4) and the following simple kinematic relation must hold

$$v_n^+ - v_n^- = \dot{\lambda}_n \, d \tag{1}$$

between this velocity 'jump' and the linear shortening rate $\dot{\lambda}_n$ of the unit cell.

On a finer scale than that of Figure 4, the nominally flat grain-to-grain contact must be visualized as a 'contact zone' of some finite thickness δ. This thickness may vary between a few nannometer for 'clean' contacts between salt single crystals of halite or quartz and tens of μm for intergranular solution seams that contain impurities shed by the dissolving grains and other less soluble minerals. We further stipulate that this contact zone tends to remain in a roughened state during active IPS, such that adjacent grains maintain solid/solid contact across insular ridges that are separated by fluid-filled grooves (Raj, 1982; Lehner & Bataille, 1984/85; Spiers & Schutjens, 1990; Cox & Paterson, 1991). In plan view such a grain boundary will then exhibit a characteristic 'island and channel structure', while in cross-section the contact zone forms a thin, porous and permeable interlayer that can be wetted and invaded by the pore fluid, i.e. water in our case. The dissolved solid material can thus be carried by diffusive transport through the grain boundary fluid phase from interior points of a grain boundary towards its periphery, i.e. into the open pore space. Thus, if J_r^{gb} denotes the radial component of the diffusive mass flux of dissolved material (in $\mathrm{kg\,s^{-1}\,m^{-2}}$), averaged over an effective grain boundary thickness δ, then the total rate of outflow across a cylindrical cross-section of area $2\pi r\delta$ and the total rate of dissolution within this contour must balance. Consequently

$$2\delta J_r^{gb}(r) + \rho^s(v_n^+ - v_n^-)r = 0 \tag{2}$$

where ρ^s denotes the density of the solid.

To stay with the simplest model of IPS, we shall assume that the diffusive flux J_r of solute mass is governed by Fick's Law as applicable to bulk diffusion through a liquid phase. We introduce this relationship in the form $J_r = -\rho^f L^{gb} d\mu/dr$, where L^{gb} is a phenomenological 'mobility' coefficient, μ is the mass-specific chemical potential (in $\mathrm{J\,kg^{-1}}$) of the solute component in the grain boundary solution phase and ρ^f is the bulk density of the latter. For a sufficiently dilute solution, Fick's law can be written in terms of the gradient in the solute mass fraction C as (see, e.g., de Groot & Mazur, 1962, Chap. 11)

$$J_r = -\rho^f D^{gb} \mathrm{d}C/\mathrm{d}r \tag{3}$$

Here $D^{gb} = L^{gb}(\mathrm{d}\mu/\mathrm{d}C)_{p,T} \approx L^{gb}kT/\rho^s\Omega^s C$ is an approximately constant mass diffusivity (in $\mathrm{m^2\,s^{-1}}$) that controls solute diffusion through the intergranular fluid phase, where k is the Boltzmann constant and Ω^s denotes the molecular volume of the solid. This relationship follows from the standard expression

$$\mu = \mu^*(p,T) + (kT/\rho^s\Omega^s)\ln\gamma C \tag{4}$$

for the chemical potential of the solute component of a binary solute-diluent solution as long as the concentration dependence of the activity coefficient γ can be disregarded.

We may now combine Eqs. (1)–(3) and integrate the resulting differential equation in $C(r)$ to obtain the radial distribution of the solute mass fraction along the intergranular contact

$$C(r) = C(a) - \frac{\rho^s \dot{\lambda}_n d}{4\rho^f D^{gb}\delta}(a^2 - r^2) \tag{5}$$

As expected for IPS ($\dot{\lambda}_n < 0$), the solute concentration and therefore also its chemical potential are higher inside the grain contact than on its periphery, i.e. in the open pore space. By taking a surface average of this distribution over the grain contact, one obtains the following expression for the rate of grain convergence

$$\dot{\lambda}_n = -\frac{8\rho^f D^{gb}\delta}{\rho^s a^2 d}[\tilde{C} - C(a)] \tag{6}$$

in terms of the average concentration $\tilde{C}$ along the grain boundary and the concentration $C(a)$ at the periphery of the contact, i. e. in the open pore space where μ and hence C are taken to be uniform.

To progress towards the desired creep law, the main task that remains is to determine the value of unknown average concentration $\tilde{C}$ in the above relation. The key to taking this step lies in the thermodynamic theory of heterogeneous and nonhydrostatic systems as developed by Gibbs (1878) and in certain additional arguments which must be brought in from nonequilibrium thermodynamics, as will now be discussed.

4. Chemical equilibrium at a stressed solid/solution interface

It was Sorby, who in 1863 remarked that an effect similar in principle to the lowering of the freezing temperature of water with increasing pressure, first established experimentally by W. Thomson (Lord Kelvin) in 1850, must hold true with respect to the solubility of salts in water and might thus explain pressure solution phenomena such as 'pitted pebbles'; and indeed, in 1861 J. Thomson had already demonstrated the stress-enhanced solubity of salt crystals in water experimentally. But it remained for Gibbs (1878) to furnish a full theoretical explanation of the effect of stress on the chemical equilibrium of a an arbitrarily stressed solid with its own solution phase. In particular, Gibbs showed that the condition for local chemical equilibrium at a flat interface between a stressed pure solid and its solution is given by (Gibbs, 1878, eq. 387):

$$u^s - Ts^s + p/\rho^s = \mu \tag{7}$$

where u^s and s^s are the specific internal energy and specific entropy of the pure solid phase, p is the pressure in the solution phase and μ is the chemical potential of the solute component in the solution phase (in the present context always a binary aqueous solution) as given by (4). Since $-p$ must equal the normal component of stress σ_n at the solid side of the interface, the left-hand side of condition (7) may be expressed solely in terms quantities associated with the solid phase. Chosing to replace the first two terms by the specific Helmholtz free energy $f^s = u^s - Ts^s$, we may thus write equivalently

$$f^s - \sigma_n/\rho^s = \mu \tag{8}$$

Condition (7) characterizes a strictly local state of equilibrium, pertaining only to points along a solid/solution interface. For curved interfaces a interfacial energy term of the form $(1/R_1 + 1/R_2)\gamma^{sf}/\rho^s$ must be added to the left-hand-side of this condition, in which R_1 and R_2 denote finite principal radii of interfacial curvature and γ^{sf} the interfacial energy (cf. Gibbs, 1878, eq. 661; see also: Heidug, 1991).

Leaving aside the question (in part already answered by Gibbs) of the stability of such equilibria, one can picture a variety of situations involving a nonhydrostatically stressed solid in global equilibrium with its own solution. In each case, however, the criterion of global interfacial equilibrium will amount to nothing less than the requirement that condition (7) be satisfied pointwise along the entire solid/fluid interface in the system under consideration. It is only in the special case in which a homogeneous solid is completely surrounded by a solution phase at uniform chemical potential μ and pressure p that the term $f^s + p/\rho^s$ becomes a constant independent of position, thus appearing in the role of a specific Gibbs free energy characterizing the bulk of the solid phase—on equal footing with the chemical potential μ of the dissolved solid. In general, however, and on leaving the territory of textbooks on chemical thermodynamics which restrict attention to hydrostatic states of stress, there exists no useful concept of a scalar Gibbs free energy or chemical potential which could serve to define the heterogeneous equilibrium of a solid composed of different phases or of a system comprising a stressed solid in contact with a liquid solution. As J. Cahn put it in a recent review of research into the physical chemistry of stressed solids (Cahn, 1989), "These results of Gibbs have upset many people who believe that the chemical potential of an equilibrated system is not only definable, but must be constant throughout. One can only surmise that Gibbs ... did not believe that a constant chemical potential was a necessary condition for heterogeneous equilibrium."

It has nevertheless become clear in recent years that the formal development of Gibbs' thermodynamics of heterogeneous systems could be clarified in this point and in a sense made more complete through the explicit recognition of the concept of a 'chemical potential tensor' (Bowen, 1976; Grinfeld, 1982, 1991; Truskinovskiy, 1984; Heidug & Lehner, 1985). It will be noticed indeed, that the condition of interfacial equilibrium (8) can be written in the form

$$\mu_n^s = \mu \tag{9}$$

in terms of the normal component $\mu_n^s = \boldsymbol{n} \cdot \boldsymbol{\mu}^s \cdot \boldsymbol{n}$ at the solid/solution interface (with unit normal $\boldsymbol{n}$) of the tensor

$$\boldsymbol{\mu}^s = f^s \mathbf{1} - \boldsymbol{\sigma}/\rho^s \tag{10}$$

where $\mathbf{1}$ is an isotropic unit tensor and $\boldsymbol{\sigma}$ is the symmetric stress tensor. Under hydrostatic conditions, when $\boldsymbol{\sigma} = -p\mathbf{1}$, $\boldsymbol{\mu}^s$ is also isotropic and its orientation-independent normal component becomes identical with the specific Gibbs free energy $f^s + p/\rho^s$ or scalar chemical potential of a single-component solid.

Applying these ideas to our model of IPS, we first note that the normal stress σ_n at the grain contact shown in Figure 4 actually represents a normal traction that acts upon the two imagined parallel surfaces through the solid grains which mark the boundaries of the grain contact zone. As will be seen presently, the normal component μ_n^s of the chemical potential tensor (10), defined from the values of σ_n, ρ^s and f^s along these contact zone margins, can be given a similarly fundamental significance as the truly superficial quantity on the left-hand-side of (9), as if the phase boundary were transferred from its actual position to the smooth margins of the intergranular contact zone (Lehner, 1990).

5. Phenomenological rate laws for free-face precipitation and intergranular dissolution

We wish to derive a model IPS that is free of any *a priori* bias towards one or the other rate-limiting process. We must therefore reject the frequently made assumption that the intergranular solution phase is in chemical equilibrium with the stressed solid phase. Similarly, we shall admit a state of disequilibrium between the two phases along the hydrostatically loaded pore walls, making allowance for precipitation from a locally supersaturated pore solution at such sites. Following earlier work by Lehner & Bataille (1984/85), and Lehner (1990, 1995), we shall employ thermodynamic arguments that will enable us to set up a more complete phenomenological description of IPS under closed-system conditions.

Consider first the solid-fluid interface comprising the pore walls or 'free faces' of our system or unit cell. We have already seen that Gibbs' condition (9) must hold along this phase boundary, wherever the transport of the soluble solid substance across it—be it by dissolution or precipitation—is to be prevented from taking place. Implied by this statement is the view that any violation of condition (9) in the form of an imbalance between the terms on either side of (9) will generate a local thermodynamic driving force $\mu_n^s - \mu (\neq 0)$. Depending upon its sign (positive or negative), this force may drive soluble matter into or out of solution across the phase boundary, although in general a certain threshold value may have to be attained by the force before it can initiate the process.

Let $\dot{m}_{sf}$ denote the rate (in kg s^{-1} m^{-2}) at which mass is transferred across the pore wall at any point. Then the product $\dot{m}_{sf} (\mu_n^s - \mu)$ can be shown to equal the rate at which work is dissipated per unit area as the material passes through the phase boundary (see Lehner & Bataille, 1984/85). This rate is expressible as the product of the absolute temperature T, which we assume to remain continuous across the phase boundary, and a nonnegative interfacial entropy production rate $\dot{\sigma}$. Accordingly one has

$$\dot{m}_{sf} \, (\mu_n^s - \mu) = T \, \dot{\sigma} \geq 0 \tag{11}$$

An important observation is now that this inequality must hold at any value of the driving force—beyond a possible threshold—by virtue of the requirement $\dot{\sigma} \geq 0$ of the second law. It follows that the flux $\dot{m}_{sf}$ and the force $\mu_n^s - \mu$ must be functionally dependent. The simplest 'phenomenological relation' of this kind, obtained upon truncating a Taylor expansion after the linear term, takes the form

$$\dot{m}_{sf} = \rho^s L^{\pm}(\mu_n^s - \mu) \tag{12}$$

where $L^{\pm} > 0$ is a phenomenological rate constant which may assume different values L^+ and L^- for precipitation and dissolution, respectively. In the following, we shall find it convenient to use the rate constant $K^{sf} = L^+(kT/\rho^s\Omega^s)$ in lieu of L^+.

A phenomenological relation of the same type has been introduced long ago by Machlin (1953) in an analysis of the growth of precipitates in metals. Since then, several authors have obtained expressions for the driving force or 'driving traction' along different routes. The advantage of this thermodynamic approach to a question of interfacial kinetics lies in the clarification of the nature of the driving force itself (in the present case it is found to represent the normal component of a certain chemical potential tensor). This can be of great conceptual help in the development rational theories of deforming two-phase materials. Although one must not loose sight of the fact that the experimental evidence will often suggest a nonlinear kinetic relation, the definition of the full thermodynamic driving force that phenomenological theory has to offer remains of interest to the experimentalist. We shall not pursue these questions any further at this point, but instaed refer the interested reader to the work cited in the above, in particular to the work of Grinfeld (1982, 1991), Truskinovskiy (1984), Lehner & Bataille (1984/85), Heidug & Lehner (1985), Heidug (1991), and Leroy & Heidug (1994).

Turning now to the kinetics of intergranular dissolution and diffusive transport, let us denote by $\dot{m}_{gb} = -\rho^s \dot{\lambda}_n d$ the uniform dissolution rate along a given intergranular contact, i.e., the flux of solid material (in kg s^{-1} m^{-2}) from both grains into the contact zone and across the actual solid/solution phase boundary lying within that zone (cf. Fig. 4 and Eq. 1). Also, let $\mu_n^s(r)$ denote the normal component of the solid-phase chemical potential tensor along the smooth outer boundaries of the contact zone and $\mu(r)$ the chemical potential of the dissolved solid component in the grain boundary fluid. The product $\dot{m}_{gb}\ [\mu_n^s(r) - \mu(r)]$ then equals the rate at which work is dissipated per unit nominal contact area in isothermal irreversible processes as the material passes through the contact zone and the phase boundary.

Furthermore, the work dissipated per unit area in the isothermal diffusive transfer of material, driven by the potential drop $\mu(r) - \mu(a)$ at the rate $\dot{m}_{gb}$ from its solution site at r to the contact periphery at a, is equal to $\dot{m}_{gb}\ [\mu(r) - \mu(a)]$. If $\dot{\Delta}_{gb}$ denotes the total rate of dissipation associated with irreversible processes in an intergranular contact zone, it then follows that $\dot{\Delta}_{gb}$ must be equal to the sum of the integrated dissipative work rates for a contact, i.e.,

$$\dot{\Delta}_{gb} = \dot{m}_{gb} \int_{S_{gb}} [\mu_n^s(r) - \mu(r)]\,\mathrm{d}A + \dot{m}_{gb} \int_{S_{gb}} [\mu(r) - \mu(a)]\,\mathrm{d}A \tag{13}$$

where the integrals are taken over the nominally flat grain-to-grain contact S_{gb} of Figure 1. Note that on carrying out a formal integration over S_{gb} and cancelling terms, this becomes

$$\dot{\Delta}_{gb} = \pi a^2\ \dot{m}_{gb}\, [\tilde{\mu}_n^s - \mu(a)] \tag{14}$$

from which it is apparent that the potential difference $\tilde{\mu}_n^s - \mu(a)$ represents the average total driving force for the combined (serial) processes of intergranular dissolution and grain boundary diffusion.

If the first integral in (13) can be neglected against the second, grain boundary diffusion acts as the rate-limiting process and the driving chemical potential difference associated with the easy deformation/dissolution step may equivalently be set equal to zero, i. e., $\mu_n^s \approx \mu$. It is this Stefan-type approximation that allows (8) to be treated as an 'equilibrium condition' as in the work of Kamb (1961) and Paterson (1973).

The rate of dissipation associated with material transfer through the contact zone and the phase boundary is given by the product $\dot{m}_{gb}\ [\mu_n^s(r) - \mu(r)]$. As in the above, we may now invoke the second law to assert that this product must remain nonnegative at any value of the driving force and to conclude that the force and flux must therefore be functionally related. If dissipation in the solid phase within the grain boundary contact zone remains negligible in comparison with the dissipation associated with the dissolution reaction, then the force $\mu_n^s(r) - \mu(r)$ will be available in full to drive the dissolution reaction. The simplest force-flux relation to be postulated for this case is a linear relation of the form:

$$\dot{m}_{gb} = \rho^s K^{gb}(\rho^s \Omega^s / kT)[\mu_n^s(r) - \mu(r)] \tag{15}$$

where $K^{gb} > 0$ is phenomenological rate constant.

The use of relation (15) can be justified, as long as the solid material may be assumed to remain in the elastic range while traversing the contact zone, except for the immediate vicinity of load-bearing islands. Islands are thus imagined to be undercut by dissolution, but to deform by dissipative inelastic micromechanisms only in the final phase of this undercutting process, when the load is about to be transferred from such a deforming contact point onto neighbouring stiffer contacts. Dissipation in the solid portion of the contact zone could then be neglected against the dissipation associated with the dissolution reaction. While the local undercutting and removal of islands is essential to achieving grain convergence by IPS, inelastic solid deformation can be viewed as an associated phenomenon, made possible only by the former and progressing therefore at the same rate. The most important effect of significant dissipation by inelastic deformation within the contact zone is likely to be a consumption in dissipated work of part of the available driving force $\mu_n^s(r) - \mu(r)$, implying a certain reduction in the force available for the dissolution reaction. In the following we shall apply (15) without any further correction for work dissipated in deforming solid contact zone material, treating K^{gb} as a phenomenological constant governing the dissolution rate at a stressed grain boundary. This then enables us to evaluate the first integral in (13).

The second integral in (13) represents the rate of dissipation associated with solute diffusion along the grain boundary. The relevant kinetic relation is Fick's law (3), which has already been used in (6) to obtain the rate of grain convergence in terms of the unknown average concentration $\widetilde{C}$ a nominally flat circular contact S_{gb}. Making use of (4), equation (6) may be cast in terms of the mass flux $\dot{m}_{gb} = -\rho^s\ \dot{\lambda}_n\ d$ and the average thermodynamic force $\tilde{\mu} - \mu(a)$ that drives grain boundary diffusion. To do so, we first introduce a reference concentration C_0, representing the solubility of the solid in the aqueous pore fluid at a fixed pressure p_0 and temperature T_0. An appropriate choice of reference state will usually suggest itself in experimental studies, but a general rule to follow is to choose the pair (p_0, T_0) as representative as possible for the average process conditions encountered. Under the assumptions of dilute solute concentrations, it then follows from (4) that $[C(r) - C(a)]/C_0 \approx \ln C(r)/C(a) \approx (\rho^s\Omega^s/kT)[\mu(r) - \mu(a)]$ and, after averaging over S_{gb},

$$[\widetilde{C} - C(a)]/C_0 \approx (\rho^s\Omega^s/kT)[\widetilde{\mu} - \mu(a)]$$

so that (6) may be written

$$\dot{m}_{gb} = (8\rho^f C_0 D^{gb}\delta/a^2)(\rho^s\Omega^s/kT)[\widetilde{\mu} - \mu(a)] \tag{16}$$

Similarly, one has

$$\dot{m}_{gb} = \rho^s K^{gb}(\rho^s\Omega^s/kT)[\widetilde{\mu}_n^s - \widetilde{\mu}] \tag{17}$$

for the surface-averaged form of (15).

6. The driving force of intergranular pressure solution

The chemical potential $\widetilde{\mu}$ may now be eliminated from (16) and (17), to obtain the relation

$$\dot{\lambda}_n = -\,\dot{m}_{gb}\,/\rho^s d = -[\tau_S + (4a^2/d^2)\tau_D]^{-1}(\rho^s\Omega^s/kT)[\widetilde{\mu}_n^s - \mu(a)] \tag{18}$$

for the rate of grain convergence. Here we have defined the characteristic time constants

$$\tau_S = d/K^{gb}, \qquad \tau_D = \rho^s d^3/(32\rho^f C_0 D^{gb}\delta) \tag{19}$$

that provide measures for the rates at which intergranular solution and grain boundary diffusion occur.

In essence, the above result already represents the desired 'creep law' for our simple model system. We notice, in particular, that the difference $\widetilde{\mu}_n^s - \mu(a)$ between the average solid-phase chemical potential in the grain-to-grain contact and the (uniform) solute chemical potential in the pore space appears in the role of the thermodynamic force that drives intergranular pressure solution—a result that is consistent with expression (14) for dissipation associated with this process.

We must now seek an expression for the driving force, making use of definition (10), that will render explicit the role of the intergranular contact stress and having done so, we shall finally wish to relate this intergranular stress to an appropriate macroscopic average stress.

For this purpose, we shall find it convenient to introduce as a reference chemical potential the potential $\mu_0 = \mu^*(p, T) + (kT/\rho^s\Omega^s)\ln\gamma C_0$ of a solution which is in equilibrium with the solid phase everywhere along a flat interface, when both phases are subjected to the same hydrostatic pressure and temperature p and T, respectively. Gibbs' condition (9) then demands that $\mu_n^s = f_0^s + p/\rho_0^s = \mu_0$ evrywhere along the solid/solution interface. With this condition in mind, we now consider the potential difference $\mu_n^s(r) - \mu(a)$ locally along an intergranular contact and write it as

$$\begin{aligned} \mu_n^s(r) - \mu(a) &= [\mu_n^s(r) - \mu_0] - [\mu(a) - \mu_0] \\ &= (f^s - f_0^s) + (1/\rho^s - 1/\rho_0^s)p - (\sigma_n + p)/\rho^s - [\mu(a) - \mu_0] \end{aligned} \quad (20)$$

Before making use of this expression in (18), we introduce the approximation

$$\mu_n^s - \mu_0 \approx -(\sigma_n + p)/\rho^s \quad (21)$$

implying the neglect of the first two terms on the right-hand-side of (20) against the third term. This approximation is frequently used in combination with the Kamb/Paterson equilibrium assumption $\mu_n^s \approx \mu$ (Paterson, 1974; Rutter, 1976; Lehner & Bataille, 1984/85). Its validity has been discussed by Lehner (1995).

If (21) is assumed, then the local driving force in the kinetic relation (15) becomes

$$\mu_n^s(r) - \mu(r) \approx -(\sigma_n + p)/\rho^s - [\mu(r) - \mu_0] \quad (22)$$

while the overall average driving force in (18) becomes

$$\widetilde{\mu}_n^s - \mu(a) \approx -(\widetilde{\sigma}_n + p)/\rho^s - [\mu(a) - \mu_0] \quad (23)$$

We thus arrive at an expression for the rate of grain convergence wherein the driving force is expressed in terms of an effective intergranular normal stress and a change in solute chemical potential from the hydrostatic equilibrium level μ_0:

$$\dot{\lambda}_n = -\ \dot{m}_{gb}\ /\rho^s d = [\tau_S + (4a^2/d^2)\tau_D]^{-1}(\rho^s\Omega^s/kT)[(\widetilde{\sigma}_n + p)/\rho^s + \mu(a) - \mu_0] \quad (24)$$

7. Compaction creep in a cubic packing of spheres

Let us focus now on the aggregate compaction behaviour of a simple cubic packing of spherical grains, with a cube-shaped unit cell that truncates a single grain along six pairwise orthogonal grain-to-grain contacts. Our aim is to derive an appropriate expression for the creep law (24) in terms of the macroscopically applied effective stress $\overline{\sigma}_{ij} + p\delta_{ij}$.

Let $\langle\sigma_{ij}\rangle_s = V^{-1}\int_{V_s}\sigma_{ij}\,\mathrm{d}V$ denote the macroscopic partial stress or 'phase averaged' stress of the solid phase. Here $V = V_s + V_f$ is the volume of the unit cell, which consists of the solid grain and fluid or pore volume portions V_s and V_f. Since equilibrium requires $\sigma_{ik,k} = 0$, the phase averaged stress is also given by the surface integral $V^{-1}\int_{\partial V_s}\sigma_{ik}n_k x_j\,\mathrm{d}A$ taken over the entire (closed) boundary $\partial V_s = S_{sf} + S_{gb}$ of the grain, where n_k and x_j denote components of the outward unit normal to ∂V_s and of a position vector in some fixed Cartesian frame, S_{gb} comprises six (in the present example) intergranular contacts, and S_{sf} stands for the solid/fluid interface of the pore walls. Using the fact that the only traction along S_{sf} is supplied by the uniform pore fluid pressure p, we split the integral over ∂V_s and write

$$\begin{aligned}\langle\sigma_{ij}\rangle_s &= -p\frac{1}{V}\int_{S_{sf}} n_i x_j\,\mathrm{d}A + \frac{1}{V}\int_{S_{gb}} \sigma_{ik} n_k x_j \mathrm{d}A \\ &= -p\frac{1}{V}\int_{\partial V_s} n_i x_j\,\mathrm{d}A + \frac{1}{V}\int_{S_{gb}} (\sigma_{ik} + p\delta_{ik}) n_k x_j \mathrm{d}A \end{aligned} \tag{25}$$

and since the first integral on the right equals $(1-\phi)p\delta_{ij}$, where $\phi = V_f/V$ denotes the aggregate porosity, the relationship

$$\overline{\sigma}_{ij} + p\,\delta_{ij} = \langle\sigma_{ij}\rangle_s + (1-\phi)p\,\delta_{ij} = \frac{1}{V}\int_{S_{gb}} (\sigma_{ij} + p\,\delta_{ij}) n_k x_j \mathrm{d}A \tag{26}$$

exists between the integrated effective grain boundary traction and the macroscopic effective stress. Note that the total stress is defined as the sum of the partial or phase averaged stresses of the solid and fluid phases according to $\overline{\sigma}_{ij} = \langle\sigma_{ij}\rangle_s + \langle\sigma_{ij}\rangle_f = \langle\sigma_{ij}\rangle_s - \phi p\delta_{ij}$.

In the example considered here, we assume the macroscopic principal stress directions to remain perpendicular to the three orthogonal pairs of grain-to-grain contacts, and we place the origin of the position vector $\boldsymbol{x}$ at the grain center. Under these circumstances one only has to consider a relation in the principal stresses of the form (in which no summation is implied over a repeated index ν)

$$\overline{\sigma}_\nu + p = \frac{1}{V}\int_{S_{gb}} (\sigma_\nu + p) n_\nu x_\nu \mathrm{d}A, \qquad (\nu = 1,2,3) \tag{27}$$

Using the fact that $n_\nu x_\nu = \lambda_\nu d/2$ together with the expressions $V = \lambda_1\lambda_2\lambda_3 d^3$ and $a_\nu^2 = (1-\lambda_\nu^2)d^2/4$ for the current volume of the unit cell and the contact radius, the integral on the right-hand-side may be expressed in terms of the average effective normal stress on a contact, which is then found to be given in terms of the macroscopic principal effective stress by

$$\tilde{\sigma}_\nu + p = \tfrac{4}{\pi}[\lambda_1\lambda_2\lambda_3/(1-\lambda_\nu^2)\lambda_\nu](\overline{\sigma}_\nu + p), \qquad (\nu = 1,2,3) \tag{28}$$

For a simple cubic packing undergoing isotropic compaction under an effective all-round compressive stress $\overline{\sigma} + p$, we must have $\lambda \equiv \lambda_1 = \lambda_2 = \lambda_3 \leq 1$, so that $a_\nu^2 = a^2 = (1-\lambda^2)d^2/4$ and relation (28) specializes to

$$\tilde{\sigma} + p = \tfrac{4}{\pi}[\lambda^2/(1-\lambda^2)](\overline{\sigma} + p) \tag{29}$$

Using this result in (24), we now obtain the truly macroscopic relation

$$\dot{\lambda} = \frac{-1}{\tau_S + (1-\lambda^2)\tau_D}\left[\frac{4\lambda^2}{\pi(1-\lambda^2)}\,P_e - \frac{\rho^s\Omega^s}{kT}\,\Delta\mu\right] \tag{30}$$

in terms of the dimensionless effective pressure $P_e = -(\overline{\sigma} + p)(\Omega^s/kT)$ and the potential difference $\Delta\mu = \mu(a) - \mu_0$. This 'creep law' displays a number of interesting properties that will now be briefly discussed.

First, it will be noticed that the factor within the brackets, i.e., the driving force, comprises a mechanical and a chemical term. Typically, although not necessarily, the two terms will have opposite signs. Relation (30) therefore predicts that IPS should cease to operate above a certain level for the chemical potential (or solute concentration).

Secondly, although linear in P_e and $\Delta\mu$, the relation displays a strong geometrical nonlinearity. In other words, at fixed values of P_e and $\Delta\mu$, the strain rate will depend strongly on the already accumulated 'pressure solution strain'. Since the source of this nonlinearity lies in the dependence of the grain-to-grain contact size on the amount of dissolved material, this particular feature of the creep law (30) may be described differently as a strong dependence of the creep rate on the current size (or size distribution) of the intergranular contacts. The latter might for example be directly obtainable from thin sections. The same geometric nonlinearity has been taken into account for some time in experimental studies of compaction creep behaviour caused by IPS. For example, care has been taken in such studies, to determine the dependence of creep rates on stress at a fixed strain (see, e.g., the work of Spiers & Schutjens, 1990, which also contains a theoretical analysis of this nonlinearity).

A third observation relates of course to the appearance of the characteristic times τ_S and τ_D in the leading factor of relation (30). Clearly, the relatively larger time constant will tend to govern the rate of the overall process, as it should in the case of two serial processes. As may be expected, however, τ_D is weighted by the grain contact size, becoming less effective at smaller relative contact sizes. A further important feature of these time constants, apparent from their definitions, is the different dependence on grain size which they exhibit. It follows, that grain-boundary diffusion controlled creep rates should display a dependence on the third power of the grain size, while dissolution controlled creep rates should vary linearly with d. This theoretically expected result has been exploited successfully by experimentalists to determine the rate-controlling process in experimentally compacted grain aggregates of various salts (e.g., NaCl, KCl, $NaNO_3$) that served as rock analogues (Spiers & Schutjens, 1990; Spiers *et al.*, 1990; Spiers & Brzesowsky, 1993).

Finally, we come to chemical term in the driving force, a term that is missing in most published pressure solution creep laws. The first thing to notice is its role in ensuring a physically consistent behaviour at vanishing effective pressure, when $\dot{\lambda}$ need not vanish (as is usually assumed) and compaction creep can continue as long as the solution phase remains undersaturated ($\Delta\mu < 0$). Secondly, as has been mentioned already, the presence of the chemical term entails the possibility of a vanishing creep rate at a certain level of supersaturation. Although qualitatively in accord with the expected behaviour, this prediction of a critical supersaturation—if tested experimentally—may nevertheless turn out too inaccurate, primarily because the complex process of grain boundary healing is not addressed by the present model.

Compaction creep in porous and permeable sediments will generally occur under 'open system' conditions: a layer affected by this process will permit the passage, through any volume element, of material dissolved in a percolating aqueous pore fluid. Thus, depending on the question a geologist may ask, the compacting volume of rock may be a single stratum embedded in a large-scale hydrologic system such as a whole sedimentary basin, or it may represent an imagined vertical prism of sediments, comprising an entire sedimentary column from some impermeable base up to the sea floor. In each case, the transport of fluids through the system entails a dependence of the large-scale solute distribution on the prevailing flow regime, as is evident from the equations governing the fluid flow and solute transport through the rock. The value of $\Delta\mu$ in (30) is therefore to some extent controlled by the flow regime and so must be the compaction creep rate due to IPS. In principle therefore, the prediction of local pressure solution creep rates requires the solution of a coupled system of equations, governing a deformation and a transport problem. Once developed in full, a theory for these coupled processes should enable geologists to explore a range of phenomena combining thermal, chemical, and mechanical effects in the diagenesis (i.e., lithification and chemical as well as mineralogical alteration) and creep deformation of rocks.

REFERENCES

Bowen, R.M.: Theory of mixtures, in: Continuum Physics Vol. 3 (Ed. A.C. Eringen), Academic Press, New York (1976).

Cahn, J.W.: The physical chemistry of stressed solids, Ber. Bunsenges. Phys. Chem., 93 (1989), 1169–1173.

Coble, R.L.: A model for grain boundary diffusion controlled creep in polycrystalline materials, J. Appl. Physics, 34 (1963), 1679–1682.

Cox, S.F. and M. Paterson Experimental dissolution-precipitation creep in quartz aggregates at high temperatures, Geophys. Res. Lett., 18 (1991), 1401–1404.

Durney, D.W.: Early theories and hypotheses on pressure-solution-redeposition, Geology, 6 (1978), 369-372.

Gibbs, J.W.: On the equilibrium of heterogeneous substances. Trans. Connecticut Academy, III: 343–524. In: The Scientific Papers of J. Willard Gibbs, Vol. 1. Longman, Green, and Co., Toronto 1906; Dover, New York 1961.

Grinfeld, M.: Phase transitions of the first kind in nonlinear elastic materials. Mechanics of Solids, 17/1 (1982), 92–101. Engl. transl. of Izv. AN SSSR Mekhanika Tverdogo Tela, 17/1 (1982), 99–109.

Grinfeld, M.: Thermodynamic Methods in the Theory of Heterogeneous Systems, Longman, Greens. Toronto Ontario 1991.

De Groot, S.R. and P. Mazur: Non-Equilibrium Thermodynamics, North-Holland, Amsterdam 1969.

Heidug, W.K.: A thermodynamic analysis of the conditions of equilibrium at nonhydrostatically stressed and curved solid-fluid interfaces. J. Geophys. Res., 96/B13 (1991) 21,909–21,921.

Heidug, W.K. and F.K. Lehner: Thermodynamics of coherent phase transformations in nonhydrostatically stressed solids, PAGEOPH, 123 (1985), 91–98.

Houseknecht, D.W. and L.A. Hathon: Petrographic constraints on models of intergranular pressure solution in quartzose sandstones, Applied Geochemistry, 2 (1988), 507–521.

Kamb, W.B.: The thermodynamic theory of nonhydrostatically stressed solids. J. Geophys. Res., 66 (1961), 259–271.

Lehner, F.K. and J. Bataille: Nonequilibrium thermodynamics of pressure solution, PAGEOPH, 122 (1984/85) 53–85.

Lehner, F.K.: Thermodynamics of rock deformation by pressure solution, in: Deformation Processes in Minerals, Ceramics and Rocks Eds. D.J. Barber & P.G. Meredith), Unwin Hyman, London (1990), Chap. 11.

Lehner, F.K.: A model for intergranular pressure solution in open systems, Tectonophysics, 245 (1995), 153–170.

Leroy, Y.M. and W.K. Heidug: Geometrical evolution of stressed and curved solid-fluid phase boundaries 2, Stability of cylindrical pores, J. Geophys. Res., 99/B1 (1994), 517–530.

Machlin, E.S.: Some applications of the thermodynamic theory of irreversible processes to physical metallurgy, Journal of Metals, Tans. AIME (March 1953), 437–445.

Mosher, S.: Pressure solution as a deformation mechanism in Pennsylvanian conglomerates from Rhode Island, Journal of Geology, 84 (1976), 355–363.

Merino, E.: American Scientist, 80 (1992), 466.

Mullis, A.M.: The role of silica precipitation kinetics in determining the rate of Quartz pressure solution, J. Geophys. Res., 96/B6 (1991), 10,007–10,013.

Paterson, M.S.: Nonhydrostatic thermodynamics and its geologic applications, Rev. Geophys. Space Phys., 11/2 (1973), 355–389.

Raj, R.: Creep in polycrystalline aggregates by matter transport through a liquid phase, J. Geophys. Res., 87/B6 (1982), 4731–4739.

Rutter, E.H.: The kinetics of rock deformation by pressure solution. Phil. Trans. R. Soc. London, A 283 (1976), 203–219.

Rutter, E.H.. Pressure solution in nature, theory and experiment. J. Geol. Soc. London, 140 (1983), 725–740.

Spiers, C.J. and P.M. Schutjens: Densification of crystalline aggregates by fluid-phase diffusional creep, in: Deformation Processes in Minerals, Ceramics and Rocks, (Eds. D.J. Barber and P.G. Meredith), Unwin Hyman, London (1990), Chap. 12.

Spiers, C.J., P.M. Schutjens, R.H. Brzesowsky, C.J. Peach, J.L. Liezenberg, and H.J. Zwart: Experimenmtal determination of constitutive parameters governing creep of rock salt by pressure solution, in: Deformation Mechanisms, Rheology and Tectonics (Eds. R.J. Knipe & E.H. Rutter), Spec. Publ. No. 54, The Geological Society, London 1990, 215–227.

Spiers, C.J. and R.H. Brzesowsky: Deformation behaviour of wet granular salt: Theory versus experiment, Seventh Symposium on Salt, Vol. I, Elsevier Sci. Publishers, Amsterdam 1993, 83–92.

Tada, R. and R. Siever: Pressure solution during diagenesis, Ann. Rev. Earth Planet. Sci., 17 (1989), 89–118.

Truskinovskiy, L.M.: The chemical-potential tensor, Geochem. International, 21/1 (1984), 22–36.

Weyl, P.K.: Pressure solution and the force of crystallization–a phenomenological theory, J. Geophys. Res., 64/11 (1959), 2001–2025.

APPLICATION TO SHAPE MEMORY DEVICES

E. Patoor, Y. Gillet and M. Berveiller
CNRS URA 1215, Metz, France

Abstract

Shape memory alloys play a large role in the development of intelligent systems. Design of these systems needs to know the global relationship between the applied forces and the conjugated kinematical variables for shape memory elements. Such a relation is strongly non linear and temperature dependent. Aim of this work is to derive these relations starting from the definition of a macroscopic criterion for stress induced transformation. This transformation criterion is deduced from micromechanical modelling and takes into account the dissymmetry observed between tensile and compressive tests in these materials. Structure calculations aspects are taken into account using the framework of beam theory. To illustrate these problems two loading cases are solved. First example deals with the analytical solution for pure torsion of a cylindrical beam. Second example deals with more complex loading conditions applying the Bresse integrals technique to superelastic structures. Numerical results obtained in that way well agree with experimental determination performed on superelastic beam in bending and on a helical spring.

1. Introduction

In chapter 5 of this book, the constitutive equations for shape memory behavior were derived. The aim of the present chapter is to define the relationships between the thermomechanical loading conditions (force, torque and temperature) and the conjugated kinematical variables (displacement or rotation) for shape memory devices.

Such relations depend on the material behavior of the shape memory alloys itself (considering the different material parameters involved in the flow rule used) and the geometrical characteristics of the device. In most case non uniform loading conditions are applied on the structure (some parts are in tension while other parts are in compression for instance). So it is very important to be able to determine with a good accuracy the stress field inside a loaded structure. In shape memory alloys, in order to take into account the dissymmetry presented in superelasticity [1], the existence of an hysteresis with sub-loop trajectories [2] and the large influence of the thermomechanical loading history (training effect [3], fatigue degradation [4] and aging phenomena [5]), the determination of this local stress state is of first importance.

For a long time, due to the lack of valuable relation devoted to shape memory structures the thermomechanical response of such elements was deduced using charts, experimentally defined on products having similar characteristics [6, 7]. First determinations of calculation rule taking account the non linear behavior of shape memory alloys date from the very beginning of the 90's. In these first works, very simplified behavior rules were used and simple loading cases were studied (pure bending, pure torsion) [8, 10]. These first attempts were improved in two directions, more accurate behavior laws were used and more complex calculation methods were applied [11, 15]

This chapter is focussed on the superelastic behavior of beams. In a first section a transformation criterion and its associated flow rule are defined for this kind of behavior. In the second section the framework of strength of materials is applied to superelastic beam.

2. Transformation criterion and associated flow rule

It is well known that the superelastic behavior of shape memory alloys exhibits a large dissymmetry between tensile test and compressive one ([1] and [16, 18]). This phenomenon was reported on several systems of alloy and originates from crystallographical aspects of the martensitic transformation itself [19]. Such dissymmetry is an important feature for structure calculation. For instance in a cantilever beam one part of the material is stressed in tension when the other part is in compression (fig.1). Symmetrical transformation criterion (like those derived from the Von Mises one) can no longer be applied for these alloys. From micromechanical modelling developed in chapter 5, an analytical criterion taking account of this phenomenon, can be derived [20].

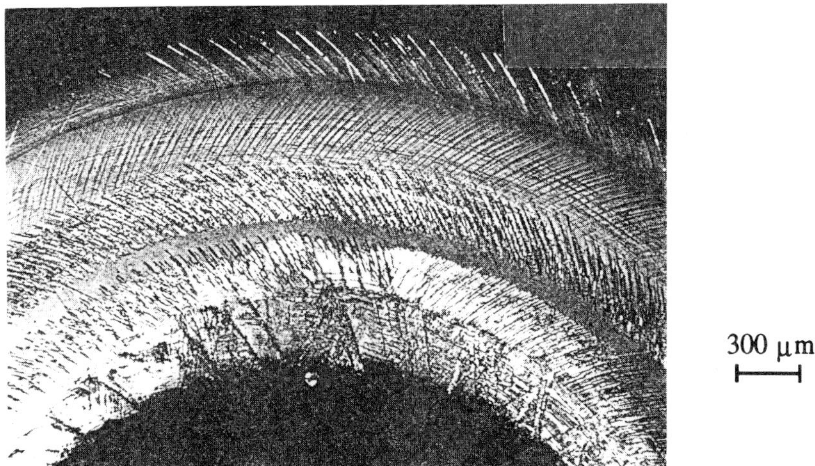

Figure 1 : Micrography of a superelastic Cu-Al-Be single crystal in bending (M_s = -20°C). Transformed area is larger in the tension region than in the compression one. Shift of the neutral axis (indicated by the untransformed domain) is very pronounced [21].

2.1. Transformation criterion

Using the crystallographical approach developed in chapter 5 a transformation surface is numerically obtained in the stress space (fig.2) [22]. This surface can be analytically described using a criterion $\mathcal{F}$ depending of the applied stress σ_{ij}, the temperature T and a set of internal variables denoted by V_k, such as :

$$\mathcal{F}(\sigma_{ij}, T, V_k) = 0 \tag{1}$$

As this surface presents an isotropic evolution [23], the macroscopic equivalent transformation strain ε^T constitutes a convenient internal variable. Considering the material as isotropic one can write (1) using eigenvalues σ_1, σ_2 and σ_3 of the stress tensor.

$$\mathcal{F}(\sigma_1, \sigma_2, \sigma_3, T, \varepsilon^T) = 0 \tag{2}$$

In shape memory alloys as the volume change associated to the transformation is weak, so that the hydrostatic pressure has little effect on the transformation temperature and criterion (2) can be expressed using the scalar invariants of the deviatoric stress tensor S (resp. $J_2 = \frac{1}{2} S_{ij} S_{ij}$ and $J_3 = \det(S_{ij})$). The simplest expression of this kind is obtained using the Prager equation [22] :

$$\mathcal{F}(J_2, J_3, T, \varepsilon^T) = J_2 \left(1 + b \frac{J_3^{2n}}{J_2^{3n}}\right) - (K(T, \varepsilon^T))^2 = 0 \tag{3}$$

In this criterion, parameters b and n and the function $K(T, \varepsilon^T)$, which depends on the temperature and on the equivalent transformation strain, have to be identified using particular loadings. The transformation surface numerically defined is well-captured setting n equal to 0.5 in equation (3) (fig 2). This gives the following macroscopic transformation criterion for superelasticity.

$$\mathcal{F}(J_2, J_3, T, \varepsilon^T) = J_2 \left(1 + b \frac{J_3}{J_2^{3/2}}\right) - (K(T, \varepsilon^T))^2 = 0 \tag{4}$$

Two different loading conditions must be performed to identify parameter b and function K. From tensile and compressive tests it comes :

$$b = \frac{\sqrt{27}}{2} \frac{\sigma_c^2 - \sigma_t^2}{\sigma_c^2 + \sigma_t^2} \tag{5}$$

where σ_t denotes the transformation stress observed in tension and σ_c the same quantity in compression. For simplicity first stage of the stress-strain tensile curve in polycrystalline material can be described using a parabolic approximation (fig.3) and decoupling the effect of the temperature on the influence of the progress of the transformation [11].

$$\sigma(T, \varepsilon^T) = \sigma_0(T) + k(\varepsilon^T)^m \tag{6}$$

Parameters m and k are material constants to be identified on the tensile curve of the alloy considered. In relation (6), the transformation stress $\sigma_0(T)$ is linearly related to the temperature according to M_s the transformation temperature, that is a material characteristic.

$$\sigma_0(T) = B(T - M_s) \tag{7}$$

In these conditions, function $K(T, \varepsilon^T)$ is identified as

$$K(T, \varepsilon^T) = \frac{1}{\sqrt{3}} (B(T - M_s) + k(\varepsilon^T)^m)(1 + \frac{2b}{\sqrt{27}})^{1/2} \tag{8}$$

Relations (5) and (8) totally define the criterion (4). This expression depends on five material parameters : k, m, M_s, B, b. From criterion (4) the superelastic behavior can be described in the framework of associated plasticity [24] using the normality rule.

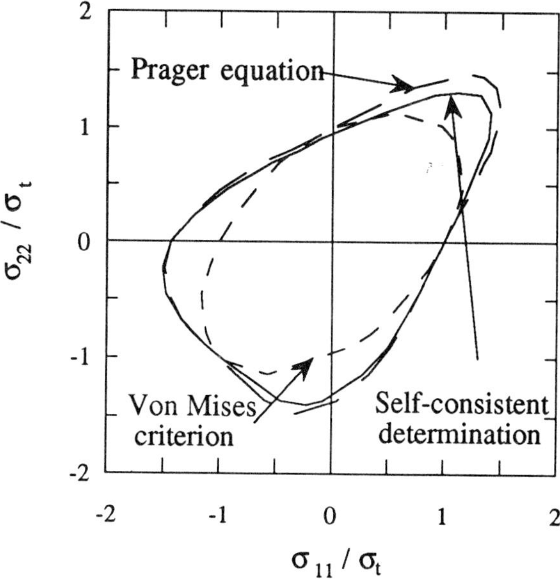

Figure 2 : Transformation surface numerically determined using the micromechanical approach presented in chapter 5. The shape of this surface is approximated using the non symmetrical criterion (4). Result obtained using a Von Mises criterion is given for comparison [22]

2.2. *Associated flow rule*

According to the general assumptions made in chapter 5 (infinitesimal strain, homogeneity of the elastic constants, no plastic deformation) the macroscopic total strain is additively composed with an elastic and a transformation part.

$$\varepsilon_{ij} = \varepsilon^{e}_{ij} + \varepsilon^{T}_{ij} \tag{9}$$

Elastic strain is derived using the Hooke's law. Transformation strain is obtained from the associated flow rule derived to criterion (4). Assuming the validity of the normality rule we obtain

$$d\varepsilon^{T}_{ij} = \frac{\partial \mathcal{F}(\sigma_{ij}, T, \varepsilon^{T})}{\partial \sigma_{ij}} \, d\lambda \tag{10}$$

In this expression, transformation multiplier $d\lambda$ is derived using the consistency condition.

$$d\mathcal{F}(\sigma_{ij}, T, \varepsilon^T) = \frac{\partial \mathcal{F}}{\partial \sigma_{ij}} d\sigma_{ij} + \frac{\partial \mathcal{F}}{\partial T} dT + \frac{\partial \mathcal{F}}{\partial \varepsilon^T} d\varepsilon^T = 0 \tag{11}$$

For simplicity we use in the following the notation $\mathcal{F}_{,ij}$ to represent the partial derivative of function $\mathcal{F}$ according to the component σ_{ij} of the stress tensor. Considering criterion (4) it comes:

$$\frac{\partial \mathcal{F}(\sigma_{ij}, T, \varepsilon^T)}{\partial \sigma_{ij}} = \mathcal{F}_{,ij} = \frac{\partial J_2}{\partial \sigma_{ij}} - \frac{b}{2 J_2^{3/2}} \left(J_3 \frac{\partial J_2}{\partial \sigma_{ij}} - 2 J_2 \frac{\partial J_3}{\partial \sigma_{ij}} \right) \tag{12}$$

and relation (11) turns into

$$\mathcal{F}_{,ij}\, d\sigma_{ij} - 2 K \left(\frac{\partial K(T, \varepsilon^T)}{\partial T} dT + \frac{\partial K(T, \varepsilon^T)}{\partial \varepsilon^T} d\varepsilon^T \right) = 0 \tag{13}$$

Introducing (10) into the definition of the equivalent transformation strain ε^T gives:

$$d\varepsilon^T = \left(\frac{2}{3} d\varepsilon_{ij}^T d\varepsilon_{ij}^T\right)^{1/2} = \left(\frac{2}{3} \mathcal{F}_{,ij} \mathcal{F}_{,ij}\right)^{1/2} d\lambda \tag{14}$$

Expression (14) allows to derive the multiplier $d\lambda$ using relation (13) and the following flow rule is obtained considering (11):

$$d\varepsilon_{ij}^T = \mathcal{F}_{,ij} \frac{1}{H} \left(\mathcal{F}_{,kl}\, d\sigma_{kl} - 2 K \frac{\partial K(T, \varepsilon^T)}{\partial T} dT \right) \tag{15}$$

In this equation parameter H is defined like

$$H = 2 K \frac{\partial K(T, \varepsilon^T)}{\partial \varepsilon^T} \left(\frac{2}{3} \mathcal{F}_{,kl} \mathcal{F}_{,kl}\right)^{1/2} \tag{16}$$

From definition (8) and from parabolic approximation (6) used, it comes :

$$\begin{aligned} \frac{\partial K(T, \varepsilon^T)}{\partial T} &= \frac{B}{\sqrt{3}} \left(1 + \frac{2 b}{\sqrt{27}}\right)^{1/2} \\ \frac{\partial K(T, \varepsilon^T)}{\partial \varepsilon^T} &= \frac{n\, k}{\sqrt{3}} (\varepsilon^T)^{m-1} \left(1 + \frac{2 b}{\sqrt{27}}\right)^{1/2} \end{aligned} \tag{17}$$

Flow rule (15) is then expressed as :

$$d\varepsilon_{ij}^T = \frac{\mathcal{F}_{,ij} (\varepsilon^T)^{1-m}}{m\, k \left(\frac{2}{3} \mathcal{F}_{,kl} \mathcal{F}_{,kl}\right)^{1/2}} \left(\frac{3}{2} \frac{1}{1 + \frac{2 b}{\sqrt{27}}} \frac{\mathcal{F}_{,pq}\, d\sigma_{pq}}{B(T - M_s) + k(\varepsilon^T)^m} - B\, dT \right) \tag{18}$$

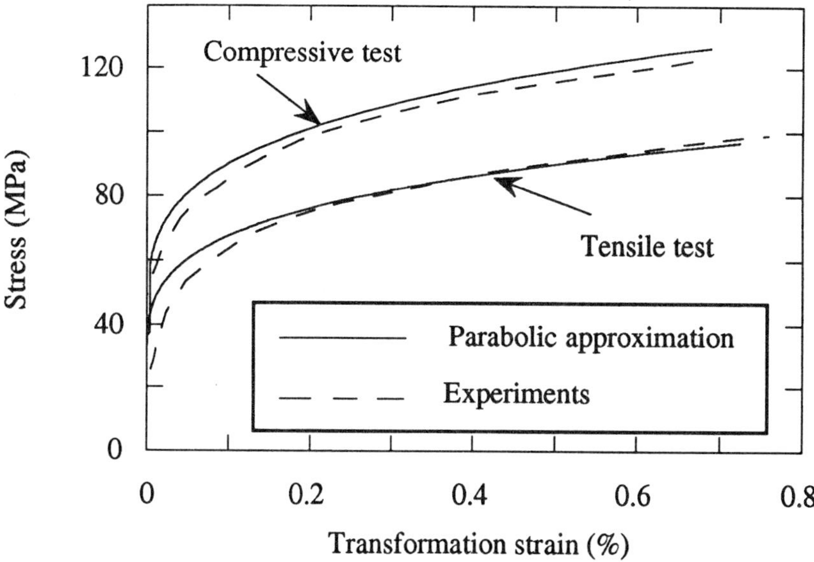

Figure 3 : Description of a stress-strain curve using the transformation criterion (4) and the parabolic approximation given in relation (6) [11].

where the $\mathcal{F}_{,ij}$ expressions are obtained using relation (12). Equation (18) defines a relationship useful to realize structures calculations under complex loading conditions [11]. According to the parabolic approximation (6) used, validity of flow rule (18) is limited to the first stage of transformation (ε^T smaller than 2 or 3% in a Cu-based alloy). This limitation has no consequence for a structure calculation point of view, because, due to fatigue life limitation, total strain must be limited too in each part of the device [25].

3. Structure calculation

Previous section deals with the material behavior. No geometrical aspect was taken into account and the stress tensor is considered as uniform. This is no longer the case in a real device. Geometrical characteristics become very important and the stress field is in most case strongly non uniform. In the framework of the beam theory adopted in this chapter, geometrical considerations are reduced to the description of plane sections and mean line. Due to the non linearity of the material behavior analytical expressions are scarcely obtained and numerical resolution must be used. To point out the different problems to be solved, this presentation is splitted into two parts. In the first one, aspects related to the stress heterogeneity inside a plane section is only considered. To illustrate this class of problem, the design of a cylindrical beam under a torsion loading is presented. In the second part, heterogeneity is now considered along the beam length. Bresse integral equations are used to solve this kind of problem.

3.1. Cylindrical beam in pure torsion, a simplified approach

One consider a cylindrical beam bounded by an external surface of radius R. This beam is loaded in pure torsion while temperature T is kept constant in the high temperature phase domain ($T > A_f$). In this part, to derive simple analytical expressions the material behavior law (defined from criterion (4) and flow role (18)) is strongly simplified considering transformation takes place at a constant shear stress level $\tau_S(T)$ only depending on the temperature. The fundamental hypothesis of beam theory (straight sections remain straight and normal to the axis) that satisfy the compatibility relationships is kept. In this scheme, due to both geometrical and loading symmetries, the strain field is simply described considering a rigid body rotation of the straight section characterized by an angle ψ per unit length. If r denotes the distance from the point considered to the center of the section, shear strain γ (r) is expressed as

$$\gamma(r) = r\ \psi \tag{19}$$

According to this relation, maximum strain is observed on the outer fiber of radius R. From this consideration wire begins to transform when a critical rotation Ψ_S is reached. In the same way critical rotation Ψ_f characterized the completion of the transformation on the outer part of the wire. These characteristic rotations are defined from shear strains γ_s and γ_f that are respectively associated to the beginning and to the end of the transformation. These quantities are temperature dependent.

$$\gamma(R) = \gamma_S = R\ \psi_S \qquad \text{and} \qquad \gamma(R) = \gamma_f = R\ \psi_f \tag{20}$$

This analysis defines three loading domains [26]. In the first one, rotation imposed to the wire is lower than Ψ_S and the alloy does not transform and behaves elastically (denoting by μ_a the elastic shear modulus). Shear stress is defined as :

(i) for $\Psi < \Psi_S$ $\qquad \tau(r) = \mu_a\,\gamma(r) = \mu_a\,\psi\ r$ (21a)

In the second one, when the rotation is between the two critical values ψ_s and ψ_f, some part of the material transforms. Due to geometrical and loading symmetries, elastic and transforming zones are assumed to be circular and concentric. They are limited by a radius denoted as R_a . For r smaller than R_a, the shear stress is smaller than the transformation stress and the material behavior is elastic. For r larger than R_a, the transformation occurs.

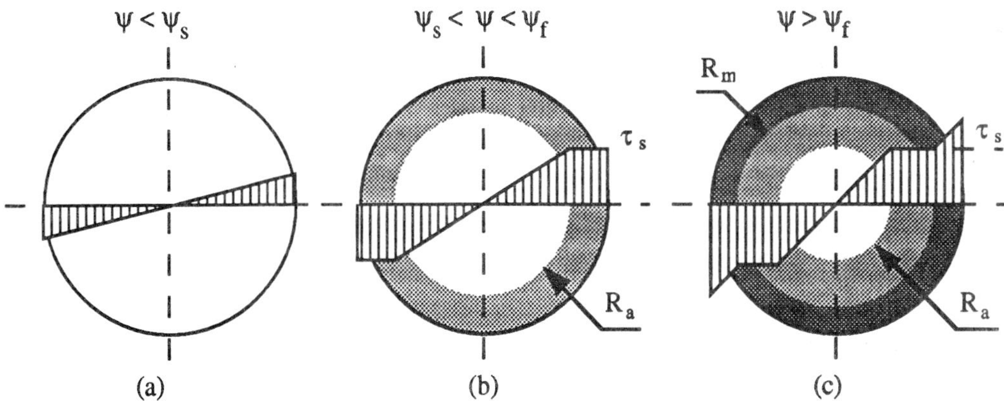

Figure 4 : Stress distribution in a straight section for a circular beam in pure torsion: (a) elastic behavior of austenite; (b) during the phase transformation (c) taking account the elastic behavior of the fully transformed domain

(ii) when $\psi_s < \psi < \psi_f$ if $r < R_a$ $\tau(r) = \mu_a \psi r$

if $r \geq R_a$ $\tau(r) = \tau_s(T)$ (21b)

In the last loading domain, for rotation, larger than ψ_f outer region is fully transformed and this martensitic domain behaves elastically (denoting by μ_m the elastic shear modulus of martensite). R_m is the characteristic radius delimiting the fully transformed area. For r smaller to R_a the material keeps the elastic behavior of austenite. For r larger than R_m, the material is totally transformed in the new phase and its behavior is again elastic.

(iii) when $\psi > \psi_f$ if $r < R_a$ $\tau(r) = \mu_a \psi r$

if $R_a \leq r \leq R_m$ $\tau(r) = \tau_s(T)$ (21c)

$r > R_m$ $\tau(r) = \mu_m(\psi r - \psi_f R) + \tau_s(T)$

For a given section one obtain thus the stress profile described by figure 4. During the loading path, evolution of radii R_a and R_m are obtained considering :

$$\gamma(R_a) = \gamma_s = \psi R_a \quad \text{and} \quad \gamma(R_m) = \gamma_f = \psi R_m \tag{22}$$

Taking account relations (20) and (22) gives :

$$R_a = \frac{\psi_s}{\psi} R \qquad \text{and} \qquad R_m = \frac{\psi_s}{\psi} R \tag{23}$$

For any value of ψ, considering the particular expression (21) for the shear stress, the applied torque M^t needed to induce this rotation, is deduced using the classical expression :

$$M^t = \int_S \tau(r)\, r\, dS \tag{24}$$

Relation between the torque M^t and the rotation ψ becomes:

(i) when $\psi < \psi_s$

$$M^t = \frac{S\, \mu_a\, R^2}{2}\, \psi \tag{25a}$$

The usual relationship of the strength of materials is found.

(ii) when $\psi_s < \psi < \psi_f$

$$M^t = S\, [\frac{\mu_a R^2}{2}\, \psi\, (\frac{\psi_s}{\psi})^4 + \frac{2}{3} \tau_s R\, (1 - (\frac{\psi_s}{\psi})^3)] \tag{25b}$$

(iii) when $\psi > \psi_f$

$$M^t = S\, [\frac{\mu_a R^2}{2}\, \psi\, (\frac{\psi_s}{\psi})^4 + \frac{2}{3} \tau_s R\, (1 - (\frac{\psi_s}{\psi})^3) + \frac{\mu_m R^2}{2}\, \psi\, (1 - (\frac{\psi_s}{\psi})^4) - \frac{2}{3} \mu_m R^2\, \psi_f\, (1 - (\frac{\psi_f}{\psi})^3)] \tag{25c}$$

Relations (25) describe the superelastic behavior of a cylindrical beam in pure torsion. In these relations the influence of the temperature is taken into account through the evolutions of τ_s, γ_s and γ_f. Despite large assumptions used to describe the shear behavior of the material, one notice that the response of the beam is complex. The unloading response can be obtained using the same framework but complexity is increased due to the existence of sub-loop trajectories inside the hysteresis loop [2]. Despite the large assumptions made on the behavior law in that part (transformation under constant stress level), this approach deals with complex non linear relation. So, numerical methods can not be avoid to solve structure calculation problems for shape memory elements as soon as loading conditions are not uniform along the beam (furthermore if considering now equation (18) a more realistic material behavior is used). Bresse integral equations appear as a method easy to use in such case.

3.2. Bresse integrals applied to shape memory structure

Results obtained in the previous section are restricted to beam submitted to pure torsion (M^t is kept constant along the beam length). Same kind of result can also be derived for bending [8] but heterogeneity of the loading state along the beam length must be taken into account. Two methods can be used to account with the non uniformity of the loading state (or with complex loading condition like in bending-torsion test). Application of energetic theorem (like the Castigliano one) is one of this method [11], using of Bresse integral equations is an other one.

Our attention will be focussed on the second one due to its great versatility and its ability to be implemented in a computer code.

Lets us consider a beam described considering its mean line $\overset{\frown}{AB}$ and plane sections (S) having inertia center G. Bresse equations relate the generalized displacement (displacement of the mean line and rotation of the plane sections) with the generalized forces applied on this structure. In each section an external loading can be decomposed between normal and shear forces (resp. N, T_2 T_3) and twisting and bending moment (resp. M_t, M_{f2},M_{f3}). Assuming classical Navier-Bernoulli and De Saint Venant assumptions these generalized forces allow to define elementary displacement $\vec{u}$ and elementary rotation $\vec{\omega}$ for each section (S) like :

$$\left\{\begin{array}{lll} u_1 = \dfrac{N}{\int_S L_{11}(r)dS} & u_2 = \dfrac{T_2}{\int_S L_{44}(r)dS} & u_3 = \dfrac{T_3}{\int_S L_{44}(r)dS} \\ \omega_1 = \dfrac{M_t}{\int_S L_{44}(r)\rho^2 dS} & \omega_2 = \dfrac{M_{f2}}{\int_S L_{11}(r)z^2 dS} & \omega_3 = \dfrac{M_{f3}}{\int_S L_{11}(r)y^2 dS} \end{array}\right. \tag{26}$$

where $L_{ij}(r)$ denotes the secant modulii of the material for the considered plane section (S). These modulii are to be determined considering, like in the previous section, that elastic and transforming domain coexist if the criterion (4) is satisfied in some part of this section. In such case, moduli $L_{ij}(r)$ are non uniform and must be determined using flow rule (18). Kinematical variables u_i and ω_i are both functions of the external loading through the generalized forces, on the shape of the plane section and on the material behavior.

In this framework, displacement of a section B is obtained as a function of the displacement and rotation observed in a given section A (resp. $\vec{U}_A$ and $\vec{\omega}_A$) and of the contribution of elementary displacement and rotations of all the section lying between A and B.

$$\vec{U}_B = \vec{U}_A + \vec{\omega}_A \wedge \vec{AB} + \int_{AB} \vec{U}\, ds + \int_{AB} \vec{\omega} \wedge \vec{GB}\, ds \tag{27}$$

In the same way, rotation of section B is defined as :

$$\vec{\omega}_B = \vec{\omega}_A + \int_{AB} \vec{\omega}\, ds \tag{28}$$

Due to the inelastic behavior of shape memory alloys, an incremental formulation of equations (26), (27) and (28) is more adapted. In this case tangent modulii are used instead of the secant one in equations (26) [11]. This scheme of resolution was applied to several loading conditions and extended to anisothermal cases [20]. Examples of results obtained in that way are presented in the next section.

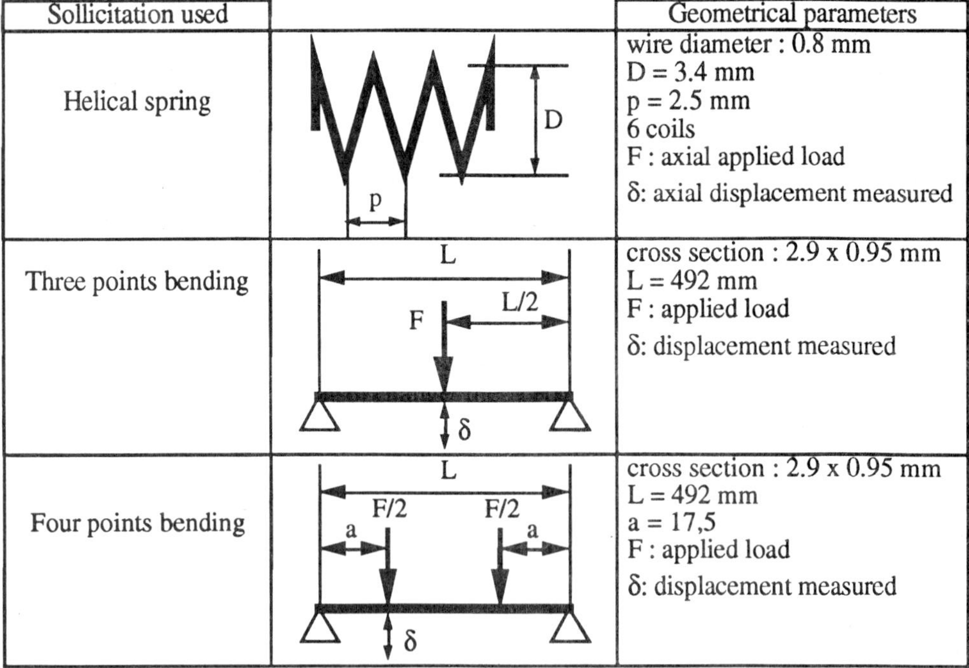

Sollicitation used		Geometrical parameters
Helical spring		wire diameter : 0.8 mm D = 3.4 mm p = 2.5 mm 6 coils F : axial applied load δ: axial displacement measured
Three points bending		cross section : 2.9 x 0.95 mm L = 492 mm F : applied load δ: displacement measured
Four points bending		cross section : 2.9 x 0.95 mm L = 492 mm a = 17,5 F : applied load δ: displacement measured

Table 1 : Geometrical characteristics of the superelastic element used in this study.

3.3. *Examples of application*

Material behavior described using criterion (4) and associated flow rule (18) is now applied to simulate the load-displacement curve of several structures. These curves are obtained using the framework of Bresse integral equations (27) and (28). Results obtained on prismatic beams in bending and helical spring are presented (geometrical characteristics are given on table 1).

The different material parameters used to compute the load-displacement response of these superelastic devices are deduced from tensile test experiment, calorimetric or resistivity measurement performed on the same alloys (Table 2) [27].

	Alloy A (helical spring)	Alloy B (bending tests)
Nominal composition (weight %)	Cu - 26.2 Zn - 3.7 Al - 1.1 Ni	Cu - 11.4 Al - 0.6 Be
Thermal treatment	15' at 843°C, water quenched, 1 hour annealing at 100°C	As received (hot rolled, water quenched)
Elastic modulus	34 GPa	73.2 GPa
M_s	- 18°C	- 105°C
A_f	- 6°C	- 90°C
K	8.3 GPa	12.7 GPa
m	1	1
B	1.2 MPa·K^{-1}	2.8 MPa·K^{-1}
b	0.84	0.84

<u>Table 2 :</u> Value of the different material parameters involved in flow rule (18) to describe the superelastic behavior of alloys used in this study.

For the helical spring design, considering the thread p is small in comparison with the mean spring diameter $\mathcal{D}$, it is assumed that the wire cross sections are only submitted to pure torsion. The load-displacement curve obtained is in good agreement with measurement performed on the real device (figure 5) [28].

Simulation of several four points and three points bending tests are performed at different temperature on the same alloy and for the same beam geometry (see table 1). To account with the large displacement involved in these loading conditions, the beam geometry is reactualized after each calculation step. Influence of the shear forces is neglected. The different simulated load-displacement curves such obtained are all in good agreement with the experimental ones (figure 6) [27]. From these results it can be concluded that this structure calculation framework is able to take into account evolution in the loading conditions, both about the test temperature and the nature of the solicitation.

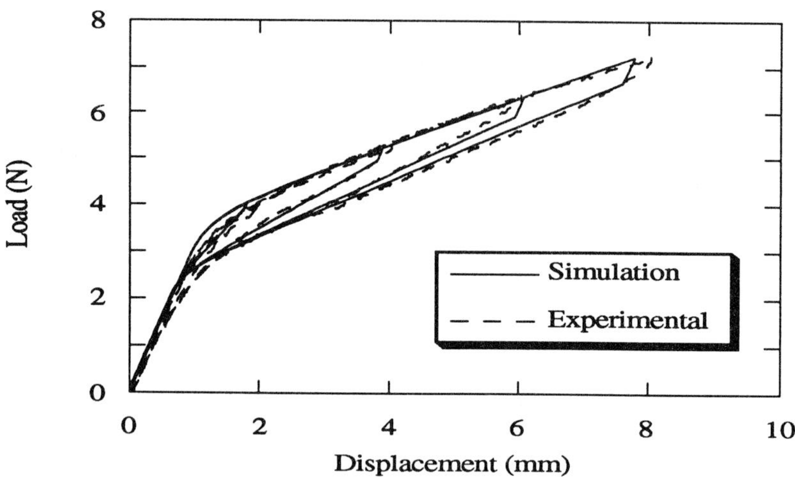

Figure. 5 : Simulation of the load-displacement curve obtained on a superelastic helical spring at T = 20°C°. Material data and geometric characteristics used are listed in tables 1 and 2. Comparison with experimental result performed on the real device [28].

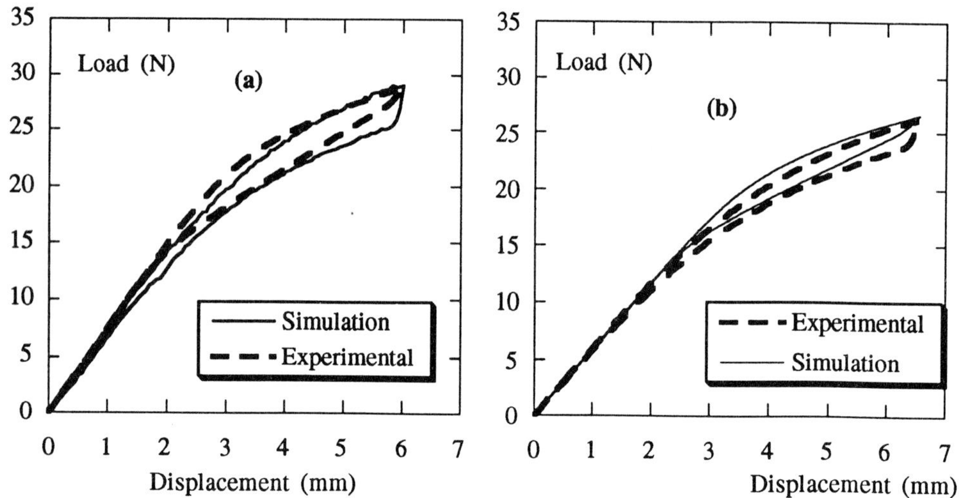

Figure 6 : Different bending tests performed on alloy B.
Comparison between experimental measurement and numerical simulation.
(a) Four points bending at T = - 0.3°C ; (b) Three points bending at T = 20°C

Unloading part simulated in figures 5 and 6 and sub-loop trajectories described in figure 5 are derived in the same way as the loading part from the definition of a reverse transformation criterion and its associated flow rule. Details about modelling of unloading process are given in [11]. Differences observed in the beginning of the unloading process in bending (figure 6) come from some wear appearing during the experiment between the beam and the support.

4. *Conclusion*

Beam theory framework is well adapted to describe the non linear behavior of superelastic structure. Influence of the test temperature and geometrical aspects are well taken into account. Hysteresis behavior related to the martensitic transformation and sub-loop trajectories can also be described using such description. Limitations comes to the general beam theory assumptions. Thus this method is restricted to simple loading conditions and simple geometrical shape but present the advantage of very fast computation (few minutes on a PC computer).

Nevertheless a great number of superelastic applications can be described in that way and and as more and more applications using shape memory alloy elements involved the superelastic properties of these alloys [29, 30], calculations presented in this chapter can give however useful informations to engineers in charge to design shape memory superelastic devices.

To deal with more complex structure or complex loading conditions, finite element method must be used. This is realized implementing the transformation criterion (4) and its associated behavior law (18) into the industrial finite element code ABAQUS [31].

References

[1] MELTON K.N., "*Ni-Ti based shape memory alloys*" in DUERIG T.W., MELTON K.N., STÖCKEL D., WAYMAN C.M., "Engineering aspect of shape memory alloys", 1991, p.21.

[2] MÜLLER I., HUIBIN XU, "*On the Pseudo-elastic Hysteresis*", Acta metall. mater., Vol. 39, No. 3, pp. 263-271 (1991).

[3] CONTARDO L., GUÉNIN G., "*Training and Two Way Memory Effect in Cu-Zn-Al Alloy*", Acta metal. mater., Vol. 38, No. 7, pp. 1267-1272 (1990).

[4] THUMANN M., HORNBOGEN E., "*Thermal and mechanical fatigue in Cu-base shape memory alloys*", Z. Metallkde, Bd. 79, H2, pp. 119-126 (1988).

[5] AHLERS M., "*Martensite and equilibrium phases in Cu-Zn and Cu-Zn-Al alloys*" Prog. in Materials Science, Vol. 30, pp. 135-186 (1986).

[6] STALMANS R. "*CADSMA*" Procs COMETT Course EEC 87/2/C-2/00863, Barcelona (1989).

[7] WARAM T., "*Design principles for Ni-Ti actuators*", in "Engineering aspects of shape memory alloys", T.W. Duerig et al. Eds, Butterworth-Heinemann (1990).

[8] ATANACKOVIC T, ACHENBACH M.,"*Moment-curvature relations for a pseudoelastic beam*", Continuum Mech. Thermodyn., p.73 (1989).

[9] PATOOR E., EBERHARDT A., BERVEILLER M., "*Comportement pseudo-élastique et effet mémoire de forme double sens. Application à la torsion de barreau cylindrique*", Rev. Traitement Thermique, Vol. 234, 43 (1990).

[10] TOBUSHI H.,TANAKA K., "*Deformation of a Shape Memory Alloy Helical Spring*", JSME Int. Journal, 1991, Series 1, Vol. 34, No. 1, p. 83.

[11] GILLET Y., PATOOR E., BERVEILLER M., "*Structure calculation applied to Shape Memory Alloys*", Journal de Physique IV, Vol. 5, pp. C2-343-348 (1995).

[12] BRINSON L.C., LAMMERING R., "*Finite element analysis of the behavior of shape memory alloys and their applications*", Int. J. Solids Structures, Vol. 30, No 23, pp 3261-3280 (1993).

[13] LECLERC S., LEXCELLENT C., GELIN J.C.,"*A finite element calculation for the design of devices made of shape memory alloys*", Journal de Physique IV, Vol. 6, pp. C1-225-234 (1996).

[14] MANACH P.Y, FAVIER D., RIO G.,"*Finite element simulations of internal stresses generated during the ferroelastic deformation of NiTi bodies*", Journal de Physique IV, Vol. 6, pp. C1-235-244 (1996).

[15] AURICCHIO F., TAYLOR R.L., "*Shape memory alloy superelastic behavior: 3D finite-element simulation*", Procs. 3rd Int. Conf. on Intelligent Material, Lyon, France, pp.487-492 (1996).

[16] ROUMAGNAC P., "*Étude du comportement en phase martensitique et de l'effet mémoire simple d'alliages à mémoire de forme Ni Ti en fonction du mode de sollicitation*", Thèse de doctorat N° D599, UTC, Compiègne, France, 1993.

[17] VACHER P., LEXCELLENT C., "*Study of Pseudoelastic Behaviour of Polycrystallin Shape Memory Alloys by Resistivity Measurement and Acoustic Emission*", Procs. ICM 6, Kyoto, Japan, pp. 231-236 (1991).

[18] MANACH P.Y., "*Etude du comportement thermomécanique d'alliages à mémoire de forme NiTi,*", Thèse de doctorat, Institut National Polytechnique de Grenoble, France (1992).

[19] PATOOR E., EL AMRANI M., EBERHARDT A., BERVEILLER M., "*Determination of the origin for the dissymmetry observed between tensile and compressive tests on shape memory alloys*", Journal de Physique IV, Vol. 5, pp. C2-495-500 (1995).

[20] HAUTCŒUR A., to appear

[21] GILLET Y., "*Dimensionnement d'éléments simples en alliage à mémoire de forme*", Thèse de doctorat, Université de Metz, France, 1994.

[22] EL AMRANI ZIRIFI M., "*Contributions à l'étude micromécanique des transformations martensitiques thermoélastiques*", Thèse de doctorat, Université de Metz, France, 1994.

[23] EL AMRANI M., BENSALAH M.O., PATOOR E., EBERHARDT A., BERVEILLER M.,"*Détermination des surfaces de début et de fin de transformation pour un alliage à mémoire de forme*", 1° Congrès de Mécanique, Rabat (Maroc), pp. 59-66 (1993).

[24] LEMAITRE J., CHABOCHE J.L., Mécanique des matériaux solides, CH. 5, Plasticité, Edition Dunod, 1985.

[25] SAKAMOTO H., "*Fatigue behavior of monocrystalline Cu-Al-Ni shape memory alloys under various deformation modes*", Trans. of the Japan inst. of Metals, Vol.24, No. 10, pp. 665-673 (1983).

[26] GILLET Y., PATOOR E., BERVEILLER M., "*Elements of stucture calculation for shape memory device*", Journal de Physique IV, Vol. 1, pp. C4-151-156 (1991).

[27] GILLET Y., PATOOR E., BERVEILLER M., "*Beam theory applied to shape memory alloys*", Procs. SMST '94, Asilomar conference center, Pacific Grove, California , USA, pp 169-174-416 (1994).

[28] GILLET Y., PATOOR E., BERVEILLER M., "*Mechanical engineering for shape memory alloys*", ICOMAT 92, Monterey (USA), 1992, p. 1241.

[29] MORAN S. "*Flexible instruments in minimal access surgery*", Procs. SMST '94, Asilomar conference center, Pacific Grove, California , USA, pp 411-416 (1994).

[30] PAINE J.S.N., ROGER C.A., "*Review of multi-functional SMA hybrid composite material and their applications*", Procs. ASME WAM '94, Chicago, AD- Vol. 45 / MD- Vol. 54, p. 37-45 (1994).

[31] SEGARD E., GILLET Y, MARTINY M. PATOOR E, "*Finite element simulation of superelastic elements*", to appear.

APPLICATION TO HEAT TREATMENT AND SURFACE TREATMENT

F.D. Fischer
University of Mining and Metallurgy, Leoben, Austria

ABSTRACT

An attempt is made for the calculation of residual stresses due to nitriding in a component. The knowledge of the determinating processes and the evaluation of their influence on the residual stress state may assist the selection of the nitriding paramters for the optimization of the nitriding process. Special interest is laid on the description of the micromechanical effects occuring during nitriding. The following processes are considered within the model: * The creep behavior and plasticity, * the diffusion of nitrogen into the material, * the kinetics of the growth of nitrides, * the volumetric strains due to the interstitial nitrogen and the nitrides, * the transformation induced plasticity (here precipitation induced plasticity, PRIP) caused by the plastic accommodation process around the growing nitrides. The influence of all these processes on the residual stress state is studied for a cylindrical specimen.

The results are quantatively compared with measurements from the literature.

INTRODUCTION

Residual stresses and strains which are developed in heat treated components may lead to a distortion or fracture of this component, see e.g. the contribution by Denis and Gautier on residual stress calculations in this book. On the other hand a residual stress state may be produced to improve the usability of an component leading to an increased fatigue resistance or better tribological and anticorrosion properties. One technique is based on the diffusion of carbon or nitrogen into the material which may be stored in form of interstitial atoms or in precipitations as carbides or nitrides. In both cases a local volume change occurs which may lead to a macroscopic residual compression stress state near the surface of the component. Considering the nucleation and growth of precipitations one may think on a diffusive phase transformation process accompanied by a significant volume change δ.

Although the transformations fraction f is usually not more than 3 % a TRIP-effect happens at the mesoscopic level. Let us speak from here on from "precipitation induced plasticity" (PRIP) instead of TRIP.

With respect to the technological aspects of carburizing and nitriding the reader is referred to the recent conference proceedings [1] or the conference report [2].

In the following context we concentrate on the nitriding process. However, the outlined procedure is quite general and not restricted to nitriding. Only the actual data are taken for the nitriding process.

Estimations and calculations of residual stresses in nitrided steels are published very rarely. Mittemeijer [3] gave some relations for the residual stresses on the surface under consideration of the development of lattice dislocations. Lesage et al. [4] calculated the residual stresses due to carbonitriding. He considered the phase proportions after quenching and the local volume variations due to the carbon and nitrogen content. In the work of Oettel et al. [5] calculations were made including volume changes during nitriding and an assumption for some relaxation stress. Recently, in a work of Schreiber et al. [6] an attempt was made to calculate the residual stresses using a model including volume changes during nitriding and an assumption of a creep law for the material. Within this work a method is reported for the calculation of nitriding stresses. Here we consider the diffusion of nitrogen into the material, the kinetics of the growth of nitrides, the volumetric dilatation, the creep and PRIP due to the growth of nitrides.

1. METALLURGICAL ASPECTS OF NITRIDING

Details on the metallurgical processes taking place during nitriding in iron-based alloys can be taken from Daves, [7], chpt. 4. One must distinguish between nitrogen diffusion either in pure iron and unalloyed carbon steels or in alloyed steels. If we consider the first case (pure iron and unalloyed carbon steels) the nitrogen concentration c (z) can be calculated by the simple relation, see [3],

$$c(z) = c_o(t)\operatorname{erfc}(z / (2\sqrt{D_N t})), \tag{1}$$

c_o is the nitrogen concentration on the surface, D_N the coefficient of nitrogen diffusion, t is the time and z the distance from the surface. Equation (1) is only valid for $\sqrt{D_N t}$ being small in relation to the thickness of the component. The nitrogen is mainly stored as interstitial nitrogen or in α - or γ - nitrides. In the investigation at hand the relation $c(z)$, equ. (1), takes into account both effects together. In the case of alloyed steels the main mechanism is the nucleation and growth of precipitations, represented by a dimensionless parameter $X_G, 0 \leq X_G \leq 1$ describing the precipitated fraction of the component G (e.g. G being chromium, Cr, and the precipitations being mainly chromium nitrides, CrN). G = 0 if no G-precipitations exist. $X_G = 1$ if all G (e.g. the whole chromium contents of a Cr-steel) has formed G-N-compounds at a distance z from the surface. Following standard physical chemistry, see e.g. Christian [8], we can then assume a time law X_G in the form

$$\frac{dX_G}{dt^*} = BX_G^{1/3}(1 - X_G). \tag{2}$$

B denotes a constant, see Bulgach et al. [9], $B \sim 10^3 - 10^4$ $[s^{-1}]$, which depends on the amount of precipitations per volume, the concentration of the alloying element in the precipitations and the solid solution and the diffusion coefficient of the alloying element. The growth of nitrides is only possible within a layer with the thickness d on the surface of the structure, where enough nitrogen is diffused into the material. The thickness d is determined by the parabolic time law using the parameter β, $(\beta \sim 1)$,

$$d = \beta\sqrt{D_N t}\,. \tag{3}$$

The time t*, within the nitrides can grow, is then given by the relation

$$t^* = \begin{cases} 0 & : z \geq d \\ t - \dfrac{z^2}{\beta^2 D_N} & : z < d. \end{cases} \tag{4}$$

The solution of (2) leads to the transcendental equation

$$\frac{1}{B}\left[\frac{1}{2}\ln\frac{1+X_G^{1/3}+X_G^{2/3}}{\left(1-X_G^{1/3}\right)^2} - \sqrt{3}\arctan\frac{2X_G^{1/2}+1}{\sqrt{3}} + \frac{\pi}{2\sqrt{3}}\right] - t^* = 0\,. \tag{5}$$

The equation (5) can be solved easily by inversion. That means one calculates to a given value of $X_G, 0 \leq X_G \leq 1$ the value t*B and then looks for a given value of t*B to the corresponding X_G. Experimental data, see Fig. 1, from Heger [10] have been taken to compare the prediction of the analysis (5). Considering a 4.5 wt. % Cr-Steel the nitrogen concentration X_G in wt % can easily be calculated for CrN-type nitrides from $X_G = X_{Cr}$, (5), with $D_N = 6.4 * 10^{-6}$ [$mm^2 s^{-1}$],

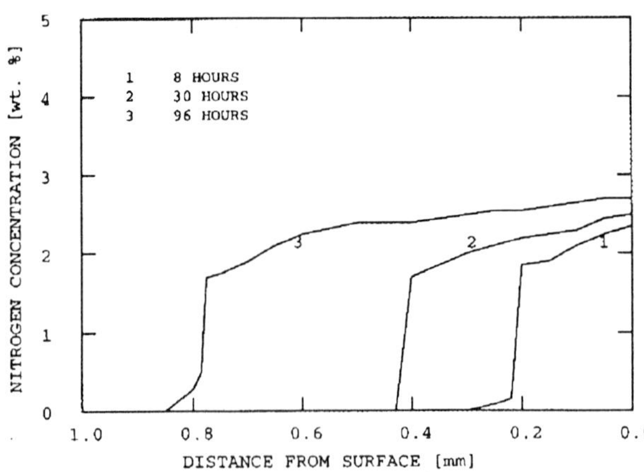

Fig. 1: Nitrogen concentration profiles for a 4.5 wt% Cr-Steel measured by Heger [10].

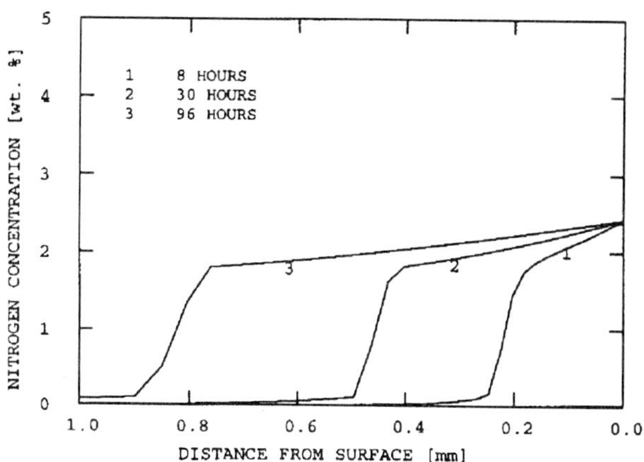

Fig. 2: Nitrogen concentration profiles for a 4.5 wt% Cr-Steel calculated due to (5).

$B = 7200\ [s^{-1}]$, $\beta = 1$. The result is depicted for various diffusion times t in relation to the z-coordinate in Fig. 2.

One can see that a typical nitrided profile possesses a depth in the order of a mm. Surprisingly good agreement with the experimental data can be achieved by this rather simple concept.

Mittemeijer presented in his work [3] a linear relation between the contents of G (in at.%) in an iron based alloy and the volumetric dilatation $\left.\frac{\Delta V}{V}\right|_{max}$ (in %), if the whole amount of G has precipitated as nitrides ,

$$\left.\frac{\Delta V}{V}\right|_{max} = \kappa_G G, \ G\ [\text{at. }\%]; \tag{6.1}$$

κ_G is ~ (1,75/4) for G being chromium; watch that 1 at.% Cr corresponds to 0,93 wt. % Cr.

The actual volumetric dilatation $\frac{\Delta V}{V}$ follows now as

$$\frac{\Delta V}{V} = \left.\frac{\Delta V}{V}\right|_{max} X_G . \tag{6.2}$$

The volume strain δ of a specific nitrid is very high and amounts from $0.2 < \delta < 0.5$. Knowing δ allows to calculate the volume fraction f of the nitrides as

$$f = \frac{\Delta V}{V}\frac{1}{\delta} = \left.\frac{\Delta V}{V}\right|_{max} \frac{X_G}{\delta} . \tag{6.3}$$

2. DERIVATION OF THE PRIP-STRAIN RATE

The following derivations are based on the contribution of Fischer on modelling and simulation of TRIP in this book, expecially on the chapters 1.1 Kinematics and 2.2. „The deformation process", subchapter 2.2.1.3 for diffusive transformation. However, the current model is specifically adapted to a configuration, see Fig. 3, with a small volume fraction f of precipitations, $f << \alpha_P$. The factor α_P, $0.524 \leq \alpha_P \leq 0.740$, describes the arrangement of the nuclei of the precipitations which are thought as spheres growing with a radius R inside the original parent phase sphere with the corresponding volume fraction α_P and the radius $\tilde{R}$. Since the precipitations are very small, only a part of the surrounding parent phase shell is assumed to plastify with the corresponding volume fraction $(1-\kappa)(\alpha_P - f)$, $0 \leq \kappa \leq 1$. Both the remaining part of the parent phase within the sphere with radius $\tilde{R}$ and the remaining gap between the spheres with the volume fraction $(1 - \alpha_P)$ are in the elastic state. From this consideration the relative radii follow as

$$\frac{R_P}{R} = \{[(1-\kappa)(\alpha_P - f) + f]/f\}^{1/3}, \tag{7.1}$$

$$\frac{R_P}{\tilde{R}} = \{[(1-\kappa)(\alpha_P - f) + f]/\alpha_P\}^{1/3}. \tag{7.2}$$

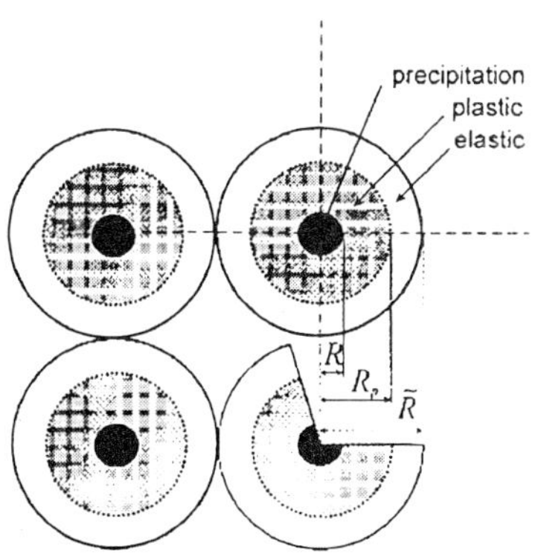

Fig. 3: Micromechanical model for the derivation of the PRIP-strain rate.

The hydrostatic pressure p in a sphere with radius R surrounded by a plastic shell with radius R_P can be taken from the literature, e.g. [11], p. 270 - 278, as

$$p = -\frac{2}{3}\left[1 + \ln\left(\frac{R_P}{R}\right)^3 - \left(\frac{R_P}{\tilde{R}}\right)^3\right]\sigma_{y0}, \tag{8.1}$$

σ_{y0} being the yield stress of the parent phase. Introducing of (7.1), (7.2) into (8) leads to

$$p = -\frac{2}{3}\left[\frac{\kappa(\alpha_P - f)}{\alpha_P} + \ln\frac{(1-\kappa)\alpha_P + \kappa f}{f}\right]\sigma_{y0}. \tag{8.2}$$

Lėt us first consider a uniaxial load stress $\Sigma_{33} = \Sigma$. As a simplification only the stress component $\Sigma_{33} = \Sigma$ is assumed to exist in the elastic part of the parent phase. Further we assume that in the transformed sphere only the load stress Σ_{33} as well as the hydrostatic pressure $-p\delta_{ij}$ are active. If we denote with $\overline{\sigma}^P_{33}$ the average of the stress component σ_{33} over the plastified zone, the global equilibrium in load direction delivers

$$[(1-\alpha_P) + \kappa(\alpha_P - f)]\Sigma + f(\Sigma + p) + (1-\kappa)(\alpha_P - f)\overline{\sigma}^P_{33} = \Sigma$$

or

$$\overline{\sigma}^P_{33} = -\frac{\alpha_P p}{(1-\kappa)(\alpha_P - f)} + \Sigma \tag{9.1}$$

and with (8.2)

$$\overline{\sigma}^P_{33} = \frac{2}{3}\frac{f}{(1-\kappa)}\left(\frac{\kappa}{\alpha_P} - \frac{1}{\alpha_P - f}\ln\frac{f}{(1-\kappa)\alpha_P + \kappa f}\right)\sigma_{y0} + \Sigma. \tag{9.2}$$

However, $\overline{\sigma}^{p}_{33}$ can also be expressed by the strain rate $\dot{\varepsilon}_{33}$ due to the concept explained in detail in the a.m. chapter 2.2 of Fischer's contribution in this book. Here we take a result which has been published earlier by Fischer, [12], equ. (30a), relating $\dot{\varepsilon}_{33}$ linearly with $\overline{\sigma}^{p}_{33}$ in the case of a "Greenwood and Johnson relation"

$$\dot{\varepsilon}_{33} = \frac{5}{6}\delta\frac{\overline{\sigma}^{p}_{33}}{\sigma_{yo}}\frac{2f}{\alpha_P + f} . \tag{10}$$

The PRIP-strain rate $\dot{\varepsilon}^{PP}_{33}$ of the specimen follows from $\dot{\varepsilon}^{PP}_{33} = \dot{\varepsilon}_{33}(1-\kappa)(\alpha_P - f)$ and insertion of (9.2) into (10) as

$$\dot{\varepsilon}^{PP}_{33} = \frac{5}{3}\delta\frac{\alpha_P - f}{\alpha_P + f}(1-\kappa)\left[\frac{\Sigma}{\sigma_{y0}} + \frac{2}{3}\frac{f}{(1-\kappa)}\left(\frac{\kappa}{\alpha_P} - \frac{1}{\alpha_P - f}\ln\frac{f}{(1-\kappa)\alpha_P + \kappa f}\right)\right]\dot{f} . \tag{11.1}$$

A generalization with respect to a triaxial load stress state Σ_{ij} with its deviator S_{ij} is proposed as

$$\dot{\varepsilon}^{PP}_{ij} = \frac{5}{2}\delta\frac{\alpha_P - f}{\alpha_P + f}(1-\kappa)\left[\frac{\|S_{ij}\|}{\sigma_{y0}} + \frac{2}{3}\frac{f}{(1-\kappa)}\left(\frac{\kappa}{\alpha_P} - \frac{1}{\alpha_P - f}\ln\frac{f}{(1-\kappa)\alpha_P + \kappa f}\right)\right]\dot{f}\frac{S_{ij}}{\|S_{ij}\|} \tag{112}$$

$\|S_{ij}\|$ is any proper norm of S_{ij} . It should be mentioned that due to the development of precipitations an addidtional hardening may happen leading to a substitution of σ_{y0} by $\sigma_{y0} + \tilde{\sigma}_{y0}(f)$, for details see [7], chpt 2.2.6.

For the volume fraction f relation (6.3) can now be inserted into (11.2).

3. MATERIAL DESCRIPTION

Within the material model we use an additive decomposition of the total strain rate as

$$\dot{\varepsilon}_{ij} = \dot{\varepsilon}_{ij}^{e} + \dot{\varepsilon}_{ij}^{th} + \dot{\varepsilon}_{ij}^{p} + \dot{\varepsilon}_{ij}^{PP} + \dot{\varepsilon}_{ij}^{cr} . \tag{12.1}$$

In addition to the elastic strain rate $\dot{\varepsilon}_{ij}^{e}$, the thermal strain rate $\dot{\varepsilon}_{ij}^{th}$ and the plastic strain rate $\dot{\varepsilon}_{ij}^{p}$ both the PRIP strain rate $\dot{\varepsilon}_{ij}^{PP}$, see (11.2), and the creep strain rate $\dot{\varepsilon}_{ij}^{cr}$ are applied. $\dot{\varepsilon}_{ij}^{cr}$ is expressed as

$$\dot{\varepsilon}_{ij}^{cr} = \frac{3}{2}\left[\frac{c\left(a^{1/c}\right)}{\Sigma_{eq}}.\Sigma_{eq}^{b/c}.\left(\dot{\varepsilon}_{eq}^{cr}\right)\frac{c-1}{c}\right]S_{ij}, \tag{12.2}$$

Σ_{eq} is the von Mises equivalent stress, $\varepsilon_{eq}^{cr} = \int_0^t \left(\frac{2}{3}\dot{\varepsilon}_{ij}^{cr}\,\dot{\varepsilon}_{ij}^{cr}\right)^{1/2} d\tilde{t}$.

Of course, the creep law contains (at least) three parameters a, b, c. One should be careful by taking those data for the parent phase material since just in the area where the nitriding stresses will arise, the precipitations (even in the nanoscale) may prevent creeping compared with creep in the "pure" parent phase. Setting the parameters a, b, c is a difficult job and, more or less, an open question. The reader is referred here e.g. to a paper by Arzt, [14].

The material formulation above is implemented as a user supplied material subroutine (UMAT) into the finite element code ABAQUS, [15].

4. SIMULATION OF NITRIDING STRESSES

The calculations are performed for a cylindrical specimen with a radius r = 3.5 mm , which is nitrided at 540°C for 5 hours. After the nitriding process the specimen has been cooled down slowly to room temperature so no additional eigenstresses occur due to an inhomogeneous temperature field. The difference between the thermal expansion coefficients of the parent phase and the precipitations has been ignored in relation to the significant volume strain δ.

This means finally that the residual stress state at nitriding has been "frozen" into the specimen. Therefor, "classical" plasticity is not activated, which means finally that in addition to the elastic strain rate only the PRIP strain rate as well as the creep strain rate are considered. The material data used are as follows:

*Nitriding

$D_N = 6.4 * 10^6$ [$mm^2 s^{-1}$] ,

B= 7200 [s^{-1}] ,

β= 1 ;

*Elastic behavior

E = 210000 [N mm^{-2}], Young's modulus,

ν= 0.3, Poisson's ratio ;

*PRIP strain rate

$\left.\frac{\Delta V}{V}\right|_{max}$ = 0.005 (1 wt. % Cr), δ = 0.39 (CrN), σ_{y0} = 400 [MPa] ;

*Creep strain rate

a = 1.25 *10^{-17} [sec^{-c} N^{-b} mm^{2b}], b= 4.3 , c = 1.

The creep material data are based on those for AISI 4340 steel (0.6 - 0.8 wt.% Cr), see [16]. However the value a here is 1/10 of that given in [16] based on the assumption that creep is slowed down by the existence of precipitations. To the knowledge of the author no experimentally assured data on creep in a nitrided material exist!
For the sake of comparison the longitudinal stress σ_{33} is calculated for various data sets and depicted in Fig. 4.,

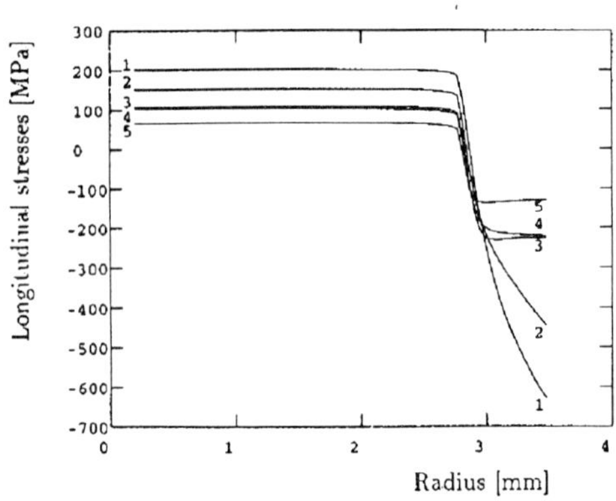

Fig. 4: Calculated longitudinal stress profiles for various material data sets 1 to 5, see the text.

Curve 1: elastic, no PRIP, no creep,
Curve 2: elastic, PRIP, no creep,
Curve 3: elastic, no PRIP, creep,
Curve 4: elastic, PRIP, creep. It can be seen that a pure elastic behaviour would lead to unrealistically high stresses σ_{33}. The PRIP-effect reduces the stress level significantly! However, if PRIP and creep are competing, the creep-effect would dominate.

Curve 5 shows a changed situation for a 2wt.% Cr-steel with a double of $\left.\frac{\Delta V}{V}\right|_{max}$ as above. An "overdominant" creep-effect with 10 times the parameter a is assumed to be operative leading to a further reduction of the stress level.
Finally the experimental data on σ_{33} by Heger and Bergner, [17] have been veryfied in a further study, (1 wt.% Cr, 4 hours nitriding), see Fig. 5. It is interesting to note that

σ_{33} shows a "hook"-behavior near the surface. This could only be reached by reducing the β value to 0.5. The applied data set allows a really good prediction of the measured longitudinal residual stresses

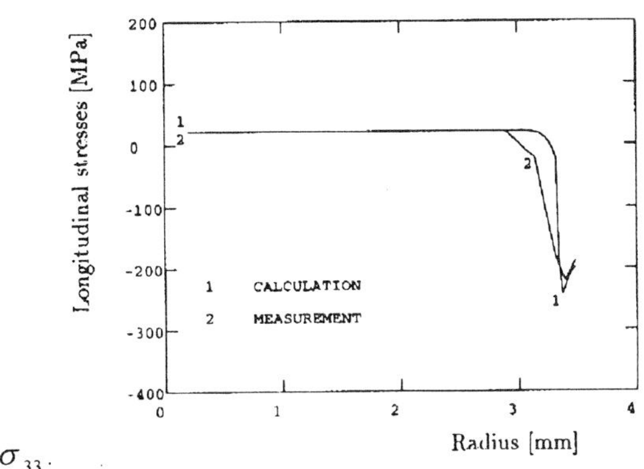

σ_{33} .

Fig. 5, Longitudinal stress profiles: (1) calculated, (2) measurement by Heger and Bergner [17].

CONCLUSIONS

For the calculation of nitriding residual stresses in structures one has to know a couple of material properties for each alloy under investigation which are not published yet. For a good quantitative description of nitriding residual stresses additional measured or calculated data are neccessary as the creep rates of the material as a function of the nitrogen concentration, the diffusion coefficient of nitrogen for the specific alloy, data for the kinetics of precipitation growth and data for the PRIP-effect.

Based on the caluculations at hand we can make following conclusions:

* The intensity of the residual stresses depend mainly on the creep behaviour of the steel.
* In the case of creep resistant steels, however, PRIP has to be considered.
* The distribution of the nitriding stresses is strongly related to the kinetics of the growth of precipitations, especially the decrease of the longitudinal and

circumferential stresses near the surface in the direction orthogonal to the surface.

ACKNOWLEDGEMENT

The author expresses his thanks to Dr. W. Daves, senior researcher at the Christian Doppler Laboratory for Micromechanics of Materials, Leoben/Vienna, who developed the theoretical framework and provided the computations.

REFERENCES

1. Carburizing and Nitriding with Atmospheres: Conference Proceedings (Eds J.Grosch, J. Morral and M. Schneider), ASM International, Materials Park 1995.
2. Conference Report: Carburizing & nitriding, Adv. Mat. Processes, 149 (1996), 63-71.
3. Mittemeijer, E.J.: Gitterverzerrungen in nitriertem Eisen und Stahl, Härt.-Techn. Mitt., 36 (1981), 57 - 66 .
4. Lesage, J., Degallaix, G. and J. Barralis: Calculation of residual stresses in case carbonitrided steel, in: Fatigue & Stress (Ed. M.P. Lieurade), ITT-International, France 1989, 265 - 276.
5. Schreiber, G., Oettel, H., Darbinjan, W. and R. Wiedemann: Residual stresses in the precipitation zone of nitride layers, in Residual Stresses (Eds V. Hauk, M.P. Hougardy, E. Macherauch and H.-D. Tietz), DGM, Oberursel 1993, 965-974.
6. Oettel, H. and G. Schreiber: Eigenspannungsbildung in der Diffusionszone, in: Nitrieren und Nitrocarburieren (Eds E.J. Mittemeijer and J. Grosch), AWT, Darmstadt 1991, 139 - 151.
7. Daves, W.: Mikro- und makromechanische Simulation des Deformations-verhaltens von Stählen unter Berücksichtigung von Umwandlungs- und Diffusionsvorgängen, Fortschrittsberichte VDI, Reihe 18, Nr. 141, VDI-Verlag, Düsseldorf 1994.

8. Christian, J.W.: The Theory of Transformation in Metals and Alloys, Pergamon Press, Oxford 1965.

9. Bulgach, A.A., Solodkin, G.A. and L.A. Gliberman: Computersimulation of the kinetics of the growth of nitrides, Met. Sci. Heat Treat. (USSR) 26 (1984), 41 - 48.

10. Heger, D.: Die mathematische Modellierung des Stickstoffkonzentrationsprofiles der Ausscheidungsschicht nitrierter Eisenlegierungen am Beispiel von Fe - Cr - Legierungen in Abhängigkeit der Nitrierparamter , Ph.D. Thesis, Bergakamie Freiberg 1990.

11. Reckling, K.A.: Plastizitätstheorie und ihre Anwendung auf Festigkeitsprobleme, Springer-Verlag, Berlin et al. 1967.

12. Fischer, F.D.: A micromechanical model for transformation plasticity in steels, Acta metall. mater. 38 (1990), 1535 - 1546.

13. Daves, W. and Fischer F.D.: Finite element simulation of the development of residual stresses during nitriding under consideration of the micromechanical and metallurgical processes, in: Materials Science Forum, Vols 163-165, Trans Tech Publications, Switzerland (1994), 713-718

14. Arzt, E.: Creep of dispersion strengthened materials: a critical assessment, Res Mechanica 31 (1991), 399-453

15. Hibbitt, Karlsson and Sorensen Inc., ABAQUS User Manual, Version 5.4, Pawtucket, R.I. 1994

16. Rothman, M.F.: High Temperature Property Data: Ferrous Alloys, ASM International, Metals Park (1980), 316 ff.

17. Heger, D. and D. Bergner: Berechnung der Stickstoffverteilung in gasnitrierten Eisenlegierungen, Härt.-Techn. Mitt., 46 (1991), 331-338

CONSIDERING STRESS-PHASE TRANSFORMATION INTERACTIONS IN THE CALCULATION OF HEAT TREATMENT RESIDUAL STRESSES

S. Denis

CNRS URA 159, Nancy, France

Abstract

The models for describing stress-phase transformation interactions (transformation plasticity and the effect of stress on transformation kinetics) are reviewed. It is shown that the macroscopic constitutive law that is generally used in the calculation of heat treatment residual stresses in steels allows to describe well the thermomechanical behaviour of the material in its transformation range. An analysis of the role of these interactions on the stress/strain evolutions during the heat treatment of a piece and on the residual stress states is proposed based on numerical simulations.

1. INTRODUCTION

The heat treatment of metallic alloys generally involves thermal, metallurgical and mechanical phenomena and their couplings. Indeed, during a heat treatment (quenching, surface hardening, thermochemical treatment) the alloy undergoes temperature variations and phase transformations. The resulting changes in density give rise to internal stresses in the piece. Only first order stresses will be considered here. In addition, the phase transformation affects the temperature evolutions through transformation enthalpy and changes in thermophysical properties and the stresses affect the kinetics of the phase transformation. Moreover, the phase transformation modifies the thermomechanical behaviour of the material through transformation plasticity and through the change of mechanical properties.
As heat treatment processes are difficult to control and optimize, big efforts have been made during the last fifteen years in order to predict the results of the heat treatment (microstructures and hardness, residual stresses and distorsions of the piece). The above mentioned phenomena have to be modelled and further introduced in mechanical calculations by finite elements. This work has been particularly developed for steels.

Considering the great complexity of the metallurgical and thermomechanical behaviour of steels, it has been considered that the only way to achieve a quantitative prediction of internal stresses and strains in a steel part during heat treatment is to establish macroscopic models based on experimental data.
The aim of this paper is firstly to review how the stress-phase transformation interactions (the effect of stress on transformation kinetics and the transformation plasticity phenomenon) have been modelled for the purpose of residual stress calculations. Then, we will analyse how these models allow to describe the thermomechanical behaviour of the material during phase transformation as it can be measured in a specimen (during a tensile test). Finally, from numerical simulations, we will analyse how the stress-phase transformation interactions affect the development of internal stresses and strains in a piece during heat treatment.

2. MODELLING OF STRESS-PHASE TRANSFORMATION INTERACTIONS

When phase transformation takes place under stress different effects are observed : phase transformation kinetics is modified (it will be called "metallurgical" interaction), a transformation plasticity deformation (or transformation induced plasticity) occurs (it will be called "mechanical" interaction) and the transformation mechanism can be modified. These effects have been reviewed for different phase transformations which occur in steels (diffusion dependent and martensitic ones) both from the point of view of the mechanisms involved, the experimental determination and the modelling [1-5].
It should be underlined that the material in a piece undergoing a heat treatment is generally submitted to triaxial stress states and to small plastic strains (typically less than 1%). Thus, in the following we shall only recall the phenomena that may occur under such conditions and give a synthesis of the models used presently for the calculation of internal stresses.

2.1 Metallurgical interaction

2.1.1 Phenomena

The present knowledge concerns essentially the effect of hydrostatic pressure or uniaxial stresses on the phase transformation :
- under hydrostatic pressure, the equilibrium diagram is modified and the transformation kinetics are slowered. For the martensitic transformation, a decrease in Ms temperature is observed with increasing pressure. During the austenite decomposition, the isothermal transformation curves (IT diagrams) and the continuous cooling transformation curves (CCT diagrams) are shifted towards lower temperatures and longer times.
- under uniaxial stresses (tension or compression), the kinetics of diffusional decomposition of austenite (ferritic and pearlitic transformation) are accelerated (due to an increase in nucleation rate). The same is true when plastic strain has occured in austenite prior to the transformation. Concerning martensitic transformation, for stresses lower than the yield stress of austenite, only Ms temperature increases. A compressive stress is less effective than a tensile stress (this behaviour is explained by the Patel and Cohen model). A plastic strain in austenite (in the range 0-5%) before martensitic transformation leads to a decrease in Ms temperature which is more significant as the deformation occurs at a temperature near Ms (this effect has to be related to strain hardening of austenite).

2.1.2 Modelling

Only a few models have been developed in order to take into account the effect of the stress/strain states generated during cooling in a piece on the kinetics of the phase transformations [6, 7, 8].
In the calculation of internal stresses, the progress of martensitic transformation is generally described by a Koistinen Marburger type law :

$$y_m = y_\gamma\,(1\text{-}\exp(\text{-}k(M_s\text{-}T))$$

where y_m is the volume fraction of martensite, y_γ the volume fraction of austenite, T the temperature and k a constant.
It has been assumed, on one hand that the effect of the stress/strain state only results in a variation in M_s temperature and on the other hand that alone the internal stress state affects the M_s temperature. Indeed, it can be estimated that an accumulated plastic strain in austenite during cooling of about 1% would lead to a decrease of M_s temperature of a few degrees whereas a stress state close to the yield stress of austenite can lead to an increase of 30°C [3].
In order to take into account that both shear and normal components of the stress affects M_s temperature, Inoue [6] proposed a model in which the change in M_s temperature is related to the mean stress σ_m and to the second invariant of the stress deviator tensor J_2 :

$$\Delta M_s = A\sigma_m + BJ_2{}^{1/2}$$

A and B are material dependent coefficients that are determined experimentally [1].

Concerning diffusion dependent transformations, the effect of stress/strain states on the isothermal kinetics has been described through a shift of the IT curves in the time scale. As before, it has been assumed that the effect of plastic strain is negligible when compared to the effect of the stress. Moreover, the effect of the mean stress can be neglected when compared to the effect of the stress deviator (at least for middle size pieces in which the mean stress does not exceed a few hundred MPa during quenching for example) [7]. Considering that the isothermal transformation is divided into an incubation period followed by the progression of the transformation that is described by a Johnson-Mehl-Avrami law [9], we have taken into account the effect of stress in the following manner [7] :

for the incubation period : $\tau_{IT\sigma} = \tau_{IT}(1+D_k)$ with $D_k = g_k(\sigma_e)$

for the growth period : $y_k = y_{maxk}\,(1\text{-}\exp\,\text{-}b_k(t/(1+D_k))^{n_k})$

τ_{IT} is the isothermal incubation period (subscript σ denotes "with effect of stress"), y_k is the volume fraction of constituent k (y_{maxk} the maximum value), n_k and b_k are temperature dependent coefficients,
t the time (t=0 is the end of incubation period), D_k is the shift of IT curves, g_k is an experimental function, σ_e is the Von Mises equivalent stress.
A different assumption has been proposed by Inoue [6] in which the shift of IT curves is only a function of the mean stress.
For bainitic transformation, a model taking into account a variation of Bs temperature with the stress state (in the same way as for Ms temperature) and a modification of the transformation kinetics (in the same way as for pearlitic transformation) can be proposed [8].

2.2 Mechanical interaction

2.2.1 Phenomenon

Transformation plasticity is a deformation that appears for a transforming material under an applied stress even for stresses lower than the yield stress of the phases. This deformation is generally attributed to two basic mechanisms

- the anisotropic plastic accommodation of the transformation strain. For transformations that occur with a volume change alone (ferritic and pearlitic transformation in steels for example), this mechanism is the only one to be considered.
- the orientation of the product phase by the stress state. This mechanism intervenes when transformation strain has a shear component.

Both mechanisms are effective for martensitic transformation in steels [5, 10].

Most of the studies concern constant applied stresses during the transformation. They generally show a linear relationship between transformation plasticity and the applied stress for stresses lower than the yield stress of the parent phase and for a fully transformed specimen. The variation of transformation plasticity with the progress of transformation has been found as linear for pearlitic and ferritic transformations of steels [11,12]. For martensitic transformation, this variation is generally found to be nonlinear with a high slope at the very beginning of the transformation due to the orientation effect (mechanism 2) and to plastic accommodation (mechanism 1) and a decrease of this slope as the orientation effect decreases [5, 10]. Moreover, the nonlinear relationship between transformation plasticity and the applied stress at given martensite contents has been clearly analysed [5]. The behaviour during bainitic transformation is found similar to the one during martensitic transformation as shown in some recent studies [13, 14].

2.2.2 Modelling

In the past, different models have been developed for describing transformation plasticity.They were based on either one or the other above mentioned mechanisms. A review can be found in [4]. More recently, micromechanical approaches in which both mechanisms are considered for martensitic transformation have been developed [15-18].

For the purpose of calculating internal stresses during heat treatment, mainly a phenomenological approach has been used in which the evolution law for transformation plasticity has been written from experiments.

Under uniaxial stresses (tension or compression) transformation plasticity strain has been assumed of the form :

$$\varepsilon^{tp} = K\sigma f_k(y_k)$$

where constant K and function f_k are both determined experimentally.

For triaxial stress states, it has been assumed that the same relations hold for transformation plasticity strains as for classical plastic strains (Von Mises associated flow rule) i.e. the transformation plasticity strain rate is proportional to the stress deviator [19, 20]. But since transformation plasticity appears as soon as transformation starts even though the stress is very small, it has been supposed that no yield criterion needs to be verified. The following expression [20] is now used by many authors :

$$\dot{\varepsilon}_{ij}^{tp} = 3/2 K f'_k(y_k) \dot{y}_k s_{ij}$$

s_{ij} are the components of the stress deviator tensor.

(Let us point out that the progress of transformation y_k has to be considered stress state dependent (see 2.1.2)).
As an alternative, Hamata et al. [21] proposed to use a power law of the stress in the expression of ε^{tp} and derived a transformation "viscoplasticity" strain by using an analogy with the theory of viscoplasticity.
More recently, Videau et al [22] obtained an expression of transformation plasticity by using also viscoplasticity theory. The authors introduce strain hardening effects and propose a proportionality of the transformation plasticity strain with an effective stress i.e. $(s_{ij} - X_{ij})$ where X_{ij} represents an "internal" stress (it is analogous to the introduction of a back stress in classical plasticity).
It must be underlined that all these expressions are supposed to be valid for the different transformations occuring during a heat treatment (diffusion dependent and martensitic ones) whatever the mechanisms involved.
It has to be mentioned that Leblond [23] has brought a theoritical justification for proportionality between transformation plasticity and stress deviator (or effective stress by considering kinematic hardening) by developing a model of the thermomechanical behaviour of a two phase transforming material. The author considers only mechanism 1 (plastic accommodation) to be at the origin of transformation plasticity and only the volume change associated with the transformation. In this model, explicit formula are obtained for constant K and function f. This model takes also into account a deviation from linearity for high applied stresses. In a later work, Fischer [24] formulated also the transformation plasticity strain as linearly related to the stress deviator for a martensitic transformation by considering only mechanism 1 (plastic accomodation of the volumic variation and the shear).
A micromechanical approach by finite elements for a transformation without shear [25] let also conclude that transformation plasticity can be considered as an additional strain in the macroscopic constitutive law of the material. The transformation plasticity strain was found to depend linearly on the macroscopic stress deviator (for constant or linearly increasing and decreasing applied stresses and not too small transformed fractions).
For martensitic transformation, first results obtained by a micromechanical model that takes into account both mechanisms but also formation of self accommodating plates [17], seems to show a similar relationship (under a constant non uniaxial loading).
In addition, recent experiments under multiaxial loadings [26] may bring some new elements to that discussion.

3. THERMOMECHANICAL BEHAVIOUR LAW OF THE MATERIAL

The modelling of the thermomechanical behaviour of a material that undergoes a phase transformation can be considered in three aspects :
- a metallurgical aspect : the microstructural evolutions of the material for the different thermal histories to which it will be submitted must be described
- a mechanical aspect : the elastic and plastic (or viscoplastic) behaviour of the multiphase material must be known at the different temperatures
- the interactions between stresses/strains and phase transformations must be taken into account.

These last years, different models for calculating the kinetics of phase transformations in steels during continuous cooling and heating have been developed [9, 20, 27-33] and associated with the thermomechanical behaviour law of the material in order to predict heat treatment residual stresses.
In the following we will only describe briefly the constitutive equation that is the most commonly used. Details on the formulations can be found in the references.

3.1 Modelling

The macroscopic constitutive equation that governs the behaviour of a material undergoing a phase transformation is generally written by assuming that the total strain rate is an addition of different contributions:

$$\dot{\varepsilon}_{ij}{}^{t} = \dot{\varepsilon}_{ij}{}^{e} + \dot{\varepsilon}_{ij}{}^{th} + \dot{\varepsilon}_{ij}{}^{tr} + \dot{\varepsilon}_{ij}{}^{tp} + \dot{\varepsilon}_{ij}{}^{in}$$

$\dot{\varepsilon}_{ij}{}^{e}$ is the elastic strain rate which is related to the stress rate by Hooke's law. Young's modulus and Poisson's ratio have to be taken temperature dependent and microstructure dependent. (Here, "microstructure" means "volume fractions of the different phases").

$\dot{\varepsilon}_{ij}{}^{th}$ is the thermal strain rate that takes into account the thermal expansion coefficients of the different phases and their dependence on temperature.

$\dot{\varepsilon}_{ij}{}^{tr}$ is the strain due to the volume change associated with the different phase transformations.

$\dot{\varepsilon}_{ij}{}^{tp}$ is the trasformation plasticity strain rate

$\dot{\varepsilon}_{ij}{}^{in}$ is the inelastic strain rate :
- either the plastic strain rate when no viscous effects are considered. It is calculated using the classical theory of plasticity with the associated hardening rules (isotropic and/or kinematic) [6, 20, 27, 34, 40] or obtained from a micro-macro approach [23, 35]
- or the viscoplastic strain rate [21, 22, 36-39].

All material parameters (yield stress, hardening parameters, strain rate sensitivity...) are to be considered as temperature and microstucture dependent. Mixture rules are generally assumed.

In addition, it should be mentioned that to take into account hardening is quite complex if a phase transformation occurs. Models have been proposed in order to take into account some possible "recovery" of strain hardening during a phase transformation, i.e.the fact that the newly formed phase remembers only part or even nothing of the previous hardening [22, 23, 34].

3.2 Application to a tensile test during phase transformation

The above described material behaviour law has been incorporated in various finite element programmes in order to calculate the internal stresses and strains during the heat treatment of a steel part. Eventhough numerous studies have dealt with the experimental validation of such calculations on heat treated parts, the ability of this type of law to represent the thermomechanical behaviour of a specimen in a much simpler case (for instance a tensile test) has been rarely analysed [41, 42]. This approach will be further illustrated in the following.

Figure 1 shows the mechanical behaviour of a Fe-0.2%C alloy measured during cooling (the cooling rate is constant 0.5°C/s) and during tensile tests performed at various deformation rates [41]. During cooling, the flow stress of austenite increases as temperature decreases. From 770°C, softening appears in two domains that correspond to the austenite-ferrite and austenite-pearlite transformations. Softening is quite dependent on deformation rates. The behaviour in the transformation range has been explained by analysing the possible origins of the softening and by quantifying them mainly :

- the variation of the mechanical properties (ferrite has a lower flow stress than austenite in the studied temperature range)
- the volume change associated with the transformation
- the transformation plasticity deformation

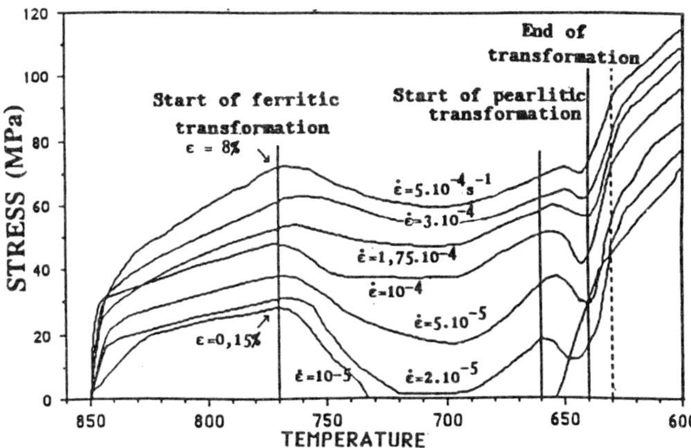

Figure 1: Stress variations versus temperature during the cooling of a Fe-0.2C steel [41].

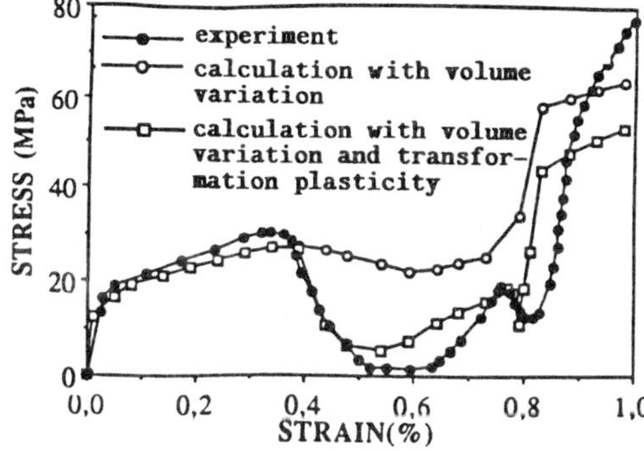

Figure 2: Calculated and measured stress variations versus deformation $\dot{\varepsilon}=2\ 10^{-5}s^{-1}$ [41].

For example, calculated results (here with an analytical model [41]) show clearly (figure 2) that for a small deformation rate, transformation plasticity is the main contribution to the softening of the alloy. In fact, it has been shown that the amplitude of softening depends largely on the ratio between deformation rate and transformation rate. For a same transformation rate, a small deformation rate will lead to a large softening; at the opposite, a large deformation rate will limit the softening and may even lead to hardening of the phase mixture. A more thorough analysis of these results can be found in [41].

More recently, a similar study has been performed for a low alloyed steel focusing on the bainitic tranformation [14]. As previously, the mechanical behaviour of the individual phases has been determined by tensile tests at different temperatures. In addition, the behaviour has been measured either during isothermal transformation or continuous cooling transformation. These results have been compared to calculated ones using different models. Figure 3 shows a tensile test during continuous cooling at constant deformation rate. The experimental curve shows two domains in which some softening appears. They are correlated to the transformation austenite-ferrite (that starts at 48s) and to the transformation austenite-bainite (at 100s). The martensitic transformation (at 160s) does not lead to stress relaxation.

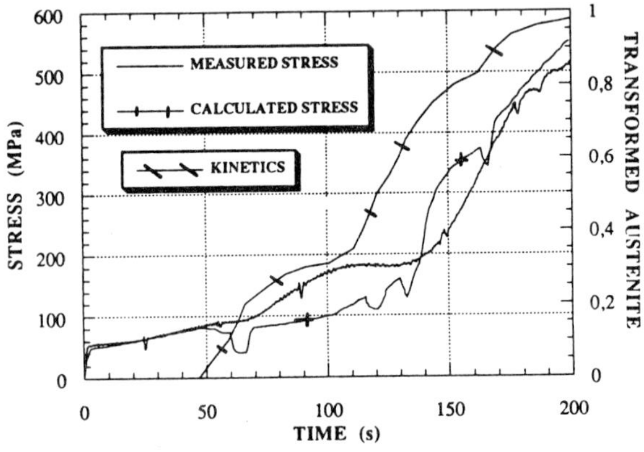

Figure 3: Stress variations versus time and phase transformation kinetics during continuous cooling
—— Experiment+ Calculation (with the finite element software described in [43])

The calculated curve has been obtained using a thermoelastoplastic behaviour law of the material with isotropic hardening and "loss of memory" during phase transformation [43]. The calculation describes very well the behaviour of austenite and shows well the stress relaxations in the transformation domains. Nevertheless, these stress relaxations are overestimated. Moreover, the stress level reached by the mixture austenite+ferrite is underestimated by the calculation.
A calculation has also been performed with the behaviour law proposed by Leblond [23]. The result is given on figure 4. Although this model gives a similar description of the material behaviour in the austenitic domain and during the austenite-ferrite transformation as the previous one, it does not show a stress relaxation associated with the bainitic transformation.

Although the complete analysis of all the results cannot be given here, our results have shown that the used macroscopic constitutive law allows a correct qualitative description of the material behaviour. Nevertheless, an accurate quantitative description is much more difficult to achieve because it depends highly on the accuracy of the input data, essentially the kinetics of the transformations and the mechanical properties of the individual phases (that are temperature dependent but also morphology dependent [41]).

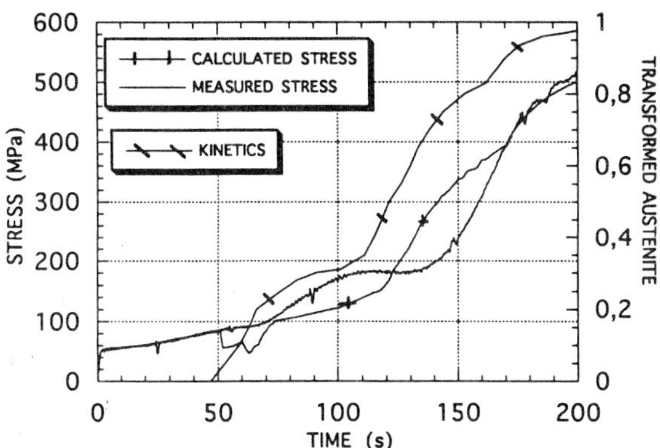

Figure 4: Stress variations versus time and phase transformation kinetics during continuous cooling
—— Experiment + Calculation (with the finite element code SYSWELD [44])

4. CALCULATION OF HEAT TREATMENT INTERNAL STRESSES

The calculation of heat treatment internal stresses includes the calculation of the temperature distributions in the piece during the treatment (solution of the heat equation) coupled with the calculation of the microstructural evolutions followed by the solution of the mechanical problem (equilibrium of stresses, compatibility of strains with suitable boundary conditions) by the finite element method. In some cases, the coupling between stress/ strain and microstructural evolutions is also taken into account.
In addition, all the input data concerning both the material behaviour (transformation kinetics, thermophysical properties, mechanical properties...) and the heat treatment process (heat flux densities...)
have to be determined. Many authors have developed such an approach. The aim was on one hand to understand better the development of internal stresses all along the treatment and on the other hand to validate experimentally the results of the calculation (mainly by comparing calculated and measured residual stress profiles). Hereafter, we will try to give some general ideas on the role of stress-phase transformation interactions in the prediction of residual stresses and illustrate them by examples.

4.1 Effect of the mechanical interaction

The most thorough analyses of the role of transformation plasticity on the development of internal stresses have been performed about ten years ago [34, 40, 45-49]. Some more recent studies can also be mentioned [50, 51]. The main points are reported below [3].

4.1.1 Effect on the stress/strain evolutions

During a heat treatment stresses arise due to the thermal gradients generated in the piece during cooling (case of quenching) or during heating and cooling (case of surface heat treatment) and to the phase transformations that occur at the different locations in the piece. Chemical gradients of the piece (as the ones generated during thermochemical treatments) will also intervene.

The stress evolution at a given location in the piece can generally be decomposed in periods where the material is loaded either in tension or in compression and periods where unloading occurs. Particularly as a phase transformation occurs, the associated volume change induces unloading. During loading, the stresses are generally high enough to produce plastic strains. For example, figure 5 shows the calculated stress (5a) and plastic strain (5b) evolutions of a cylinder during martensitic quenching in cold water. In the first stage of cooling, the surface is under tension (the center in compression) due to the thermal gradients alone and plastic strains occur in austenite.

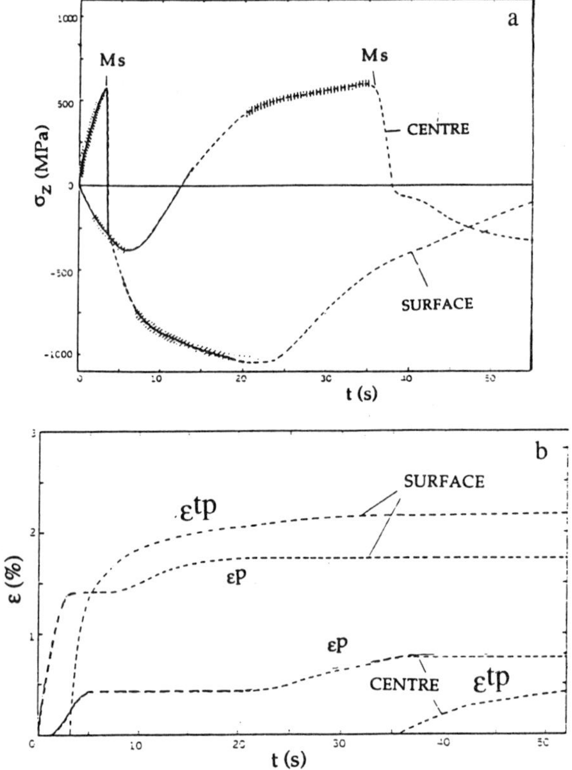

Figure 5: Quenching of a 60NCD11 steel cylinder (35mm in diameter) in cold water. a.Calculated axial stress evolutions over time at the surface and in the center. /////// Domains where plastic strains occur

b. Evolutions of the equivalent plastic strain ($\varepsilon_e{}^p$) and equivalent transformation plasticity strain ($\varepsilon_e{}^{tp}$) versus time. ($\varepsilon_e{}^p=\int d\varepsilon_e{}^p$, $d\varepsilon_e{}^p=(2/3 d\varepsilon_{ij}{}^p d\varepsilon_{ij}{}^p)^{1/2}$ and $\varepsilon_e{}^{tp}=\int d\varepsilon_e{}^{tp}$, $d\varepsilon_e{}^{tp}=(2/3 d\varepsilon_{ij}{}^{tp} d\varepsilon_{ij}{}^{tp})^{1/2}$) [47]

As martensitic transformation takes place at the surface (at Ms) immediately unloading occurs due to the volume increase. Transformation plasticity strains accompany the transformation. New plastic strains only occur as the material is submitted to high compressive stresses. In a last stage, the surface is unloaded. While the transformation progresses in the surface area, the center undergoes firstly unloading in compression.

During the further loading in tension, plastic strains are generated; a last stage of unloading due to the martensitic transformation in the center can be observed.

In order to highlight the effect of the transformation plasticity strain, comparisons have been made between calculated results obtained either by taking it into account or not. The analysis showed that :
- as the material is in an elastic regime as transformation occurs, transformation plasticity acts as an additional strain and leads to stress relaxation
- as the material deforms plastically, either transformation plasticity (whose amplitude is transformation rate and stress dependent) is sufficient to accommodate the deformations of the material (no additional plastic strain is necessary) and stress relaxation generally occurs; if transformation plasticity is not sufficient further plastic strain is necessary and no stress relaxation occurs. (The stress follows the evolution of the flow stress of the material).

This behaviour is illustrated in figure 6 that shows the "loading paths" (at the surface of a cylinder) during martensitic quenching (same case as in figure 5).This kind of representation [3, 45] shows clearly the influence of transformation plasticity both on the stress and strain states. For the calculation that includes transformation plasticity, as the martensitic transformation starts at the surface (at 247°C), the material is unloaded: the equivalent total strain becomes higher and the stress changes quicker from tension to compression. As the transformation progresses (at the surface but also in the inside of the piece) stress relaxation occurs untill 172°C (the equivalent stress remains lower than the flow stress of the material). Between 172°C and 140°C, both transformation plasticity strains (whose amplitude decreases due to the decrease in transformation rate) and classical plastic strains occur; the equivalent stress is equal to the flow stress. The unloading between 140°C and room temperature is due to the martensitic transformation in the inside of the piece.

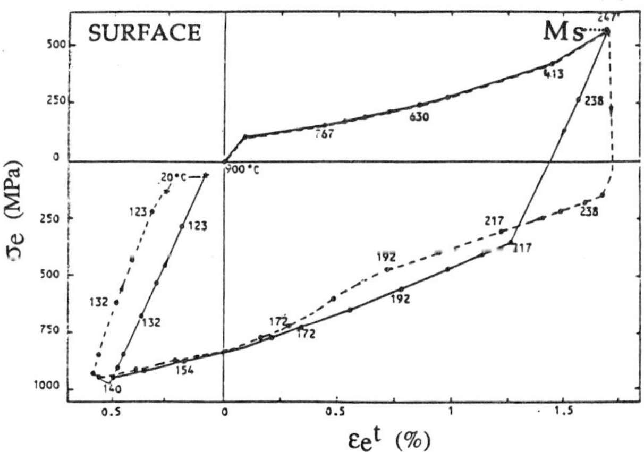

Figure 6: Quenching in cold water of a 60NCD11 steel cylinder (35mm in diameter) [3] Calculated "loading path" at the surface. ——— without transformation plasticity - - - - - with transformation plasticity

($\varepsilon_e^t = (2/3\ d\varepsilon_{ij}^t d\varepsilon_{ij}^t)^{1/2}$ is the equivalent total strain, σ_e is the equivalent stress)

4.1.2 Effect on residual stress states

Whereas the introduction of transformation plasticity will always modify the stress and strain states during the phase transformation, its effect on the residual stress states will be more or less significant depending on each particular case. Indeed, the residual stress states in a piece depend highly on the irreversible strains (plastic strains and transformation plasticity strains) that are generated all along the treatment. Particularly, the relative amplitudes of the irreversible strains that are generated as the material is in tension or in compression are determinant.

For surface heat treatments (like induction or laser hardening) in which only the surface area undergoes a martensitic transformation during cooling, transformation plasticity generally induces stress relaxations during the whole transformation process. Transformation plasticity is able to accommodate alone the deformations of the material. Finally it leads to much lower compressive residual stresses in the hardened zone.
For example, figure 7 shows the transverse stress evolution during heating and cooling at the surface of a laser hardened plate (7a) and the corresponding residual stress distribution (7b) : the stress relaxation effect due to transformation plasticity during martensitic transformation appears clearly. Moreover, a good agreement between calculated results (including transformation plasticity) and residual stress measurements was found in the hardened zone [52]. Similar results have been reported in literature either for laser hardening [53, 54] or induction hardening [55].

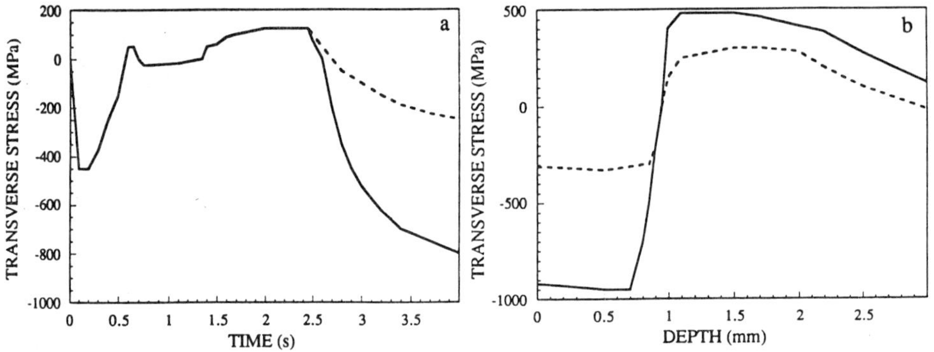

Figure 7: Laser surface hardening of a 0.42%C steel plate [52]
a. Calculated stress evolutions at the surface in the middle of the laser track
b. Residual stress distributions versus depth in the middle of the laser track
—— without transformation plasticity - - - - - with transformation plasticity
(the transverse stress is the stress perpendicular to the laser beam displacement)

The above analysis holds also in the case of thermochemical treatments that involve quenching after the diffusion treatment. Results obtained [56] for a case hardening treatment (figure 8) showed that the inclusion of transformation plasticity is of great importance for the prediction of the residual stress levels.

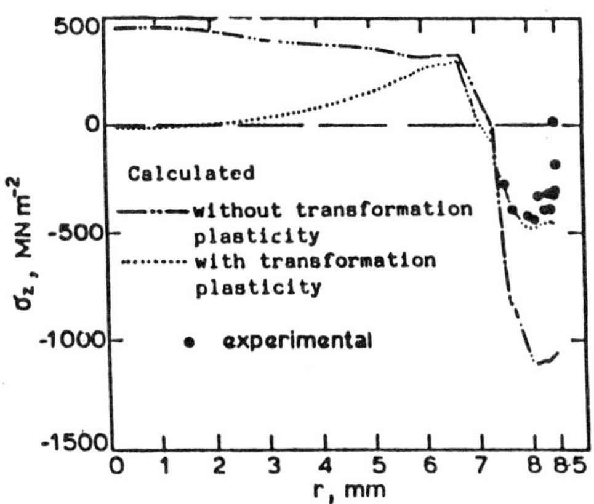

Figure 8: Carburizing and quenching in oil of a SIS2511 steel cylinder [56]
Calculated residual stress distributions

If we consider now quenching, we can find a number of results in literature where the authors have compared residual stress profiles that are obtained either by taking transformation plasticity into account or not. Most of these results concern martensitic quenching. The analysis can be the following [3]:

- firstly, if we consider the most simple case of quenching of a material in which no phase transformation occurs, only thermal stresses (due to thermal gradients) are generated and the residual stress profiles are generally characterized by compressive stresses at the surface and tensile stresses in the center (we consider here only simple shaped pieces like cylinders or plates).
- Secondly, if in addition the material undergoes martensitic transformation, the residual stress profile will tend to invert : surface stresses will tend to go to tension and the stress state in the center will tend to become compressive. This analysis was proposed very early in the qualitative approach by Rose et al. [57]. From the numerical approaches, it came out that these tendencies are highly dependent on the irreversible strain histories i.e. on the temperature evolutions in the piece (depending on the quenching medium, on the size of the piece) and on the nature of the steel (Ms temperature, mechanical properties for instance).
- Third, by the introduction of transformation plasticity strains the above mentioned tendencies will be enhanced. The enhancement will generally be larger when no classical plastic strains are generated at the location where transformation takes place.

As an illustration, figure 9 gives the calculated residual stress profiles for martensitic quenching in cold water of a steel cylinder (same case as in figure 5). By taking into account only the volume change and the variations of mechanical properties due to the transformation (full line) the residual stress levels are small. When transformation plasticity is included, the residual stress profile gains a more typical shape (tensile stresses in the surface area, compressive stresses in the center). Another example is the quenching in oil of a steel plate [45] (figure 10). Here, the inclusion of transformation plasticity leads to a complete inversion of the residual stress profile.

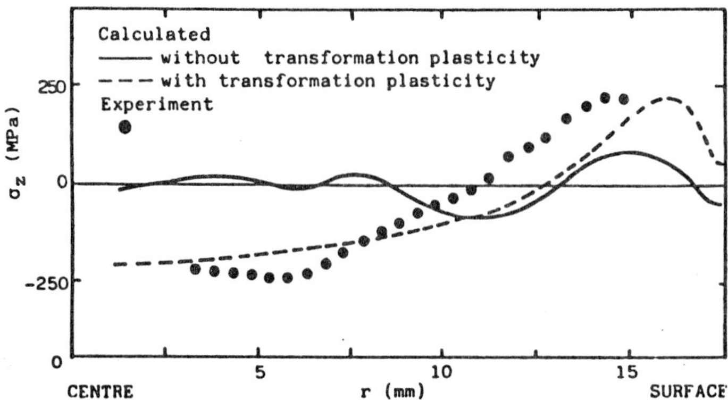

Figure 9: Quenching in cold water of a 60NCD11 steel cylinder (35mm in diameter) [3]
Residual stress distributions

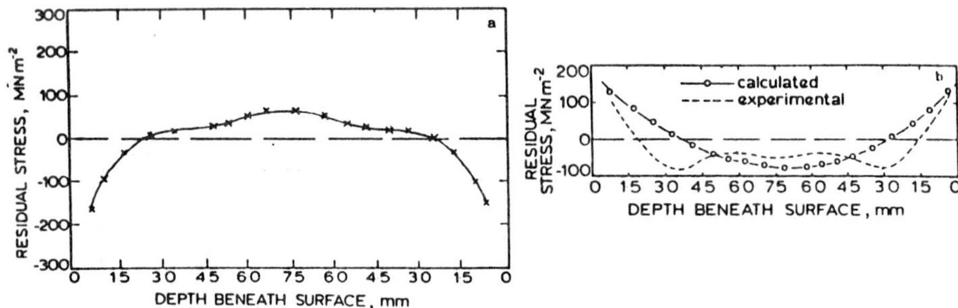

Figure 10: Oil quenching of a 835M30 steel plate (thickness 15mm) [45]
Residual stress distributions a.—— without transformation plasticity
b.- - - - with transformation plasticity

Only few results exist concerning the effect of transformation plasticity when the heat treatment involves high temperature transformations. Figure11 illustrates this effect in the case of pearlitic "quenching" of a cylinder [43]. The small plastic strains that develop during cooling, as the steel is austenitic, lead to a residual stress profile with small compressive stresses at the surface (full line). Transformation plasticity brings an additional irreversible strain and leads to the inversion of the residual stress profile (dotted line).

Finally it should be mentioned that in welding processes of steels the phenomena occuring in the heat affected zone are the same as during a heat treatment. Thus, transformation plasticity will affect the stress/strain evolutions during cooling and have an effect on the residual stress distributions that depends on the analysed conditions [58-60].

4.1.3 Effect on distortions

Of course, the final mechanical state of a piece after heat treatment is not only characterized by residual stress distributions but also deformations of the piece. The effect of

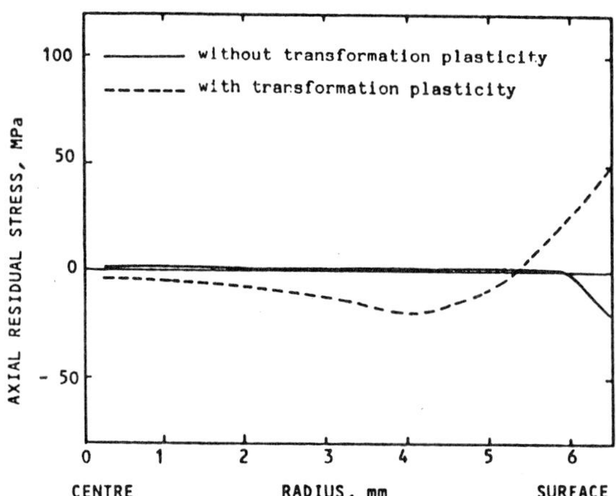

Figure 11: Cooling (14°C/s) of an eutectoid carbon steel cylinder (diameter 13mm) [43]. Residual stress profiles

transformation plasticity on the distortions of a heat treated piece has been less studied [45, 48, 51, 61]. Again, the more or less important effect of transformation plasticity on distortions will depend on each particular heat treatment process.

For martensitic quenching of cylinders it has been shown [48] that the inclusion of transformation plasticity enhances a concave-type deformation of the cylinder. It should be recalled that thermal stresses only would lead to a convex-type deformation and that the occurence of martensitic transformation will tend to give a concave type deformation of the cylinder [62]). More recently, for quenching of steel bars, only slight differences have been observed when the distortion is calculated with the transformation plasticity strain considered both for martensitic quenching and for quenching involving various transformations during cooling as shown in figure 12 [51].

From all these results, it is now well admitted that any residual stress analysis for materials that undergo phase transformations must include transformation plasticity.

In most of the studies the agreement with experimental results was found to be more satisfactory if considering transformation plasticity. Nevertheless some authors [46, 48] concluded on better simulation results in comparison with experiments (for martensitic quenching) by limiting the amplitude of transformation plasticity considered in the model It should also be mentioned that only few results have been reported on complex shaped pieces [61, 63-65].

From our point of view, the discussion on the possible reasons of discrepancies between calculated results and experimental ones is difficult : the calculated results depend on one hand on the material behaviour model and the associated input data and on the other hand on the process parameters. To obtain the data on the thermomechanical behaviour of the material is a hard experimental work and some data have to be extrapolated (for example from high temperature variations to lower temperatures). Moreover, the modelling of the process is complex, particularly the quenching process due to the heat transfer mechanisms that are difficult to control. This problem becomes even more acute for industrial pieces.

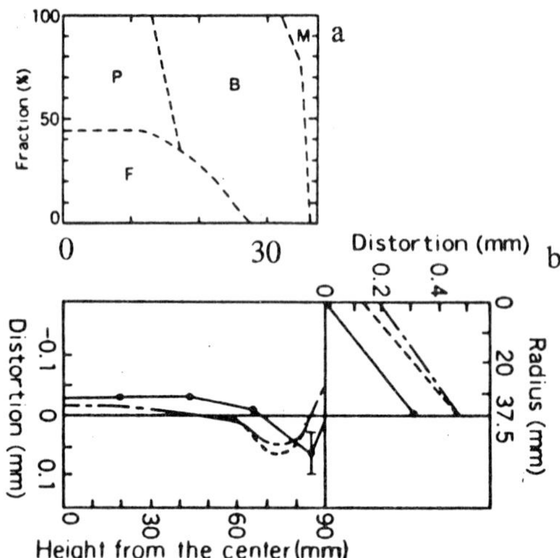

Figure 12: Quenching of a 1035 steel bar (75mm diameter)[51]
a. Radial profiles of microstructure at midlength
b. Distortions --------- without transformation plasticity
--- --- with transformation plasticity
—— experiment

4.2 Effect of the metallurgical interaction

Very few results have been obtained by taking into account in the model the coupling between the internal stress states generated in a piece during heat treatment and the transformation kinetics [6, 43]. This coupling will, of course, affect the transformation kinetics at the different locations in the piece and consequently the temperature evolutions (through the latent heat associated with the transformation) and the stress/strain evolutions. The residual stress state may be also affected.

4.2.1 Effect on transformation kinetics

If we refer to the models (2.1.2), the effect of the stresses on the transformation kinetics will depend on the type of transformation and on the stress state in the piece, at the location where the transformation occurs.
Concerning martensitic transformation, we expect that the transformation in the piece starts earlier during cooling when the material is submitted to tensile stresses (Ms temperature is increased) or starts later if compressive stresses with a relatively large hydrostatic component are present. For middle size pieces, generally an increase in Ms temperature has been calculated. During further cooling, the progress of the transformation is slowered due to the mechanical unloading of the material that is concomitant with the transformation (the mechanical driving force for the transformation decreases). For example, figure 13 shows the evolution of the volume fraction of martensite versus temperature in a quenched cylinder (case of figure 5). The effect of the internal stresses appears in the first stages of transformation (until 25% martensite formed). It is more pronounced in the center of the piece due to a higher contribution of the hydrostatic (tensile) stress. These modifications in the transformation progress have a negligible effect on the temperature evolutions in the piece.

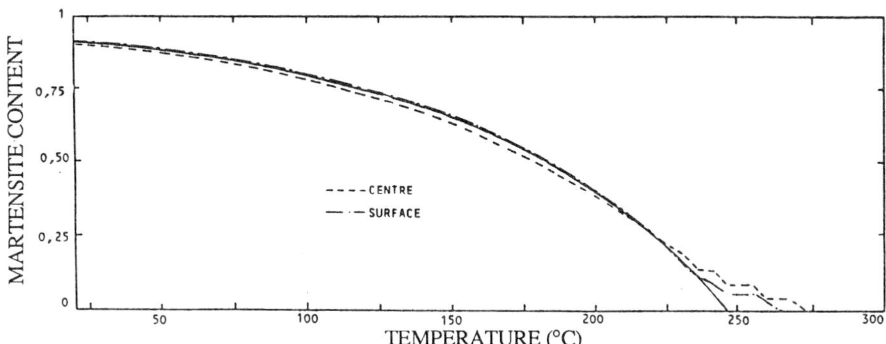

Figure 13: Quenching in cold water of a 60NCD11 steel cylinder (35mm in diameter) [1, 3]. Calculated progress of martensitic transformation versus temperature : —— without effect of internal stresses - - - - with effect of internal stresses.

For diffusion dependent transformations, the transformation kinetics are generally accelerated. In the example of the cooling down of a cylinder in which only a pearlitic transformation occurs (figure 14) a shortening of the incubation period and an increase in transformation rate is clearly observed at the surface when the metallurgical interaction is included (14a). These effects are much smaller in the center due to a lower level of the stress as the transformation starts. Due to these kinetics changes the temperature evolutions in the piece are highly modified. Particularly, the recalescence starts earlier and has a lower amplitude (figure 14b). A more thorough analysis of these results and a discussion of the proposed model (2.1.2) can be found in [7].

The effect of stresses on transformation kinetics may have important practical consequences :
- the hardness distributions in the piece can be modified. Figure15 shows a comparison between calculated hardness profiles after pearlitic quenching considering or excluding the metallurgical interaction.
- When a fully pearlitic microstructure is looked for (for example in steel wires), generally a low interlamellar spacing i.e. low transformation temperature must be obtained. The numerical simulations allowed to show that as soon as the thermal gradients become high, the increase in the cooling rate is less efficient in order to decrease the transformation start temperature. The transformation start temperature increases with the diameter of the wire for a given cooling rate (due to higher internal stresses) as can be seen on figure 16.

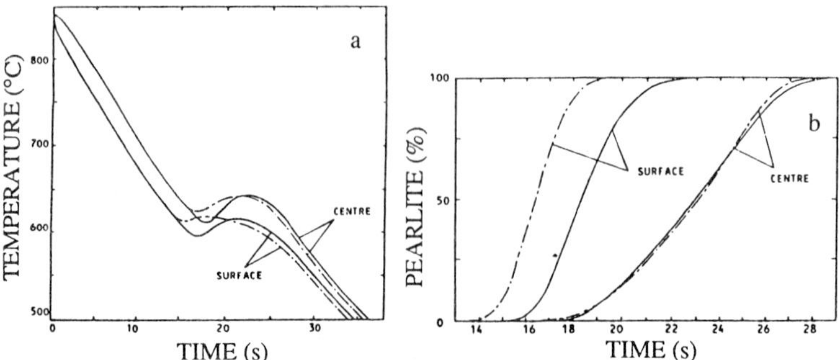

Figure 14: Cooling down (14°C/s) of an eutectoid carbon steel cylinder (13mm diameter) [43]. Calculated cooling curves (a) and pearlitic transformation kinetics (b)
—— without effect of internal stresses - - - - with effect of internal stresses

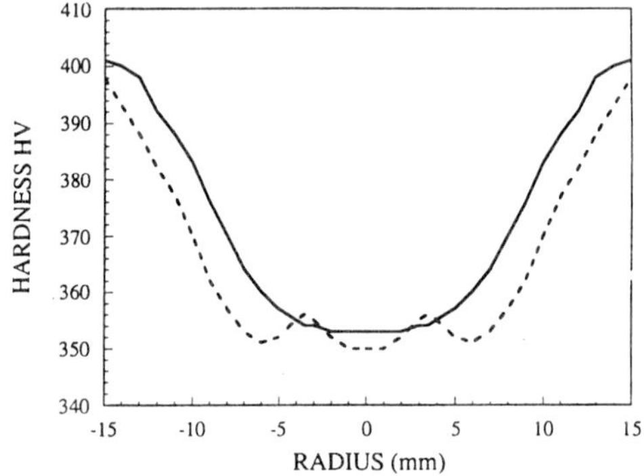

Figure 15: Quenching (in a polymer solution) of an eutectoid carbon steel cylinder (30 mm diameter) [66].
Hardness distributions —— without effect of internal stresses
- - - - with effect of internal stresses

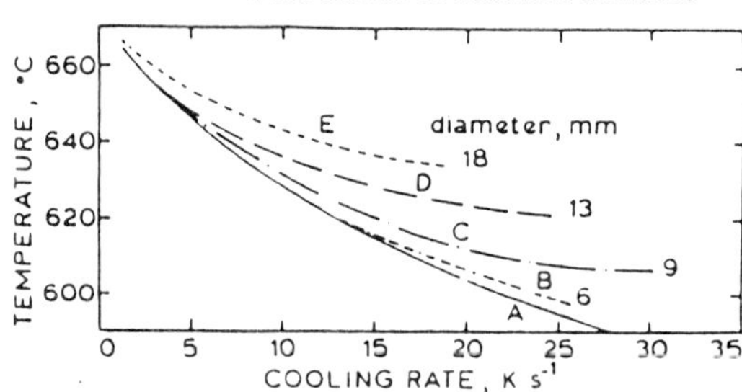

Figure 16: Calculated transformation start temperature as a function of cooling rate and wire diameter [67]
A without effect of internal stresses
B-E with effect of internal stresses

4.2.2 Effect on stress/strain evolutions and on residual stresses

As the deformations of the material during a heat treatment are highly dependent on transformation rate, we expect that the changes in transformation kinetics described above will affect the stress/strain evolutions :

- For martensitic quenching, the main effect is a shift (in the time scale) of the stress evolutions (shown in figure 5) due to the increase in Ms temperature. For the same reason, a small shift (in deformation scale) of the loading path (figure 6) has been observed [3]. The consequences on the residual stress distributions are relatively small (figure 17).

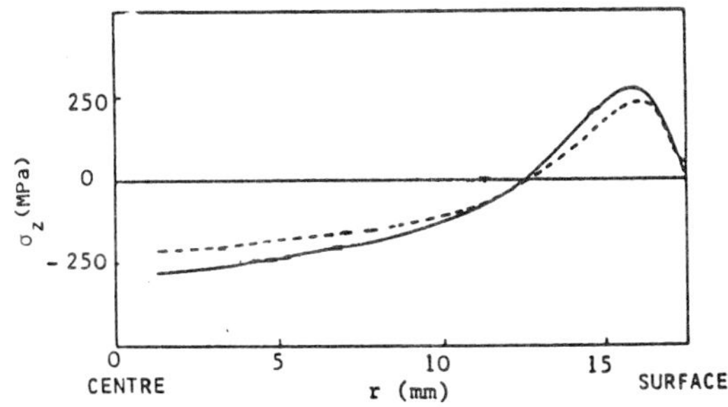

Figure 17: Quenching in cold water of a 60NCD11 steel cylinder (35mm diameter) [1,3].
Calculated radial residual stress profiles
—— without effect of internal stresses on transformation kinetics
- - - - with effect of internal stresses on transformation kinetics

- For pearlitic "quenching", the acceleration of transformation kinetics resulted in an enhancement of the variations that were already depicted by the introduction of transformation plasticity. Thus, the calculated residual stress profile (figure 18) shows higher tensile stresses at the surface and higher compressive stresses in the center due to higher amplitudes of transformation plasticity strains [43].

Results on quenching of a carbon steel cylinder [6] seems showing the same tendency (figure 19). (Note that no transformation plasticity was included in these calculations).

From these results (unfortunately very few), it is difficult to put a final statement on the importance of the metallurgical interaction in the prediction of heat treatment residual stresses.

It seems that in most cases of quenching, the metallurgical interaction does not influence significantly the residual stress states, even if it affects locally the stress evolution. Nevertheless, in some specific cases (where high temperature transformations occur) a more significant effect has been evidenced, that cannot be neglected (particularly on the transformation kinetics). This has been confirmed more recently by residual stress and distorsion calculations performed for more complex shaped pieces [64, 65].

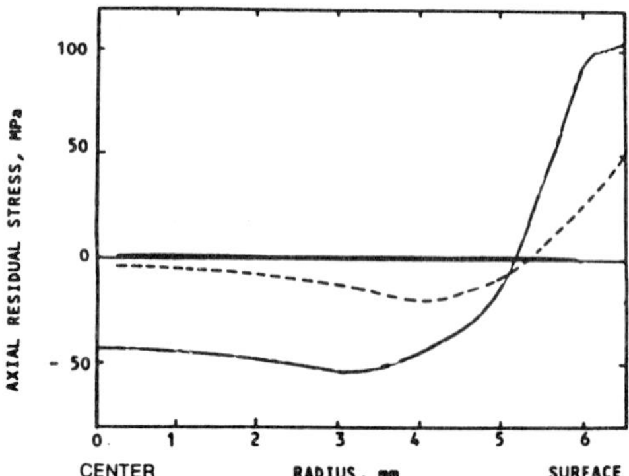

Figure 18: Cooling down (14°C/s) of an eutectoid carbon steel cylinder (13mm diameter) [43]. Calculated radial residual stress profiles
------- without effect of internal stresses on transformation kinetics
—— with effect of internal stresses on transformation kinetics

From these results (unfortunately very few), it is difficult to put a final statement on the importance of the metallurgical interaction in the prediction of heat treatment residual stresses. It seems that in most cases of quenching, the metallurgical interaction does not influence significantly the residual stress states, even if it affects locally the stress evolution. Nevertheless, in some specific cases (where high temperature transformations occur) a more significant effect has been evidenced, that cannot be neglected (particularly on the transformation kinetics). This has been confirmed more recently by residual stress and distorsion calculations performed for more complex shaped pieces [64, 65].

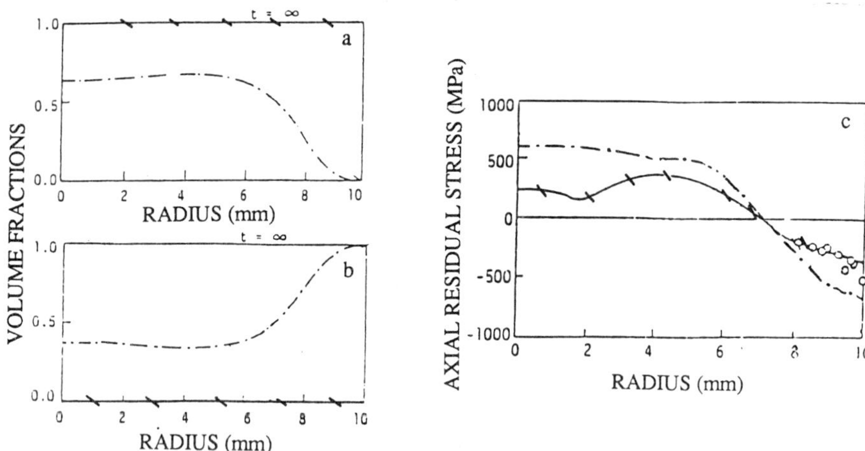

Figure 19: Water quenching of a carbon steel (S45C) cylinder (20mm diameter) [6]. Calculated distributions of microstructures a. pearlite b. martensite. Calculated residual stress distributions (c) (o experimental values)
-·-·- without effect of internal stresses on transformation kinetics
-+-+- with effect of internal stresses on transformation kinetics

5. CONCLUSIONS

The macroscopical behaviour law of a material undergoing phase transformations has been recalled. Particularly, the modelling of the stress - phase transformation interactions (transformation plasticity and the effect of internal stresses on transformation kinetics) has been described.The main conclusions are :

- the macroscopical approach that is the most commonly used for calculating heat treatment stresses and strains and that considers transformation plasticity as an additional deformation in the behaviour law of the material allows to describe the mechanical behaviour of a specimen for different types of transformations (ferritic, pearlitic, bainitic, martensitic) in anisothermal conditions.
- The effect of stress - phase transformation interactions on the residual stress distributions and on the distortions can generally not be forecasted à priori as they depend highly on the local stress states in the piece when transformation occurs. Thus, coupled temperature-phase transformation-stress/strain calculations have to be performed. It is the only way to analyse the influence of the material behaviour on the result of the treatment, for given heat treatment conditions, and eventually to neglect some minor effects for a given application.

Aknowledgments

I am very grateful to A. Simon who initiate my activity for useful improvements and comments and to E. Gautier for the years of collaboration and fruitful discussions.
Many students have contributed to this research field: C. Basso, F.M.B. Fernandes, D. Farias, F. Saliou†, L. Massicart, M. Zandona, M. Boufoussi, A. Mey.
I wish also to mention the fruitful collaboration with S. Sjöström (Linköping University).
This research field has been supported by ARBED, CETIM, PSA, RENAULT, UNIMETAL RECHERCHE...I thank especially RENAULT for providing non published results.

References

1. Denis S., Gautier E., Simon A., Beck G. : Stress-phase transformation interactions - basic principles, modelling and calculation of internal stresses, Material Science and Technology, 1 (1985), 805-814.
2. Gautier E. : Transformations perlitique et martensitique sous contrainte de traction dans les aciers, Thèse de Doctorat d'Etat, INPL Nancy 1985.
3. Denis S. : Modélisation des interactions contrainte-transformation de phases et calcul par éléments finis de la genèse des contraintes internes au cours de la trempe des aciers, Thèse de Doctorat d'Etat, INPL Nancy 1987.
4. Denis S., Gautier E., Simon A. : Modelling of the mechanical behaviour of steels during phase transformation : a review, in : International Conference on Residual Stresses ICRS2 (Ed. G. Beck, S. Denis, A. Simon), Elsevier Applied Science 1988, 393-398.
5. Gautier E., Zhang J.S., Zhang X.M. : Martensitic transformation under stress in ferrous alloys. Mechanical behaviour and resulting morphologies, Journal de Physique IV, Colloque C8, suppl. Journal de Physique III, 5 (1995), 31-40.
6. Inoue T., Wang Z.G. : Finite element analysis of coupled thermoinelastic problem with phase transformation, in : Int. Conf. Num. Meth. in Industrial Forming Processes (Ed. J.F.T. Pittmann, R.Wood, J.H. Dand Alexander), Pineridge Press 1982.
7. Denis S., Gautier E., Sjöström S., Simon A. : Influence of stresses on the kinetics of pearlitic transformation during continuous cooling, Acta Met., 35 (1987), 1621-1632.

8. Loshkarev V.E. : Interaction between quenching stresses and transformations in steel, Metal Science and Heat Treatment, 28 (1986), 3-9.
9. Denis S., Farias D., Simon A. : Mathematical model coupling phase transformations and temperature in steels, ISIJ International, 32 (1992), 316-325.
10. Gautier E., Zhang X.M., Simon A. : Role of internal stress state on transformation induced plasticity and transformation mechanisms during the progress of stress induced phase transformation, in : International Conference on Residual Stresses ICRS2 (Ed.G. Beck, S. Denis, A. Simon), Elsevier Applied Science 1988, 777-783.
11. Gautier E., Simon A., Beck G. : Plasticité de transformation durant la transformation perlitique d'un acier eutectoïde, Acta Metall., 35 (1987), 1367-1375.
12. De Jong M., Rathenau G.W. : Mechanical properties of iron and some iron alloys while undergoing allotropic transformation, Acta Metall., 7 (1959), 246-253.
13. Matsuzaki A., Bhadeshia H.K.D.H., Harada H. : Effect of stress on bainitic transformation in Fe-Si-Mn-C alloy, in : ICOMAT 92 (Ed. C.M. Wayman and J. Perkins), Monterey Institute for Advanced Studies 1993.
14. Gautier E., Denis S. : Comportement thermomécanique d'un acier au cours de la transformation bainitique, in : Report RENAULT 1995.
15. Ganghoffer J.F., Simonsson K., Denis S., Gautier E., Sjöström S., Simon A. : Martensitic transformation plasticity simulations by finite elements, Journal de Phys. Coll. C3, 4 (1994), 215-220.
16. Marketz F., Fischer F.D. : A micromechanical study of the deformation behaviour of Fe-Ni alloys and martensitic transformation, in : PTM 94 Solid Solid Phase Transformation (Ed. Johnson W.C., Howe J.M., Laughlin D.E., Soffa W.A.), TMS 1994, 785-790.
17. Simonsson K. : Micro-mechanical FE-simulations of the plastic behaviour of steels undergoing martensitic transformation, Dissertation N°362 Linköping 1994.
18. Wen Y., Denis S., Gautier E. : Computer simulation of martensitic transformation under stress, Journal de Physique IV Coll. C1, 6 (1996), 475-483.
19. Franitza S. : Zur Berechnung der Wärme- und Umwandlungsspannungen in langen Kreiszylindern, Dissertation TU Braunschweig 1972.
20. Giusti J. : Contraintes et déformations résiduelles d'origine thermique. Application au soudage et à la trempe des aciers, Thèse de Doctorat d'Etat, Univ. Pierre et Marie Curie Paris 1981.
21. Hamata N. : Modélisation du couplage entre l'élasto-viscoplasticité anisotherme et la transformation de phase d'une fonte G.S. ferritique, Thèse de Doctorat, Univ. Paris 6 1992.
22. Videau J.C., Cailletaud G., Pineau A. : Modélisation des effets mécaniques des transformations de phases pour le calcul des structures, Journal de Phys. IV Coll. C3, 4 (1994), 227-232.
23. Leblond J.B. : Mathematical modelling of transformation plasticity in steels II Coupling with strain hardening phenomena, Int. Journal of Plasticity, 5 (1989), 573-591.
24. Fischer F.D. : Transformation induced plasticity in triaxially loaded steel specimens subjected to a martensitic transformation, European Journal Mech. A/ Solids 11 (1992), 233-244.
25. Sjöström S., Ganghoffer J.F., Denis S., Gautier E., Simon A. : Finite element calculation of the micromechanics of a diffusional transformation. II. Influence of stress level, stress history and stress multiaxiality, Eur. J. Mech. A/Solids 13 (1994), 803-817.
26. Videau J.C., Cailletaud G., Pineau A. : Experimental study of the transformation induced plasticity in a Cr-Ni-Mo-Al-Ti steel, Journal de Physique IV Coll. C1, 6 (1996), 465-474.
27. Graja P., Scholtes B., Müller H., Macherauch E. : Residual stress distributions in cylindrical parts due to continuous and discontinuous hardening processes, in : International

Conference on Residual Stresses ICRS (Ed. E. Macherauch and V. Hauk) DGM 1987, 687-694
28. Hildenwall B. : Prediction of residual stresses created during quenching, Dissertation N°39, Linköping University 1979.
29. Melander M. : Computational and experimental investigation of induction and laser hardening Dissertation N° 124, Linköping University 1985.
30. Leblond J.B., Devaux J. : A new kinetic model for anisothermal metallurgical transformations in steels including effect of austenite grain size, Acta Metall., 32 (1984), 137-146.
31. Hougardy H.P., Wildau M. : Berechnung der Wärmebehandlung von Stählen-Umwandlungsverhalten, Spannungen, Verzug, Stahl u. Eisen, 105 (1985), 1289-1296.
32. Waeckel F. : Modélisation du comportement thermométallurgique des aciers, Journal de Physique C3, 4 (1994), 221-232.
33 Buchmayr B., Kirkaldy J.S. : A fundamental based microstructural model for the optimization of heat treatment processes, in : 1st Int. Conf. on Quenching and Control of Distortion (Ed. G.E. Totten) ASM Intern. 1992, 221-227.
34. Sjöström S. : The calculation of quench stresses in steel, Dissertation N° 84, Linköping University 1982.
35. Leblond J.B., Devaux J., Devaux J.C. : Mathematical modelling of transformation plasticity in steels I Case of ideal-plastic phases, Int. Journal of Plasticity, 5 (1989), 551-572.
36. Rammerstorfer F.G., Fischer F.D., Till E.T., Mitter W., Gründler O. : The influence of creep and transformation plasticity in the analysis of stresses due to heat treatment, in : Numerical Methods in heat transfer (Ed. R.W. Lewis K. Morgan B.A. Schrefler) John Wiley&Sons Ltd 1983, 447-460.
37. Assaker D., Hogge M., Dubois M., Magnee A. : Computer prediction of residual stresses in bi-metallic heat treated roll mill cylinders, Société Belge des Mécaniciens, 1404 (1989), 7.1-7.8.
38. Wang Z., Inoue T. : Viscoplastic constitutive relation incorporating phase transformation - Application to welding, Materials Science and Technology, 1 (1985), 899-903.
39. Colonna F., Massoni E., Denis S., Gautier E., Wendenbaum J. : On thermo-elastic-viscoplastic analysis of cooling processes including phase changes, Journal of Mat. Proc. Techn., 34 (1992), 525-532.
40. Rammerstorfer F.G., Fischer F.D., Mitter W., Bathe K.J., Snyder M.D. : On thermo-elastic-plastic analysis of heat treatment processes including creep and phase changes, Computers and Structures, 13 (1981), 771-779.
41. Liébaut C. : Rhéologie de la déformation plastique d'un acier Fe-C durant sa transformation de phase, Thèse de doctorat INPL Nancy 1988.
42. Gautier E., Denis S., Liébaut Ch., Sjöström S., Simon A. : Mechanical behaviour of Fe-C alloys during phase transformations, Journal de Phys. C3, 4 (1994), 279-284.
43. Denis S., Sjöström S., Simon A. : Coupled temperature, stress, phase transformation calculation model. Numerical illustration of the internal stresses evolution during cooling of a eutectoïd carbon steel, Met. Trans, 18A (1987), 1203-1212.
44. SYSWELD, Users Manual, Framasoft + CSI, Lyon (France)
45. Abbasi F., Fletcher A.J. : Effect of transformatin plasticityon generation of thermal stress and strain in quenched steel plates, Material Science and Technology, 1 (1985), 830-837.
46. Convert F., Turbat A. : Estimation and experimental determination of residual stresses and distortion in quenched bars, in : Eigenspannungen (Ed. Macherauch E., Hauk V.) DGM 1983, 251-277.

47. Denis S., Simon A. : Discussion on the role of transformation plasticity in the calculation of quench stresses in steels, in : International Conference on Residual Stresses ICRS (Ed. E. Macherauch and V. Hauk) DGM 1987, 565-572.
48. Desalos Y., Giusti J., Gunsberg F. : Déformations et contraintes lors du traitement thermique de pièces en acier, Report RE902 IRSID, May 1982.
49. Mitter W. : Umwandlungsplastizität und ihre Berücksichtigung bei der Berechnung von Eigenspannungen, Gebrüder Bornträger Berlin 1987.
50. Besserdich G., Scholtes B., Müller H., Macherauch E. : Development of residual stresses and distortion during hardening of SAE 4140 cylinders taking into account transformation plasticity, in : Residual Stresses (Ed. V. Hauk, H.P. Hougardy, E. Macherauch, H.D. Tietz) DGM 1993, 975-984.
51. Nagasaka Y., Brimacombe J.K., Hawbolt E.B., Samarasekera I.V., Hernandez-Morales B., S.E. Chidiac : Mathematical model of phase transformations and elastoplastic stress in the water spray quenching of steel bars, Met. Trans., 24A (1993), 795-808.
52. Denis S., Boufoussi M., Chevrier J.Ch., Simon A. : Analysis of the development of residual stresses for surface hardening of steels by numerical simulation: effect of process parameters, in International Conference on Residual Stresses ICRS4 (Ed. M.R. James) Society of Experimental Mechanics Bethel 1994, 513-519.
53. Yang Y.S., Na S.J. : Effect of transformation plasticity on residual stress fields in laser surface hardening treatment, J. Heat Treating, 9 (1991), 49-56.
54. Bergheau J.M., Pont D., Leblond J.B. : Three dimensional simulation of a laser treatment through steady state computation in the heat source's comoving frame, in : Mechanical effects of welding (Ed. L. Karlsson L.E. Lindgren M. Jonsson) Springer Verlag 1992, 85-92.
55. Zandona M., Mey A., Boufoussi M., Denis S., Simon A. : Calculation of internal stresses during surface heat treatment of steels, in : European Conf. on Residual Stresses (Ed. V. Hauk, H.P. Hougardy, E. Macherauch und H.D. Tietz) D.G.M. 1993, 1011-1020.
56. Sjöström S. : Interactions and constitutive models for calculating quench stresses in steels, Material Science and Technology, 1 (1985), 823-829.
57. Bühler H., Rose A. : Darstellung des Entstehens von Eigenspannungen in Werkstücken aus Stahl in ihren Umwandlungsschaubildern, Archiv. Eisenhüttenw., 40 (1969), 411-423.
58. Josefson B.L. : Effects of transformation plasticity on welding residual stress fields in thin walled pipes and thin plates, Material Science and Technology, 1 (1985), 904-908
59. Oddy A., Goldak J., McDill M. : Transformation plasticity and residual stresses in single-pass repair welds, Journal of Pressure Vessel Technology, 114 (1992), 33-38.
60. Leblond J.B., Devaux J., Devaux J.C. : Simulation de l'essai d'implant, Soudage et Techniques connexes (1988), 312-324.
61. Anastassiou M. : Distorsions lors de la trempe d'organes de boîte de vitesses, in : Internationaux de France de Traitement Thermique (Ed. ATTT) PYC Edition 1993, 11-36.
62. Toshioka Y. : Heat treatment deformation of steel products, Material Science and Technology 1 (1985), 883-892.
63. Inoue T., Wang Z.G., Miyao K. : Quenching stress of carburised steel gear wheel, in : International Conference on Residual Stresses ICRS2 (Ed. G. Beck, S. Denis, A. Simon) Elsevier Applied Science 1988, 606-611.
64. Bourdouxhe M., Denis S., Simon A. : Computation of phase changes and deformations in long products undergoing thermal treatments, in : International Conference on Residual Stresses ICRS3 (Ed. H. Fujiwara, T. Abe, K. Tanaka) Elsevier Applied Science 1992, 202-207.
65. Colonna F. : Modélisation numérique du refroidissement de rails, Thèse de Doctorat de l'Ecole des Mines de Paris 1992.
66. Saliou F., Zandona M., S. Denis, E. Gautier, Internal work LSG2M 1991.

67. Denis S., Basso C., Fernandes F., Simon A. : Contribution des contraintes internes d'origine thermique dans le calcul de l'avancement des transformations de phases en refroidissement continu d'un acier XC80, Mém. et Etud. Scient. Revue Métall. (1986), 533-542.

This paper has been already published in Journal de Physique IV, Coll. C1, Vol. 6 (1996), 159-174.